# Fireworks

# Fireworks

## Pyrotechnic Arts and Sciences in European History

SIMON WERRETT

The University of Chicago Press
Chicago and London

**Simon Werrett** is associate professor of history at the University of Washington.

The University of Chicago Press, Chicago 60637
The University of Chicago Press, Ltd., London

Printed in the United States of America

18 17 16 15 14 13 12 11 10 09 1 2 3 4 5

ISBN-13: 978-0-226-89377-8 (cloth)
ISBN-10: 0-226-89377-4 (cloth)

Publication of this book has been aided by a grant from the Bevington Fund.

Library of Congress Cataloging-in-Publication Data

Werrett, Simon.
Fireworks : pyrotechnic arts and sciences in European history / Simon Werrett.
p. cm.
Includes bibliographical references and index.
ISBN-13: 978-0-226-89377-8 (cloth : alk. paper)
ISBN-10: 0-226-89377-4 (cloth : alk. paper)
1. Fireworks—Europe—History. I. Title.
TP268.W47 2010
662'.1094—dc22

2009024802

♾ The paper used in this publication meets the minimum requirements of the American National Standard for Information Sciences—Permanence of Paper for Printed Library Materials, ANSI Z39.48-1992.

*For Anna*

CONTENTS

Map 1. Map of London ca. 1730, showing (1) Green Park; (2) St. James's Square; (3) Covent Garden; (4) Crane Court, the location of the Royal Society from 1710; (5) Cheapside, the location of the pope-burning processions; (6) the Royal Exchange (also pictured in right inset); (7) Gresham College, the location of the Royal Society from 1660 to 1666 and from 1673 to 1710. Johann Baptist Homann, *Accurater Prospect und Grundris der Königl: Gros-Britanisch: Haupt und Residetz Stadt London* (Nuremberg, ca. 1730). Art Resource, New York.

Map 2. St. Petersburg in the 1750s, showing (1) Troitskii Square; (2) the Peter and Paul Fortress; (3) the Fireworks Theater; (4) the Imperial Academy of Sciences; (5) St. Petersburg Arsenal and Fireworks Laboratory; (6) the Summer Garden; and (7) the Winter Palace. From Mikhail Makhaev, *Plan goroda Sanktpeterburga* (St. Petersburg, 1753–61). Slavic and Baltic Division, the New York Public Library, Astor, Lenox and Tilden Foundation.

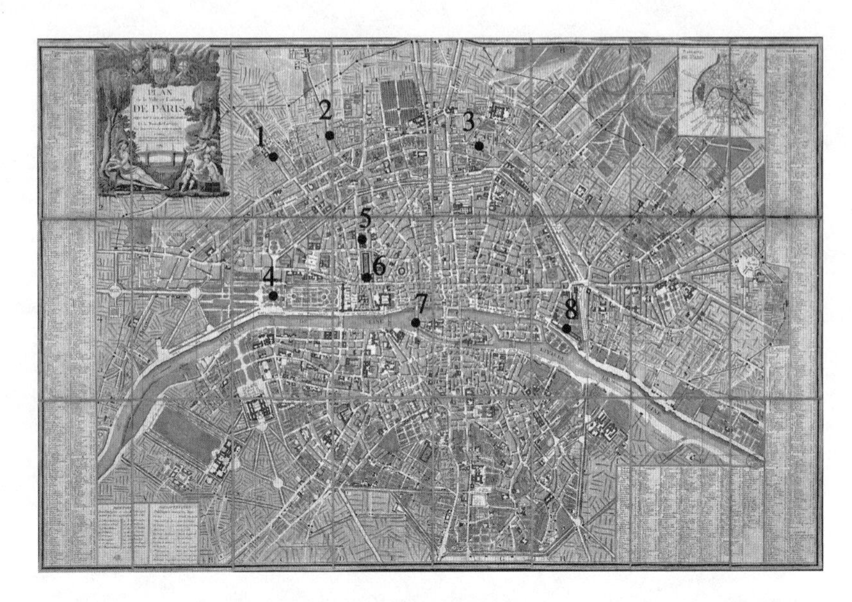


Map 3. Paris ca. 1789, showing (1) rue de Clichy, the location of Claude Ruggieri's atelier; (2) Jardin Ruggieri; (3) Torré's Waux-Hall; (4) Place Louis XV; (5) Bibliothèque du roi, location of the Paris Academy of Sciences; (6) Palais Royal; (7) Pont Neuf; (8) Grand Arsenal. *Plan de la ville et faubourg de Paris avec tous ses accroisemens* (Paris, 1789). Bildarchiv Preussischer Kulturbesitz/Art Resource, New York.

# INTRODUCTION

In January 1748, Professor Gerhard Friedrich Müller set up a new department of history in the St. Petersburg Academy of Sciences. Müller had previously returned from an extended expedition to Kamchatka that saw him travel across the Russian Empire into the far eastern reaches of Siberia, gathering natural historical and ethnographic information on the many peoples and cultures of the region. Now the historical department would assemble the material into a history of Siberia.[1] Three weeks earlier, on New Year's Eve, the imperial court of St. Petersburg, gathering in the Winter Palace on the opposite side of the river Neva to the academy, were entertained with fireworks. From a great wooden jetty extending over the water near the academy were erected huge painted wooden panels around which a great explosion of rockets, fiery wheels and bombs, shining stars, and fire fountains erupted into the night sky. In the central panel, painted in bright colors, was the image of a Siberian pine tree, growing in a garden of parterres and greenery (fig. 1). The designer of this allegorical scenery was Jacob Stählin, Müller's colleague at the Academy of Sciences, who printed a short pamphlet to explain its meaning to spectators. The tree, he said, represented growth in the prosperity of the Russian state, growth that, while continuous, was also imperceptible from day to day. Consequently, just as only a long-term view of the tree would make its growth apparent, so only a long-term understanding of Russian history would permit a true appreciation of Russia's greatness. To achieve this true appreciation, Russians needed "the understanding of those who grasp in enlightened terms the ventures of great states."[2] The message of Stählin's fireworks was, in other words, that the Russians needed a new department of history.

If, in the eighteenth century, some savants promoted enlightenment through coffeehouse lectures, salon debates, or experimental demonstra-

1. Fireworks and illuminations for the New Year 1748 in St. Petersburg. Elias Grimmel (drawer), *Izobrazhenie feierverka i illuminatsii kotoryia v novyi 1748 god, pred zimnim . . . domom predstavleny byli* (St. Petersburg, ca. 1748). Research Library, the Getty Research Institute, Los Angeles, California.

tions, others used fireworks. Stählin's was one of many such performances devised by the professors of St. Petersburg's academy in the eighteenth century, and their activities were not unique. This is a book about fireworks and the enduring, widespread, and variegated relations among pyrotechnics, art, and science over three centuries, from the Renaissance to the end of the ancien régime.

Fireworks were a significant element of early modern European life. In a world without electric light, fire was a powerful medium, a source of light and heat whose divine and magical connotations were strong.[3] Prometheus, who stole fire from heaven, presided over the human arts, by which men engaged with the natural world and turned nature to human ends.[4] Between the fourteenth century and the nineteenth, many communities displayed their authority by spectacular demonstrations of power over fire, in "artificial fireworks" that exploded around allegorical scenery and ephemeral architecture on rivers and town squares throughout Europe. Exhibiting "stars," "suns," "fiery dragons," "comets," "fizgigs," and "serpents," fireworks played with explosive forces to bring the heavens down to earth. The Catholic Church, princely courts, artillerymen, painters, architects, industrious entrepreneurs, and natural philosophers all sought out the credit such artifices could bring, and it is the history of their varied interactions and engagements with fireworks that this book recounts. The community of "artificers," the craftsmen and -women who made pyrotechnics and set them off, and the natural philosophers will be of particular interest because from their engagements and interactions emerged, not only a great variety of ingenious pyrotechnics, but also significant innovations in the sciences. Fireworks, this book suggests, contributed much to early modern science.

This book is also about the different places where fireworks, the arts, and natural philosophy came together, the distinctive regionality of these relations in history, and the consequences of location for their development. Recently, historians have highlighted the value of understanding geography for understanding the history of art, science, and enlightenment. Historians of science have shown how the sites and venues where science has been undertaken shape the nature of the knowledge and practices they produce, whose particular forms have varied historically and from region to region. Far from being a quintessentially "placeless" form of knowledge, valuable precisely because it transcends the local sites of its production, science turns out to have been fundamentally dependent on the particularities of its local contexts of production and circulation.[5] Similarly, art historians have identified the importance of place for artistic production. Works of art are often identified with national traditions and identities, yet they emerge from complex geographies crossing the physical, political, and cultural borders of different countries and continents, hybridizing variegated styles, and circulating in diverse spaces and networks.[6] Historians have also explored relations of geography and enlightenment, considering new sensibilities of landscape, empire, and the public sphere in the eighteenth century, or examining the ways in which enlightened savants took up space as a metaphor for knowledge, constructing new theories of human nature through novel geographic sciences.[7]

Drawing these diverse disciplinary arenas together, this book explores the changing geography of interactions among several arts and sciences engaged with the production of pyrotechnics. Fireworks are an ideal subject for exploring such a geography because they entailed so many diverse skills, from chemistry to poetic composition, from artillery to architecture. At no time did those involved in making displays agree on which skills were most important and how they were to be related, and, moreover, emphasis on particular skills and interactions varied from place to place. This book traces the emergence of a particular regionality in relations of the arts and sciences in pyrotechnics and how that regionality was itself transformed by the dynamic travels and activities of pyrotechnic practitioners.

Traditionally, the artist's studio and the artisanal workshop have provided important sites where historians have explored the relations of art and science. Historians of art have considered the ways painters and sculptors took up the liberal arts to raise the status of their labors during the Renaissance. Historians of science pursuing the "Zilsel thesis," after the Marxist historian Edgar Zilsel, looked to the workshop as a critical location for the shaping of a new, more empirical form of science in the seventeenth century.[8] According to this argument, natural philosophers (the term *scientist* came into use only from the 1830s) exploited the techniques and skills of artisans to reform traditional scholasticism into a more practical, utilitarian form of inquiry, able to transform nature for human ends. As Zilsel himself put it, writing of the magnetic philosopher William Gilbert in 1941: "At least a part of his laboratory must have looked like a smithy."[9] Such an account, developed most recently in Pamela H. Smith's *Body of the Artisan*, identifies a traffic of practical skills, knowledge, and epistemic values (in Smith's phrase, an "artisanal epistemology") flowing from the arts into the sciences to reform methods and approaches and, thus, contributing to the "Scientific Revolution."[10] Other historians have traced similar movements of practices, instruments, bodily gestures, and material cultures of the artisanal workshop into science, transforming knowledge in the process.[11]

The makers of fireworks also managed workshops, which they called *laboratories* many years before the term entered science, and it will be part of the argument here to show how the skills and the techniques of the fireworks laboratory, like those of other artisanal workshops, found a place in early modern science. At the same time, there is room to move beyond the workshop and sites of artisanal production to other spaces, regions, and circulations where art and science met, and it is to this goal that *Fireworks* is devoted. The book thus offers a *geography of art and science* as much as a history of fireworks and natural philosophy, using the history of pyrotechnics to map the regionality of interactions between artisans and natural philosophers in a variety of different locales. In addition to examining further the contributions of the workshop to early science, I shall examine sites such as the battlefield, the church, the arsenal, the court, the pleasure garden, and the scientific academy and consider the ways in which their knowledge, performances, and techniques were drawn together or separated to serve the ends of science and pyrotechny.

A particular focus will be on the *scenes of performance* where fireworks were played off in early modern Europe. Historians such as Paula Findlen and Mario Biagioli have pointed to the importance of festive sites for a number of scientific endeavors in the early modern period, urging close ties between the playful and carnivalesque nature of early modern social space and its science.[12] Biagioli, for example, has drawn attention to the incorporation of Galileo's discoveries of the moons of Jupiter in the pageants and spectacles of the Medici court, lending the dynasty a potent emblem and Galileo the social legitimacy that a relationship to the court secured. In tracing the history of fireworks, this book naturally explores many such theatrical and festive locales and suggests that they were just as significant as the workshop as sites where natural philosophers experienced and engaged with the arts. Early modern Europe was, after all, a theatrical culture that believed that "all the world's a stage" and took the theater and spectacle as models for nature and social order, with "all the men and women merely players."[13] In the eighteenth century, as Simon Schaffer has shown, spectacle became integral to public scientific lecturing, following Newton's advice that abstract arguments about the deity would be more convincing through demonstrations of phenomena.[14] The skills needed to exhibit science as a spectacle depended as often on the consumption of art as on the techniques of artisanal production, and the leisured and the learned who pursued science spent as much time in theatrical spaces such as opera houses, city squares, pleasure gardens, and theaters as they did in workshops. Thus, we shall often see how the techniques of *performance* in fireworks, the many changing ways in which artificers staged and fired off pyrotechnics, provided vital resources for a variety of scientific endeavors.

Appreciating the geography of art and science also entails a regional understanding. Just as art and science met in varied locations, so the nature of their interactions and the direction of their exchanges differed significantly from place to place, producing different forms of pyrotechnic and scientific knowledge and practice in different locations. To bring this out, I take a comparative approach, examining the history of fireworks and natural philosophy in a number of different urban and national settings. In some chapters, many locations serve this goal and take us variously to England, France, the German and Italian states, Russia, and China because the art of fireworks was dynamic, filled with traveling artificers and circulating techniques. Other chapters are more geographically focused, and much of the book concentrates on three particular sites: London, Paris, and St. Petersburg. Tracing the pyrotechnic and philosophical histories of these cities reveals many regional particulars in relations of art and science in European history, in the forms relations took between philosophers and artisans, in the practices or knowledge that they chose to exchange, and in the consequences of those exchanges for existing sciences and arts. Each locale constituted discernible traditions of interactions between art and science, and these might endure over several centuries.

At the same time, such relations were not essential to any particular place, and there was much change. Pyrotechnic and scientific knowledge and techniques traveled, and, as they did so, they transformed regional identities, creating new kinds of similarities or differentiation in space. These transformations in turn reformed the nature of science and fireworks. Similarly, the identity of the philosopher and the artisan was never static but negotiated, complex, and distinct at some places, integrated at others. Fireworks culture was also open to non-European practices, never limited to the borders of Europe, but absorbing skills and experiences from elsewhere. I thus trace both the more static regionalities that emerged in relations of European art and science and the ways in which such regional distinctions were dynamic and subject to change.

Another way this book moves beyond localized studies of art and science is by covering an extended period of history. Beginning in the fourteenth century, when the first gunpowder fireworks were used for war and celebration in Europe, the book ends in the 1820s, a moment when fireworks settled into a form still largely recognizable today. Covering the *longue durée* is worthwhile because it further reveals the ongoing and dynamic nature of interactions between fireworks and natural philosophy. Just as their interactions differed in space, so there was nothing fixed about the traditions that emerged in different sites over time. Observing their changes over several centuries thus demonstrates the dynamism of geography itself. If place is a determinant of knowledge and practice, then it is always being revised by the circulation of skills. Furthermore, considering the *longue durée* reveals how definitions of art and science changed over time. In the fourteenth century, *natural philosophy* (*philosophia naturalis*) referred to the study of nature within the scholastic tradition of the "liberal arts," distinct from the "mechanical arts," which were equivalent to artisanal crafts. *Science* (*scientia*) referred to any systematic or certain knowledge, natural or otherwise. Through the seventeenth century, natural philosophy became increasingly identifiable as a practical, experimental form of inquiry, and, by the end of the eighteenth century, the English and French "science" approximated the modern notion of the experimental study of nature (the eighteenth-century Russian *nauka*, however, was closer to the German *Wissenschaft*, incorporating natural and human sciences such as law and history). By 1800, the French and English *art* and the Russian *remeslo* continued to mean a craft skill or technique, distinct from the *fine arts* (*beaux arts; izobrazitel'noe iskusstvo*) of painting, sculpture, and architecture.[15] Engagements with fireworks helped constitute these changes, and we shall see how, over several centuries, artificers, architects, men of letters, and natural philosophers constantly debated the identities of these various fields of practice.

In sum, I shall argue that there was no simple relation of science and artisanry in early modern Europe, no once-and-for-all movement of craft techniques into science, and no privileged Scientific Revolution in which all this took place. Rather, pyrotechnic and philosophical techniques arose and circulated between many communities of practitioners who exploited one another's credit in collaboration or in competition to advance the status of their practices. Out of these interactions, whose nature changed continually with local circumstances, emerged numerous innovative skills, theories, and performances. We shall see how fireworks transformed disciplines as varied as meteorology and electrical physics, astronomy and navigation, while techniques drawn from rhetoric, optics, mathematics, and alchemy reconfigured pyrotechnics with spectacular new effects.

Finally, this book is not a history of fireworks per se, nor does it claim to be representative of all the arts in early modern Europe. Approaching the arts and sciences from a geographic perspective prompts the need to map the historical experiences of particular arts and their particular relations to the sciences, which will differ from one another across time and space. The map of interrelations between natural philosophy and, say, the art of turning, pottery, glassmaking, or printing will, thus, be distinct from that of natural philosophy and fireworks. Nevertheless, it is my hope that this study offers some resources for considering other arts from a regional perspective. Similarly, this is only a partial history of fireworks. Fireworks are a truly global phenomenon, used in myriad ways in cultures around the world for a great variety of religious, social, festive, and ritual purposes. That fireworks originated in China is a fact known to all. In the early decades of the twelfth century, long before there were fireworks in Europe, the Chinese fired off *pao-chang* (firecrackers), *ti lao shu* ("earth rats" or rockets), and *yen huo* (fireworks) to celebrate the new year and to ward off malignant spirits. While there will be some discussion of Chinese fireworks in this book, the focus will be on traditions in Europe. The ancient history of pyrotechnics in China is outside my scope here, as are the debts of the earliest fireworks in Europe to other non-European cultures, of which no doubt there were many, but whose nature remains to be explored.[16]

This book begins by explaining the emergence of a particular manifestation and use of fireworks in early modern Europe, and the first four chapters deal with the formation of new venues for pyrotechny and the sciences between the sixteenth century and the early eighteenth. Chapter 1 traces the creation of festive fireworks in Europe between the fourteenth century and the early seventeenth and presents them as a fusion of techniques from artillery and alchemy with the practices of church and court dramas. Considering events in several countries, it argues that such hybrid effects of "artificial fires"—imitating meteors and creating curious motions—emerged as gunners sought to raise the status of artillery and gain honor and credit with princes. In the process, they created new spaces of festive performance in European cities and made new knowledge out of pyrotechny, writing descriptions of their techniques and recipes to entice patrons to their new art. To make these techniques appeal to more noble communities, gunners exploited the credit of the liberal arts, using the idioms of mathematics, Aristotelian physics, and courtly rhetoric to articulate their skills. This knowledge, together with a remarkable repertoire of fiery effects and performance techniques, helped them earn a significant reputation in the early seventeenth century, manifested in the institutionalization of fireworks at courts and arsenals around Europe. This in turn made fireworks attractive to figures seeking reforms of natural philosophy.

Chapter 2 then shows how artificial fireworks proved significant in learned efforts to create new knowledge of nature in the sixteenth and seventeenth centuries. Both gunners and philosophers contributed to this process, synthesizing techniques and knowledge to make fireworks more scholarly and scholarship more pyrotechnic. Fireworks fed into new practices in the sciences. The example of the "fiery dragon" shows how, under the nomination of *experiments*, artificial meteors and fiery motions were adopted by natural magicians and alchemists to enhance their performances of magical effects, to gain patronage at court, and to make spectacular demonstrations of European powers in missionary and colonial contexts. Fireworks also shaped speculative knowledge of nature. Gunner-fireworkers helped establish a popular genre of "mathematical recreations" in the early seventeenth century, in which the contemplation of fireworks might lead to a deeper understanding of nature and God. Mechanical philosophers then used fireworks to model nature's operations, making pyrotechnic accounts of meteors, the heavens, and the human body. The chapter concludes by analyzing utopian schemes for science from the seventeenth century, schemes that often included fireworks and that were often modeled as spaces of performance and spectacle.

The fate of fireworks in institutions arising from such plans is then the subject of chapters 3 and 4. These chapters bring out the regional variation of pyrotechny's relations to the sciences by comparing two institutional settings for science during the late seventeenth century and the early eighteenth. Chapter 3 examines the history of fireworks and pyrotechnic experimentation in London's Royal Society between the restoration of the monarchy in 1660 and the early years of Hanoverian rule. English science at this time is often presented as antitheatrical, avoiding overt spectacle to protect itself from accusations of religious radicalism in unstable times. The Royal Society is better seen as managing spectacle to suit changing circumstances, in dialogue with the changing politics and uses of fireworks by the English court. In times of intense religious paranoia, often linked to fiery passions and incendiary tempers, natural philosophers eschewed spectacle and presented experimental philosophy as a means to manage overheated spirits. But, at other times, and increasingly after the Glorious Revolution, experimenters embraced spectacle and mobilized fireworks in the pursuit of knowledge. Properly circumscribed as commercially useful and theologically instructive, fireworks became creditable elements of London's natural philosophy, prompting much interaction between experimenters and pyrotechnic art, exemplified here in the careers of the Newtonian lecturers J. T. Desaguliers and William Whiston.

As we have seen, the professors of the St. Petersburg Academy of Sciences also used fireworks to promote the sciences, but they did so quite differently than their counterparts in London, fitting local Russian circumstances. Chapter 4 demonstrates how dynastic politics and court patronage, rather than religious turmoil and commercial culture, shaped the uses of fireworks in science in early-eighteenth-century Russia, prompting a relationship between gunner-fireworkers and academicians quite distinct from that in London. With no scientific tradition in Russia, academicians found that experimental lectures failed to interest the Russian nobility, whose support was critical to the survival of the academy. Simultaneously, academicians learned that the design or "invention" of allegorical fireworks could improve their fortunes as spectacles appealing to the Russian court. During the 1730s, professors produced scores of such allegorical fireworks. Becoming a significant component of academic activity, fireworks designs did much to secure the place of the academy in St. Petersburg and of the sciences in Russia. However, to ensure that the gaze of the court was centered on academic designs, academicians worked hard to erase the significance of spectacular performances of fireworks by Russian gunners. While Russia's philosophers thus engaged more directly in producing courtly fireworks than did their contemporaries in London, their interactions with artificers could be more hostile and dismissive. The chapter ends with a brief comparison of fireworks and natural philosophy in Russia and in France under Louis XIV, revealing the distinctive forms that relations of science, art, and pyrotechnics could take in different absolutist cultures.

By the end of the seventeenth century, both fireworks and the sciences were institutionalized across Northern Europe, in arsenals and academies, with communities meeting at the scenes of performance, in city squares and theaters, on rivers and embankments. Chapter 5 turns from the creation of venues for science and fireworks to the circulation of pyrotechnic and philosophical skills and knowledge and explores the ways in which the travels of artificers prompted regionally distinctive responses to their performances. In the mid-eighteenth century, several Italian pyrotechnic masters traveled to Paris, London, and St. Petersburg exhibiting dramatic new fireworks centered on grand temporary edifices called *machines*. This chapter highlights the different ways in which French, English, and Russian audiences responded to the Italians' pyrotechnic innovations. Parisian artificers and men of letters debated fireworks in print, arguing over the nature of pyrotechnic labors and values of openness and secrecy in the art. English and Russian reactions were quite different, and these varied responses led to differing relations of artisans and philosophers, fitting regional traditions. Simultaneously, as the pyrotechnics of the Italians circulated about Europe, they contributed to a homogenization of European courtly fireworks, a development matched by the emergence of an increasingly cosmopolitan "polite society" and public sphere at this time. After the mid-eighteenth century, "polite" and "vulgar" divisions of space would increasingly replace regional and national contrasts in the culture of science and fireworks. There was also a widening division of pyrotechnic labor. As fireworks became fashionable, the artificers who made them were routinely subjected to derision, while architects and men of letters gained status as the designers of displays.

Chapters 6 and 7 then explore new interactions between fireworks and the sciences within this changing geography and hierarchy of pyrotechnic labors. Chapter 6 proposes that the fiery spectacles made fashionable by the Italians in the 1740s and 1750s provided an important context for two of the most significant theoretical and practical enterprises of the middle decades of the eighteenth century: the debate over art and science in Denis Diderot and Jean Le Rond d'Alembert's *Encyclopédie* and the contemporary appearance of a new culture of "public science" centered on experiments with electricity. Literary debates over fireworks in Paris provide a means to situate the opinions of the philosophes Diderot and d'Alembert on the relations of arts and sciences in the *Encylopédie,* often considered seminal in debates on art and science in the Enlightenment. Simultaneously, the fashion for fireworks also contributed to the rise of an increasingly spectacular form of public natural philosophy in the mid-eighteenth century, one whose principal commodity was electricity. The chapter shows how fireworks provided resources both for this increasingly theatrical culture of scientific demonstrations and for the new science of electricity on which it often relied. Pyrotechny offered a language for describing and theorizing the "electric fire" and performance techniques that passed into electrical shows and new theories of medicine and meteorology.

Electricity was itself an unprecedented spectacle, lending much credit to natural philosophers. Chapter 7 then turns to new venues for fireworks in the late eighteenth century and explores the ways in which artificers exploited this growing credit of the experimental sciences to produce a new genre of "philosophical fireworks," designed to appeal to an increasingly "polite" and paying audience. This period witnessed the emergence of a new geography of fireworks and natural philosophy as fireworks entered novel venues catering to polite society such as pleasure gardens, theaters, and *cabinets de physique.* These venues erased some of the regional distinctions of earlier pyrotechnic culture and saw the emergence of a remarkable array of novel effects presented simultaneously as philosophical and pyrotechnic spectacles. Performers, whose identities similarly encompassed skills in natural philosophy and pyrotechnics, appealed to polite audiences by making fireworks "inoffensive," domesticating them into commodities devoid of smoke, smell, and danger, and making them more utilitarian, improving, and educational. The chapter surveys the great variety of philosophical fireworks appearing at this time, using variously optical, hydraulic, electrical, and chemical effects to entice audiences, before focusing on the rivalry of two philosophical fireworkers. Like the gunners of the sixteenth century, these performers mobilized the sciences to reassert the status of the artificer, while their performances epitomized the genre of philosophical fireworks and shaped the form that pyrotechnics have taken ever since.

The conclusion offers a summary of the argument and briefly follows the career of philosophical fireworks into the nineteenth century to indicate the ongoing interactions of science and art these entailed. That these interactions have been erased from the history of art and science then prompts some thoughts about the nature of history in relation to the bustling traffic of skills and practices that this book recounts. I conclude by returning to the geography of art and science and considering its significance for traditional notions of the "Scientific Revolution" and the "Enlightenment."

Readers should note that, in the course of the book, they may wish to refer to maps 1–3 (preceding this introduction), which identify the locations of many of the sites discussed below.

CHAPTER ONE

# "Perfecting the Pyrotecnique story": The Ingenious Invention of Artificial Fireworks

At the end of the fourteenth century, the art of gunnery was often despised. In times of peace, artisans and laborers departed from their regular employment to man the king's guns, and, in times of war, they traveled with the trains of artillery to battle. These were lowly, anonymous craftsmen, notorious for drunkenness, and mistrusted as outsiders because gunnery was not yet part of the regular army. The gunners' status was lower even than that of the common soldier, for the nobility deeply resented the cannon's power to put them at the mercy of the vulgar, and commoners feared the terrible destruction wreaked by newfangled ordnance on their churches and villages. At best, gunners commandeered precious church bells and metal tools for their furnaces to undertake their practice of "fire work" or *pyrotechnia*, the crafting of great ordnance and construction of diabolical weapons and incendiaries.[1]

By the early seventeenth century, the reputation of gunners had improved significantly, or so gunners asserted to themselves and their patrons in images and verse. In 1635, there appeared a fine engraved portrait of the Frankfurt artillerist Christophor Suenck, for twenty-six years the man in charge of exhibiting fireworks for the court of the Danish king Christian IV.[2] The engraving, by Simon de Pas, shows Suenck amid the trappings of his craft, flanked by the paraphernalia of war and festive pyrotechnics above a shield displaying the instruments of geometry (fig. 2). A motto indicates his dual concern, announcing Suenck's skills in "Arte et Marte." In the same year, the likeness of another gunner, the Englishman John Babington, appeared within the frontispiece to his *Pyrotechnia*, the most elaborate of printed fireworks manuals produced in England up to that time.[3] Other portraits followed, indicating a new self-awareness and self-confidence among gunners, who now appeared, not as anonymous artisans, but as

2. Portrait of Christophor Suenck, artillery master to Christian IV of Denmark. Simon de Pas (engraver), *Christophor Suenck Francofurtensis, Armamentarii* (1635). The Royal Library, Copenhagen, Department of Maps, Prints and Photographs.

skilled individuals, praising each others' achievements in laudatory verses, such as those dedicated to the English gunner Francis Malthus in 1629:

> Thy Archimedean hand hath learnt to frame
> Celestiall Meteors out of Nitrous flame:
> And represents strange fires of different sorts
> Suted to Martiall vse, & Courtly sports:
> So pleasing that great Kings haue spar'd some houres
> To be spectators of thy golden showres.
>
> .................................................
>
> And sure thou hast attain'd sufficient glory,
> In perfecting the Pyrotecnique story.[4]

What had happened to improve the gunners' reputation in this period? Certainly, their fame resided in the powerful new weapons gunners offered to the European courts. But between the fourteenth century and the seventeenth gunners also constructed a new genre of spectacle for princely patrons, called *artificial fireworks* or *feux d'artifice—artificial* because they were full of art and *fireworks* because they mobilized gunpowder to produce a series of remarkable fiery effects, motions, and imitations of nature that gunners used to appeal to princes. By the 1630s, "strange fires" for "courtly sports" became one of the principal and most spectacular means of political celebration and theater, a playful but potent rendition of the raw power of the state's warmaking capacity, and a fearful yet pleasurable experience that fascinated early modern audiences.[5] Gunners made wonders, open-air parallels to the exotica of courtly cabinets of curiosities and *Kunstkammern*, exhibiting, according to one gunner, all that was "beautiful and rare in all the universe."[6]

This chapter explores how the new culture of artificial fireworks emerged in Europe and, with it, a new reputation for Europe's gunners. To make their artificial fires, gunners drew on practical and speculative knowledge, the natural philosophy, geometry, and poetics of the liberal arts, and the alchemical skills entailed in manufacturing ordnance. They also fashioned a new form of pyrotechnic knowledge, creating a literature of fireworks treatises whose articulations of craft skills were carefully designed to appeal to noble patrons. By combining pyrotechnic practices with this creditable knowledge, gunners made material the discursive niceties of courtly rhetoric, fashioning a kind of poetry in black powder, an artful craft that would enthrall audiences and secure the gunners a place at court and higher reputations.

## The Spectacular Sites of Early Pyrotechnics

Gunners did not originate the festive use of fireworks. Crackers were known around the globe, and more complex fireworks had been employed in war and celebrations since their invention by the Chinese in the twelfth century.[7] These dual purposes dictated the sites of pyrotechnic use in Europe from the time gunpowder technologies appeared there in the early fourteenth century. European battlefields thus witnessed a great variety of fireballs, fire arrows, and fire-throwing tubes deployed to inundate enemies with horrible incendiary compositions, which from the fourteenth century included gunpowder explosives, rockets, and bombs. Simple fireworks also added drama to mystery plays and holy festivals at Christian religious sites, imitating the sound and appearance of celestial and meteoric scenes that mediated between heaven and earth. It was the gunners' achievement to mix together these pyrotechnics of the church and the battlefield to produce a new genre of fiery effects, suited to a growing courtly demand for spectacle celebrating martial values and princely power. Like other artisans of the fifteenth and sixteenth centuries, gunners presented their skills as critical in the construction of new representations of power. As painters, sculptors, and architects proposed buildings and decorations that would glorify the prince, so gunners offered ephemeral scenes of festive performance in which the same glory might be displayed using fireworks. To achieve this, gunners had to fashion new forms of language and practice that allowed their skills to enter and transform the spaces where the courts resided.

Around 1379, when the first pyrotechnics in both war and festival in Europe are recorded, fireworks took place amid the devastations of battle and inside the sacred spaces of the church. In that year, the Paduans and Venetians used rockets and various gunpowder incendiaries to blast the town of Chioggia. Fortified spaces were thereafter attacked with fiery mixtures of gunpowder, bitumen, and oil, used to tip fire arrows or fill bombs and "fire tubes," which cast out flames from a handheld pole.[8] It is not clear where such fireworks were made, but gunners responsible for ordnance and fireworks probably constructed them in temporary camps and workshops on or near the field of battle. Certainly, images of pyrotechnic art of the sixteenth century usually depicted it as outside work, without specifying any locale in particular.[9]

Against this violent mobilization of pyrotechnics in contested or exterior spaces was a more tranquil, though no less dramatic, location for fireworks: inside, and immediately surrounding, the church. Communal spaces, the churches entertained people with spoken and performative displays of Christian truth, an objective to which fireworks were turned first in the city of Vicenza. Here, before the bishop's palace in 1379, a Pentacostal mystery play was staged in which an artificial dove representing the Holy Spirit flew down a string from the top of a platform, emitting a stream of sparks as it descended. The scene provoked a powerful reaction in the audience: "There was a flash and a loud thunder-clap, and straight away there descended down this rope the image of a shining dove. Almost all fell to the ground in terror and amazement, beseeching God in hymns and chants that the promised Holy Spirit should descend upon them according to the prophecies."[10] Fireworks captured something of the portentousness of heavenly phenomena and directed it to aid in the teaching of Christian morals. The interior of the church was equally amenable to such impressive pyrotechnic displays. Churches had long been locations for Christian dramas, staging divine and diabolical action with scenery depicting stars, clouds, ascending angels, demons, and the fiery mouth of hell.[11] Fireworks added drama to such decorations, transforming the space of the church into a fiery virtual environment: "In the meantime a fire comes from God and with a noise of uninterrupted thunder passes down three ropes towards the middle of the scaffold, where the Prophets were, rising up again in flames and rebounding once more, so that the whole church was filled with sparks."[12] Fireworks thus brought the ominous phenomena of the skies into the microcosm of the church interior.

Alternatively, the urban center was a location of festive pyrotechny, again connected to the religious calendar. In 1540, the Sienese metallurgist Vannoccio Biringuccio recorded in his *Pirotechnia* the colorful festive pyrotechnics of Florence and Siena on the feasts of Saint John and Our Lady of the Assumption. On summer days, great wooden and papier-mâché wheels covered in painted human and animal figures were raised up over the town square on ropes and filled with a variety of simple rockets (*razzi*), squibs (*soffioni*), fire tubes (*trombe*), and crackers (*scioppi*). Linked together by cotton-wick fuses, the fireworks burst from the wheel, or "girandola," as it rotated over the square. Biringuccio praised the fireworks as "ingenious and beautiful."[13]

Historians have traditionally divided the subsequent history of fireworks in Europe between two regional styles distinguished by religion, with

Catholics in Italy and France sending off fireworks from inside artificial castles and temples to create miraculous and unexpected effects and Protestants laying out fireworks on the ground in an antimiraculous effort to avoid hidden artifice.[14] Such distinctions fail because both types of display are evident in Catholic and Protestant countries. Regional variation hinged, rather, on different degrees of proximity to the church or the battlefield. Theatrical performances descending from mystery plays mobilized miraculous effects, while triumphal reenactments staged fireworks to mimic the appearances of the battlefield. Both styles informed fireworks across denominations and national boundaries until gunners gradually merged them to create the distinctive genre of courtly fireworks displays.

From the late fifteenth century, Italian courts began to imitate the mystery plays of the church, staging religious and, later, secular dramas in chapels and then around permanent architecture or ephemeral theaters in the city squares. Displays might be private, for a select courtly audience, or staged in public before crowds of onlookers. The Este family thus patronized mystery plays with fireworks at their court chapel in Ferrara in the 1480s, going on to stage a classical comedy outside in the courtyard of the ducal palace with rockets and a spinning girandola a few years later, a pageant witnessed by some ten thousand people.[15] Gradually, fireworks emerged as an independent spectacle, staged outside before crowds of rich and poor spectators to provide the finale to church festivals and courtly pageants. Fireworks often accompanied tournaments and jousts, where princely pomp and classical iconography supplemented or replaced biblical narratives in performances.[16] Miraculous artificial meteors continued to proliferate in these court displays. Thus, a tournament around the Este castle moat provided the location for spectacular artificial meteors in 1569: "There sprang from the shore, from the margins, the buildings, the mountaintop, blazing fires, and more girandolas began spinning with frequent bangs and roars . . . and the blaze was so terrible because of the endless multicoloured flames flickering through the air, creating strange shapes that the whole sky was alight; and almost at the same moment there came such a terrible earthquake that the earth seemed to open up."[17] Audiences marveled at the variegated effects on display: "It was wonderful to see so many different fireworks."[18] The whole environment appeared transformed by fireworks, everywhere announcing the glory of the prince.

Alternatively, "mock battles" entailed the playful reenactment of warfare. They celebrated martial values and triumph and spanned both Catholic and Protestant nations, taking place in Bologna, Rome, Florence, Ferrara, and many German states during the sixteenth century.[19] Persons acting as soldiers, usually soldiers themselves, restaged engagements with the enemy using benign military pyrotechnics to imitate and celebrate war, running about on a stage or field according to the intended unfolding of the action, and casting out rockets and fireballs. In the German lands, real and artificial castles became a focus for displays.[20] The first recorded German firework thus took place in 1506, gunners honoring the emperor Maximilian with a show of 350 rockets fired from three barrels set on a barge on Lake Konstanz.[21] Then, in 1530, the printers of Nuremberg, the city that manufactured arms for the emperor, recorded a triumph in honor of Charles V's entry into Munich.[22] Another followed five years later and celebrated the capture of Tunis from the Ottoman Turks. Erhard Schön recorded this "Freuden Feuer" or "celebratory fire" in a woodcut depicting a mock battle before Nuremberg's castle in which rockets flew in all directions and the figure of a Turk was thrown from the walls beside a giant.[23] Such playful combat defined imperial fireworks for the remainder of the century, though they increasingly combined mock battles and naumachia, that is, festive fights between ships on rivers or lakes with the theatrical scenery and artificial meteors common in Italian dramas, creating pyrotechnic "pantomimes" staged on the squares and rivers of various German cities.[24] By the seventeenth century, guilds of professional fireworkers in Augsburg, Nuremberg, and Frankfurt staged displays, examining each other in *Probefeuerwerke*, where fireworks were laid out on the ground, perhaps not so much for religious reasons as for the purposes of examination.[25]

The tournament, the mock battle, and the pantomime all constituted novel contexts where fireworks were presented as contributions to princely festival and the representation of power, bringing together the portentousness of religious drama and the pomp of the court and battlefield. Perhaps the most spectacular instance of this synthesis was the Girandola festival in Rome, which became a model for courtly fireworks across Europe from the late sixteenth century. The Girandola festival was staged at the Castel Sant'Angelo on the river Tiber to commemorate the election of a new pope and the feast days of Saints Peter and Paul. Recorded as early as the fifteenth century, the Girandola had by Biringuccio's time evolved into a spectacular display of papal power.[26]

The castle itself was the focus of the display, which began with illuminations of the castle walls, an effect later to become ubiquitous on occasions of celebration in early modern Europe. Biringuccio explained how lanterns of paper and tallow candles were set in "many rows as far as the eye can reach," creating "a very beautiful thing to see."[27] Then came cannon salutes and the lighting of burning tar barrels positioned around the castle walls.

The centerpiece was a vast explosion of thousands of rockets above the castle tower rising up into the sky, their spreading "wheatsheaf" shape giving the spectacle its title by analogy to "girandola" fountain sprays (not to be confused with the Florentine and Sienese *girandole*, or "wheels"). To judge from a print of 1579 by the Roman engraver Giovanni Brambilla, crowds of mostly male spectators thronged about burning tar barrels on the shores of the Tiber to watch the display (fig. 3). The spectacle was immersive, the spectators' entire environment being eclipsed by fireworks, transformed momentarily into an apocalyptic scene. Biringuccio likened the explosion of rockets over their heads to the "fire imagined in Hell," a familiar image in smaller religious dramas.[28] Brambilla alternatively identified the firework with a meteoric tempest, describing how rockets "burst in the air in the shapes of stars." It "[seems] as if the sky has opened . . . it seems as if all the air in the world is filled with fireworks, and all the stars in the heavens are falling to earth—a thing truly stupendous and marvelous to behold."[29] Artifice here supported the symbolic and ritual significance of the event, allying the micro- and macrocosms to figure the power of the prince with an artificial version of a portentous event. The dramatic image of the fall of the firmament evoked Gospel descriptions of stars falling from heaven presaging the return of Christ after the apocalypse.[30] In Christomimetic festival, this was an appropriate allusion for the coming of a new pope, manifesting the restoration of order after chaos.

The Girandola was probably staged by gunners and certainly brought together the space of warfare, a heavily fortified castle, with the space of religious drama.[31] It linked military architecture and pyrotechnics with heavenly and meteoric imitations of biblical portentousness, producing what Brambilla and Biringuccio evidently considered an immensely powerful spectacle. Indeed, in his *Pirotechnia*, Biringuccio went on to equate passion itself with "a fire that devours and burns."[32] The grand effect of the Girandola was quickly imitated elsewhere, with the result that, with the rise of the courts, gunners' fireworks proliferated across Europe.

In France, for example, gunners transformed the center of Paris and the river Seine with spectacular artificial fires reminiscent of those in Rome. The king's firework masters, known as the *ingenieurs*, later *artificiers du roi*, assisted by the *gardes du roi*, designed and executed displays regularly from the sixteenth century.[33] Between 1613 and the late 1620s, Horace Morel, the artillery's *commissaire ordinaire*, and several other gunners staged fireworks for the court theater and ballet and for celebrations of royal births, marriages, and triumphs.[34] Their pyrotechnics echoed the celestial spectacle of the papal Girandola, displaying stars in the sky as astrological in addition

3. Giovanni Ambrogio Brambilla (draftsman), *Castello S. Angelo con la girandola* (Rome, 1579). Research Library, the Getty Research Institute, Los Angeles, California.

to apocalyptic portents. For the entry of Louis XIII and Anne of Austria into Lyon in 1623, Morel presented a display centered on a huge artificial lion, fire bursting from its jaws amid zodiacal scenery. The object was to show "by the sun at the sign of its celestial lion our king, who, traveling through the cities of his realm like the king of the planets through the signs of the zodiac, has at last arrived at the one that . . . justly deserves to be called on

4. Perseus rescues Andromeda from the sea monster in fireworks on the Seine, 1629, designed by Morel. *Projet pour le feu d'artifice* (Paris, ca. 1629). Bibliothèque nationale de France.

earth the sign of the lion." Artificial meteors were salient: "The air was filled with rockets, some discharging stars to represent the fall of the firmament."[35] The *artificiers'* spectacles turned the river and the surrounding buildings into temporary scenes of classical allegory, princely splendor, and martial pomp. Late in 1628, Morel allegorized the king's recent military triumphs over Protestant enemies at the siege of La Rochelle through the story of Perseus saving Andromeda from the sea monster, which he displayed using grand pyrotechnic automata moving on ropes across the river (fig. 4).[36] At these and later fireworks, audiences were divided between the "people" (*les gens*), crowding the shores of the Seine to watch the fireworks ignited between the Pont Neuf and the Tour de Nesle, and the king and the court (*la cour*), who watched from the balconies and windows of the Louvre.[37]

Travel was an important means by which these new fireworks proliferated, as gunners and artificers took their skills abroad with the court. The island nation of England learned of fireworks in this way, when gunners at the Tower of London were joined by Dutch and Danish artificers during the reign of James I after Suenck's patron King Christian IV visited in 1606, presenting James with fireworks of his own invention.[38] Christian's fireworks recalled the papal Girandola and Louis XIII's zodiacal scenery, showing Leo the lion, "true nursling of the supreme Sun," and other images "burning in

Etnae's flame resembling hell's endless torments."[39] English spectators admired the orderliness of the display: "This firework very methodically, one part after another, continued burning and cracking for the space of three quarters of an hower."[40] These were not the first fireworks in England, but, soon after, in 1613, English gunners presented London's first grand firework on the Thames, "in the sight of many thousands of people," for the wedding of Princess Elizabeth and the Elector Palatine. Five of the king's gunners, under the master gunner of England, William Hammond, combined the common elements of continental displays, producing a mock battle of "friendly opposition and imaginary hurley-burley battalions" and artificial fireworks resembling meteors whose "blowes . . . fell like Thunder-clappes, enforcing lightning."[41]

## Fireworks and Alchemy

Between the fifteenth century and the seventeenth, gunners thus succeeded in transforming a variety of sacred, urban, and courtly spaces into temporary stages for a great variety of dramatic fiery effects. These effects combined the portentousness of artificial meteors and celestial apparitions with the martial pomp and classical iconography of courtly spectacle, a synthesis of "Arte et Marte" that helped dramatize the power of princes. To appreciate the nature of these new artificial environments created by gunners, it is worth examining in more detail their technical accomplishments in pyrotechny during the sixteenth century. Not only did gunners bring together the explosive power of military weapons and the scenic imitations of the heavens and meteors in their dramas, but they did so with an arsenal of novel effects in fire that hybridized the two domains. They added further ingredients to these dramas by mobilizing the commonly utilized techniques of practical alchemy, the idiom through which gunners understood their work. In combination, their alchemically inspired fireworks established the vocabulary of pyrotechnic effects that marked early European displays.

When spectators of the Roman Girandola claimed that the stars were falling from the heavens, this was not simply a metaphor. The "falling stars" of sixteenth-century fireworks were designed and constructed precisely to imitate this phenomenon. Recipes for "stars" were recorded from the late sixteenth century, the stars being fired from rockets that "cast forth flames of fire and in coming downe towards the ground will shew like starres falling from heauen. . . . The stuffe which is to be put into the Rocket for to flame and giue crackes is made of twelve partes of Saltpeeter refined, of Citrine Brimstone nine partes, of grosse gunpowder five partes and ¼ of a part

5. Artificial stars (*estoilles*) fired from a rocket, from François Thybourel and Jean Appier dit Hanzelet, *Recueil de plusieurs machines militaires, et feux artificiels pour la guerre, & recreation* (Pont-a-Mousson, 1620), bk. 5, p. 11. Bibliothèque nationale de France.

mingled togeather with your hand."[42] The mixture was to be packed into a ball and wound tightly round with packthread, given a fuse, and placed in the head of a rocket (fig. 5). When the rocket exploded, the stars would stream down in the air. Recipes for stars proliferated in books on fireworks thereafter, and in later recipes gunners moistened the mixture slightly before it was fashioned into nut-sized balls and rolled in dry powder.[43] This was the first kind of firework invented exclusively for festive use, prior displays using simple squibs (paper rolls rammed with gunpowder) or rockets and other pyrotechnics that were adaptations of military fireworks. Stars captured in artifice the capacity of the heavens to display supernatural powers in the apocalyptic fall of the firmament and the stellar influence of astrology. Gunners saw their work as directly connected to such powers, one of them writing an astrological calendar of propitious times to carry out pyrotechnic labors.[44]

However, stars were by no means the only significant bodies imitated in pyrotechnics. Gunners made many other "artificial" imitations of meteoric and celestial appearances in the late sixteenth century and the early seventeenth as courtly fireworks grew increasingly elaborate. Contemporary treatises on meteors listed many forms of weather and fiery exhalations in the higher regions of the atmosphere that were imitated by gunners. Most of the fiery exhalations described in William Fulke's *Meteors*, published at London in 1563, thus found artificial equivalents in pyrotechny, including "torches," "shooting stars," "burning candles," "burning spears," "shields, globes or bowles," "lamps," "flying dragons," "the pyramidal pillar," "fire scattered in the aire," and "comets, or blazing stars."[45] According to the gunner Malthus, if powder-filled goose quills were used instead of stars in rockets, a "glorious and pleasing raine" would result.[46] Alternatively, one could make "streames of Gold much like the falling downe of Snow being agitated by some turbulent winde."[47] In 1630, the Lorraine engraver and *maitre des feux artificiels de Son Altesse* Jean Appier, known as Hanzelet, published *La pyrotechnie*, which explained how to make "a sun by means of two whale ribs laid in a cross, tying stars to the four ends with brass wire."[48] Attaching rockets to the ribs made them spin to produce a swirling circle of light.

Engravings of fireworks at this time depicted artificial clouds, rocks, celestial bodies, volcanoes, and lightning bolts traditionally figured in church and court theater. In *La pyrotechnie*, Hanzelet gathered together many of these effects in an engraving depicting a great battle on board a ship, reminiscent of the mock battles of the Holy Roman emperors (fig. 6). Now, however, the live soldiers were replaced by more theatrical artificial automata of skirmishing warriors, filled with fireworks that exploded from pores in their wooden and papier-mâché bodies. All around them stars, rockets, small "fiery dragons" (also meteoric imitations), and figures of the sun and the moon exploded in a blaze that filled up the scene with fire.

Such diverse pyrotechnic effects reflected, not only the heavenly and meteoric iconography of church and princely power, but also the idioms through which gunners comprehended their work. That is, in creating new spectacular environments for the court, gunners presented effects that did not have precedents in earlier dramas and that literally mobilized the views they entertained of their materials. Pamela Smith has argued that alchemy provided an important idiom through which painters and sculptors articulated their practices to one another in early modern Europe.[49] Artisans approached their materials as animated matter that they needed to negotiate into forms following or contradicting that matter's peculiar characteristics. Art was a process of imitating nature, not only by representing nature's forms, as in the case of artificial meteors, but also by imitating natural *processes* of generation, works of art emerging from the synthesis of the artisan's body and materials in much the same way that living creatures or minerals were generated out of the elements in nature. Smith and others have also

6. Artificial warriors battle on a fireship. Engraving from Jean Appier, dit Hanzelet, *La pyrotechnie* (Pont-a-Mousson, 1630), 253. Courtesy of the Hagley Museum and Library, Wilmington, Delaware.

noted how artisans distinguished this "low" alchemy, equivalent to a practical chemistry, from a high or "sophistical" alchemy of transmutation in this period, often praising the former and condemning the latter. The work of Biringuccio is often considered exemplary of this division.[50] However, while gunners shared the alchemical idiom with other artisans, they used it, not only to articulate their practices in manuscripts and printed treatises, but also to produce new fireworks that played on alchemical and generative principles in their forms and motions. Moreover, they appealed to both high and low alchemy in fireworks, which, like transmutations, were performances designed to display mystical powers appealing to princes.

Pyrotechny was a complex business, and gunners often kept manuscripts detailing recipes and procedures of manufacture. These reveal the alchemical idiom of early pyrotechny. The earliest manuscript work on pyrotechny was the German *Feuerwerkbuch*, which appeared anonymously ca. 1420 for an audience of "master gunners."[51] The *Feuerwerkbuch* exhorted its readers to literacy: "The master must . . . be able to read and write because in no other way can he keep all the required knowledge in mind, which is necessary to exercise his art."[52] Purporting to serve as such an aide-mémoire, the manuscript, copied many times and elaborated over the next century, gave the basic recipes of what was then still a largely military pyrotechny. Directions appeared for making gunpowder and its ingredients, saltpeter, sulfur, and charcoal, followed by instructions for casting and shooting guns and then for making various incendiary weapons. Indicative of the rising self-confidence of gunners, many manuscripts following the *Feuerwerkbuch* were signed by their authors. Such was the *Buch von den probierten Künsten*, prepared by the "shooter, cannonier, and fireworker" Franz Helm of Cologne around 1530. Helm, who had fought the Turks with the emperor's artillery, included in his *Feuerwerkbuch* recipes of "Feuerwerk zu Ernst und Schimpfe" (pyrotechnics for war and peace), among which were fire arrows, fireballs, incendiaries, bombs, fire barrels, fire pots, and sulfurous candles. These were military recipes adaptable for use in peaceful display, and Helm gave no recipes for exclusively peaceful fireworks, though he did depict a simple platform for setting them off.[53]

Alchemical language was salient in these manuscripts. Hence, the master gunner, claimed the author of the *Feuerwerkbuch*, should be adept in separation, sublimation, and confortation, all standard alchemical operations.[54] Gunpowder was understood to operate by the mixing of contraries, following the translated works of the Arab alchemist Geber (Jabir ibn Hayyan), who himself distinguished low and high alchemy as different levels of the same art. Saltpeter was a cold and dry principle, like mercury, and, once it was brought into contact and ignited with the hot and dry principle of sulfur, the antagonism of the two materials would lead to an explosion.[55] The living qualities of one principle could alter the behavior of another. Hence, mercury was "killed" (*getötet*) by "living" sulfur (*lebendiger Schwefel*) to improve the force of ignition, the resulting fires being "wild or tame" (*wilde oder zahme Feuerwerke*), depending on the ferocity of their antagonism.[56] Mercury, the volatile alchemical principle opposed to sulfur, was frequently included in the earliest artillery recipes, as when a drop of mercury added to the touchhole of a cannon rendered its blast much louder.[57]

The higher form of transmutational alchemy was also invoked in the *Feuerwerkbuch*, in the image of Black Berthold, the legendary discoverer of gunpowder in Europe. In the seventeenth century, Berthold was represented as a monk, given the dreadful recipe for black powder by the devil. But, in earlier versions of the *Feuerwerkbuch*, he appeared as an alchemist, flanked by alembics and furnace beneath the sun and moon (fig. 7). Historians now reject the once common myth that Berthold discovered gunpowder or invented the cannon but recognize that accounts of Berthold, which may have been inserted into some manuscripts by scribes, provided a foundation story for gunnery that in some cases identified the creation of ordnance with alchemy. Thus the *Feuerwerkbuch* said of Berthold:

> He worked with the great alchemy [*mit grosser Alchemie*] like those masters who are engaged with precious and valuable things, with silver and gold, and with similar metals. These masters can separate silver and gold from other precious jewelry, and from other valuable colours which they can produce. Now this master Berthold wanted to induce a golden coloration. For this he used saltpetre, sulphur, lead and oil. Then he put these ingredients in a container made of copper, which he sealed completely, exactly as it should be done, but when he put it on the fire and the container became hot, it burst into many pieces.[58]

Hence, the character of Berthold took on a persona, likely appealing to both gunners and their noble patrons, as a master "engaged with precious and valuable things," rather than the despised engines of war. With their promises of gold, alchemists had enduring appeal to the nobility, so allusions to the high art of alchemy might ennoble gunnery and provide the gunner access to the court.[59]

Alchemy continued to shape the practice and literature of pyrotechny after the *Feuerwerkbuch*. Indeed, the term *pyrotechnia* referred to any work done by fire and so could be a synonym for transmutational alchemy in

Das · iij · Capitell

Hye wyl ich sagen wye auff-
komen ist de büxsen de konst-
hait fonden eyn meister gnte
nyger bertoldus. vnd ist gewest eyn ny-
gramaticus der myt groisser alchamey
vmb gangen hait vnd als dye selben
meyster myt groszen kostlichen vnd hoff-
lichen sachen vmb gaint myt silber
golt vnd alle · vij · metalen vnd die selbe
meister gewonlich sylber vnd golt
scheiden kommen vnd das in kostliche
farben brengthen Also wolt der selbe
meister eyn golt fuer brennen zu de
selben dan gehort salpeter swebel
bly vnd olye vnd do er disse stuck in eyn
kufferen haben wail vermacht bracht
als man thun muss vnd vber das
fuer thet als balde das warm wardt
so brach der haben zu vil kleynem

7. The earliest image of Black Berthold shows him as an alchemist. From the late-fifteenth-century manuscript *Feuerwerksbuch*. Manuscripts and Incunabula Collection, Stadt- und Universitätsbibliothek, Frankfurt am Main, Germany.

addition to artificial fireworks.[60] Similarly, both gunners and alchemists claimed the term *laboratory* for the site of pyrotechnic labors in the early modern period, and into the nineteenth century workshops for making fireworks were often called *laboratories*.[61] Gunners and artisans then exploited alchemical secrets and used alchemical notions to describe their labors. Biringuccio told how he bought the secret of a fire that burned when wetted from an alchemist, and Samuel Zimmermann of Augsburg, a self-described "Feuerwerks-Künstler" and the author of a 1573 manuscript treatise on fireworks, claimed to have owned marvelous recipes for fireballs of mercury, sulfur, and saltpeter that "resolve themselves into smoke." These he claimed to have been given in dreams or by the "Alchemists," and he recalled: "I myself worked in Alchemy."[62]

## Artifice and the Imitation of Nature

An alchemical conception of nature was also critical to many of the new fiery effects produced by gunners. While gunners followed traditions of the church in imitating the appearances of celestial and meteoric phenomena with fireworks, they also took from alchemy a notion of imitating nature's *processes*. As Smith argues, artisans replicated both the external appearances of nature and nature's underlying processes of generation and animation, the act of producing art a human reflection of nature's own creative powers.[63] Gunners followed this reasoning. As Malthus explained, it was the task of the gunner making fireworks "to imitate nature . . . who far from error seemes to haue produced and brought forth all things by a curious and speciall order; without whose beautifull disposition the whole world had still continued in the most prodigious confusion of Caos, which displeased the highest, eternall and divine powers, being but a cloud or mixture of darknesse."[64]

Gunners viewed their practices in such generative terms. The *Feuerwerkbuch* conceived materials through their visible behaviors as "living," and, in 1620, Hanzelet, writing with the surgeon François Thybourel, reminded readers that pyrotechny and life were intimately linked because fire was "necessary to human life" as "the cause of all generation."[65] He continued: "Everything is engendered and grows by heat and humidity; which is the Symbol of generation."[66] The English gunner Robert Norton repeated this claim a decade later, explaining: "Gunpowder with the soule Petre, and the life Sulpher, and the body thereof Coale, be indeed the chiefe bases and foundations . . . of Fireworkes."[67] Writing in the 1640s, the Lithuanian-born gunner in Polish service Kazimierz Siemienowicz explained how "it is observed, that Fire is so fond of Sulphur, and that reciprocally, Sulphur takes such Pleasure in being devoured by that Element, that if any Bits of it happen to lye about any Wood, so as they can feel the Heat of it, they seem to call it to them, and really attract it sometimes."[68]

Gunners then made reference to life and generation in pyrotechnic recipes and displays. Complementing artificial meteors were fireworks that evoked generation, moved in erratic ways, or resembled animated beings. Spinning wooden wheels were one of the first of these kinds of fireworks, flinging out fireballs and rockets as they whirled around over a city square, prompting Biringuccio to exclaim: "Truly it was an ingenious and beautiful thing to see it produce so many effects in fire, just as living things do by themselves."[69] Alternatively, "serpents" were packed like stars into the head of a rocket, and, when the rocket exploded, they streamed into the air in erratic trajectories, reminiscent of the movements of a snake. Made to "the bigness of one's little finger" from paper rolled several times around a quarter-inch mold, the serpent was filled half with star composition and half with gunpowder. By controlling the size of the vent at either end of the case through different degrees of choking (when a wire is wound around the case and pulled tight to close it), the serpent could be made to "wamble in the ayre."[70] Serpents were also noisy. A choked section containing stronger powder gave the serpent a loud bang after it had "plaied a while."[71]

The name *serpent* was also significant since serpents were among the creatures commonly associated with nature's alchemical generation of life in the sixteenth century, along with insects, frogs, lizards, and other creatures often featured in works of art as symbols of generation.[72] Many of these creatures appeared in fireworks displays as automata, "self-moving" mechanical figures traditionally associated with the imitation of life.[73] At the fireworks on the Thames in 1613, for example, the gunner John Tindall showed the image of a rock covered with "Adders, Snakes, Toades, Serpents, Scorpions, and such venemous vermin, from whose throates were belched many fires."[74] Also common was the salamander, a symbol of regeneration in alchemy and courtly emblem books, which figured in fireworks displays as a creature immune to fire and so emblematic of constancy.[75] There were also various human figures—of battling soldiers, monks and nuns, deities, and princes—all animated by internal pyrotechnics (plate 1).[76]

The most important of these animated motions was the rocket, whose trajectories gunners varied in remarkable ways. The technical vocabulary of festive rocketry belies its fascination with the animate. Rockets move by a powerful thrust generated inside a hollow cavity in the rocket's body that gunners called its "soul," claiming rockets were "the soul of artifice."[77] Gunners took great pride in their skill at making rockets, a procedure entailing

8. Various rocket trajectories: from left to right, a "fountain" made with "little rockets upon one great rocket"; a "shower of fierie rayne" made with powder-filled quills; stars; and serpents. From Francis Malthus, *A Treatise of Artificial Fire-Works Both for Warres and Recreation* (London, 1629), 98. Beinecke Rare Book and Manuscript Library, Yale University.

the precise ramming of powder into paper cases, which were attached to sticks giving the rocket an exact poise and balance. Motions, height of ascent, and the timing of explosions could all be carefully managed by adjusting the composition, its density in the rocket's case or "coffin," the size of the cavity, and the length of the fuse leading into the cap (fig. 8). A good rocket exploding at the apex of its trajectory then released "garnitures" of stars, serpents, and *saucissons*, that is, squibs wound tightly with cord to give a noisy report.[78]

Complementary to this animistic way of presenting fireworks, gunners organized displays around the notion, common to their understanding of gunpowder, of bringing together contrary qualities to create dramatic and explosive reactions. This expression of the artisan's way of thinking became an increasingly salient element of pyrotechnic performance in the sixteenth century. Pyrotechny was an elemental art, and gunners played on their power to force fire into confrontation with the other elements. Performances took place on lakes and rivers, close to fire's contrary element. Particular pieces then raised admiration by seeming to act in ways contrary to their nature. Some of the most enduring were "aquatic fireworks"—water balls and water rockets, which ducked and dived below the surface of a river before exploding or shooting into the air. Melted pitch, rosin, grease, or paint prevented water entering the cases, while long wooden sticks kept rockets afloat. Exotic compositions of "wildfire" made with brimstone, frankincense, linseed oil, petroleum, and quicklime also burned in the water.[79] Alternating the composition with half spoonfuls of powder dust would make balls and rockets swim "now upon the water, and now under the water," while two or three fingers' breadth of fine powder dust would make the rocket "flye farre off."[80]

These were some of the most admired inventions of early European pyrotechny and hinged on the notion of contraries. As Malthus explained: "Take fire & water being two elements of contrary qualities, the one to the other cause the rockets which worke their effects in or vpon the water to appeare to the spectators more beautifull, and seem more rare and admirable."[81] Audiences comprehended the artifice with amazement—and on similar terms. As one spectator recorded in 1575: "So waz thear abrode at night very straunge & sundry kindez of fyer wooeks, compeld by cunnyng too fly too & fro and too mount very hy intoo the ayr upward, and allso too burn vnquenchashabl in the water beneath: contrary ye wot, too fyerz kinde."[82] Such observations crystallized in a common description of fireworks in the early seventeenth century as a "battle of the elements." Now military combat was integrated with courtly cosmology to create spectacles in which the elements warred and competed to create a confusion in the skies. Thus one author described fireworks on the Thames in 1613, his account recalling the Roman Girandola: "A Racket of Fire burst from the water, and mounted so high into the Element that it dazeled the beholders eyes to looke after it. Secondly, followed a number more of the same fashion, spredding so strangely with sparkling blazes, that the skie seemed to be filled with fire, or that there had beene a combate of darting starres fighting in the ayre."[83]

Another display showing the battle of the elements was staged on the Piazza Spada in Rome at the election of Ferdinand III in 1637. Designed by the Sienese painter Niccolò Tornioli, and executed by the engineer Giovanni Andrea Ghiberto, the performance presented an artificial mountain resembling Mount Etna and showed the sun and moon, fiery stars, and dragons raging, the scenery undergoing several transformations to represent the victories of the empire over its Protestant enemies and the Ottoman Turks. The Bolognese humanist and priest Luigi Manzini adapted this image to signify the power of the prince, saying of the fireworks: "Others claimed to see the renewal of the world predicted under his happy reign, the sign of this being the competition of all the elements together: fire lit from powder (which is also earth) in order to soar skyward, while water, raining down, filled the air."[84]

The artificial meteors and celestial appearances of church and court drama and artisanal alchemy were, thus, merged together. In transforming the spaces of the city into festive scenes, gunners made spectacular demonstrations of their power over the elements, effecting transmutations, and forcing together contraries whose drama was then directed to signal the glory of the court or the prince. In the open air, gunners exhibited wonders and curiosities. Fireworks were like other marvelous artifices in the court, comparable to the exotic *artificialia* of cabinets of curiosity or the artificial waterworks that graced princely gardens at this time.[85] But they were also unique, balancing the risks and dangers of playing with fire and human art's Promethean power to bring the unruly element under control. Risk was a necessary part of fireworks' appeal for audiences, and, in a court society where advancement depended on taking risks to secure honor and status, the possibility of failure underwrote the credit to be gained by a successful performance.[86] Sixteenth-century evidence is scarce, but later gunners worked hard to minimize errors, aware, as Siemienowicz observed, that pyrotechny was a "Ticklish art," where "you may readily imagine, That you will stand in the greatest need of the Celestial Protection, when encompassed on all Sides with extraordinary Dangers."[87] Fireworks could, indeed, be unpredictable. Accidents were common, though not always to the detriment of gunners. To accept a challenge and fail at court was better than to decline a challenge altogether, and a gunner wounded by stray pyrotechnics might be rewarded for bravery. If fireworks struck "vulgar" or common spectators, the scene might be considered amusing, but, if a noble was struck or insulted by fireworks, reprimands for the artificers soon followed.[88]

The status of fireworks at court was also problematic. The flourishing of spectacular fires enamored princes of gunners' pyrotechnic abilities, but these fires needed to be presented in forms suitable to the court to succeed. The language of practical alchemy was inappropriate in this regard. While it might produce magnificent effects, those effects could not be represented as the mere products of lowly artisanry. Instead, gunners looked to the liberal arts for a language in which to express their skills and order their performances, deploying such liberal language with increasing vigor during the sixteenth century.

## Pyrotechny and the Liberal Arts

Early modern Europeans distinguished the mechanical from the liberal arts. Liberal arts were the disciplines taught in the university trivium and quadrivium and consisted of arithmetic, geometry, astronomy, grammar, rhetoric, music, and logic. With the rise of humanism and efforts to make learning an element of courtly virtue, the liberal arts gained substantial credit at court, prompting practitioners of the more lowly mechanical arts to appeal to their status. Mechanical arts, of which pyrotechny was one, entailed physical and remunerated labor, which is why they were not "free" or liberal and did violence to nature by forcing "her" out of her ordinary ways through art, as in the contrary motions of fireworks. The liberal arts, in contrast, included the study of the ordinary course of nature through the writings of the ancients and, above all, Aristotle.

The language of humanism provided an alternative idiom through which to express and shape artisanal practice in ways that might appeal to the courts. In the Renaissance, painters, sculptors, architects, and other artisans thus looked to classical precedents in rhetoric and poetry promoted by humanist scholars as a means to reinterpret art. They gave new emphasis to a personal inspiration or ingenuity (*ingegno*) in their work, identified as a creative "fantasy" or "imagination" that individualized artistic styles away from the stylized forms of the Middle Ages. A naturalistic imitation of nature (*imitatio*) and the invention (*inventio*) of novel forms based on the underlying harmonies of nature or subject matter drawn from the classics constituted the ideals of humanist aesthetics.[89] The liberal art of Euclidean geometry revealed these harmonies, which could be incorporated in painting as perspective and composition and in architecture through the orders and proper proportions. Humanism produced an emblematic culture, in which individuals might offer princes diverse visual, spoken, or written images and forms whose contemplation gave pleasure to the prince while expressing the glory of his rule. The production of suitably "ingenious" and varied forms might earn their "inventors" praise and honor according to the degree of wit and subtlety with which they were achieved.[90]

Gunners and the practitioners of other military arts also sought to raise the status of their labors during this period. As one author put it, in appropriately poetic form:

As Grammar, Musicke, and Physicke,
With high Astronomie:
And other Ars Mathematicke,
And braue Geometrie.
This Art of Gunnerie likewise,
Amongst the rest let stand,
Whose god-father this Author is,
Which tooke the same in hand.[91]

Early humanist artisans included military machines among the ingenious inventions they sought to present to princes, alongside new forms of painting and architecture. In fifteenth-century Padua, the physician Giovanni da Fontana included a number of pyrotechnic recipes in his *Bellicorum liber instrumentorum*, while Leon Battista Alberti's *Mathematical Games* described techniques of gunnery.[92] Leonardo da Vinci was the quintessential Renaissance "inventor," his varied proposals including many infernal machines, among them finned shells filled with gunpowder and a rocket fired to a great height by chaining it to a cannonball.[93] Leonardo celebrated his *ingegno* by adopting the appellation of *ingenarius* at the courts of the Sforza and Francis I. He also vigorously promoted a more liberal status for painting and, with other *ingenarii*, created spectacular theatrical machines and illusionistic scenery for the church and the courts. Alternatively, Mary Henninger-Voss and Serafina Cuomo have drawn attention to the work of Niccolò Tartaglia in raising the status of gunnery. Tartaglia utilized the idioms of Aristotelian natural philosophy and Euclidean geometry to model the art of shooting in a manner presentable to noble audiences, for whom such references to the liberal arts made an otherwise lowly and unfamiliar craft comprehensible and appealing. In mathematizing shooting, he also laid the foundations for a mathematico-physical science of ballistics since his work provided the basis for Galileo's geometrical accounts of projectile motion in *Two New Sciences*.[94]

Pyrotechny was equally represented as noble during the Renaissance using the forms of the liberal arts. This is evident above all in a growing literature on fireworks that emerged coextensively with the rise of fireworks performances and that makes apparent how gunners and interested scholars worked to represent the art of pyrotechny as one suited to courtly consumption. Books, like fireworks themselves, offered a means for gunners to cross the boundary from the workshop to the court, to gain access to the prince and entertain a noble audience. The literature of pyrotechny was, then, as important as performances in the rise of the gunner, and examining it offers a window onto the aspirations of gunners to ennoble their art. Whereas gunners used a practical and alchemical idiom when addressing one another in manuscripts, works addressed to noble patrons tended, rather, to exploit the language of the liberal arts. Authors presented their work by referring to Euclidean geometry and scholastic natural philosophy and particularly stressed the importance of ingenuity and invention in pyrotechny. As in other artisanal treatises of this period, such appeals were intended both to make artisanal labors transparent and amenable to noble audiences and to prompt other gunners to adopt knowledge of the liberal arts.[95] These efforts also underlay the proliferation of new effects in pyrotechnic performances, effects framed in classical allegory and promoting princely splendor. When gunners imitated and invented new fireworks, they did so within a context that gave much credit to these skills. The proliferation of new pyrotechnics was represented as a form of ingenious invention, while the arts of gunnery and fireworks were exhibited to princes not as a lowly form of alchemy but as a noble discourse drawing on the liberal arts. It is this courtly contextualization of the new pyrotechnics, as much as the physical contexts created using pyrotechnics themselves, that secured a glowing reputation for gunners by the beginning of the seventeenth century.

A number of gunners articulated novel fireworks practices in writing, in manuscripts and printed works intended to aid practitioners in the making of fireworks and to represent those practices to noble audiences. Printed treatises on fireworks began to appear in the sixteenth century and proliferated thereafter. Most were dedicated to a noble patron or a prince, though they might also be written for other "curious and ingenious Artists," as Malthus put it.[96] Following the brief discussion of fireworks in Biringuccio's *Pirotechnia*, the first printed book devoted exclusively to fireworks for peace was the 1560 *Künstliche und rechtschaffene Feuerwerck zum schimpff* of the Nuremberg artificer Johann Schmidlap, which listed recipes for adapted military rockets, fire trunks, and wheels.[97] Schmidlap dedicated his book to Wilhelm von Januwitz, master of ordnance to the Duke of Wittenberg. Some two dozen more works describing fireworks followed in the next century, culminating in the extensive treatise of Kazimierz Siemienowicz, whose *Artis magnae artilleriae* was published at Amsterdam in 1650. Siemienowicz's work was dedicated to Leopold Wilhelm, archduke of Austria, and translated into several languages, becoming a standard reference on making fireworks for many decades thereafter.[98]

Most treatises dealt with fireworks for both war and peace, often after discussing the techniques of shooting with cannon.[99] Not all pyrotechnic treatises were authored by gunners, and some gunners authored works on subjects other than gunnery. Leonard Fronsperger, who served in Charles V's artillery train and directed Munich's arsenal, or *Zeughaus*, in the 1560s, published books on shooting and fireworks but also wrote on wine and architecture and produced a tract "in praise of self-interest."[100] Many printed books on gunnery borrowed recipes from previous works, and gunners sought to accumulate credit by accusing other authors of plagiarism in order to highlight the originality of their own work.[101] Thus, Francis Malthus lambasted Robert Norton's 1628 *The Gunner* as nothing more than a translation of the Spanish captain Diego Ufano's *Tratado dela artilleria* of 1612.[102] Malthus noted that, while his own work set down the "true rules and mixtures" of fireworks, authors like Norton were "calumniators" who "write with serpents tongues spitting their venome upon the silent and sleeping innocents gone before us."[103]

To further bolster the credit of their works, authors turned to the liberal arts to shape their discourse. In discussions of both gunnery and fireworks, they exploited the language of geometry, natural philosophy, and the notion of ingenious invention to express and lend credit to their labors. There were differences in emphasis, however. In discussions of gunnery, geometry served as a means to model the motion of cannonballs. But, in the case of fireworks, whose motions were more complex and variable, geometry served, rather, as a means to frame and order accounts of pyrotechnic skills and recipes. Thus, in 1611, the Louvain mathematician Adriaen van Roomen, or Romano, published a small Latin treatise on fireworks that set out techniques for making different fireworks in a neat Euclidean order, presenting recipes under the headings of *definitions*, *propositions*, and *postulates*. Romano also offered the first classification of fireworks, which worked according to the Aristotelian elements, dividing fireworks into those which operated in the air, in the water, or on the ground. Subdivisions were based on whether they moved and the complexity of their construction.[104] Subsequent pyrotechnic treatises, both printed and manuscript, elaborated on these classifying and organizational conventions, often beginning with accounts of tools and ingredients, starting with gunpowder, before describing additives and recipes for fireworks of the air, earth, and water.

Gunners also used geometry to give proportions to the tools and cases or cartridges used in making fireworks. Siemienowicz was an especially enthusiastic proponent of such measures and applied the caliber scale, used to proportion different sizes of cannonball, to propose different standard sizes for rockets. As we shall see in chapter 6, this was a controversial technique rejected by some fireworks masters, and the degree to which gunners actually learned and used geometry in the everyday work of pyrotechny is unclear, though it certainly became part of a gunner's training, used in shooting and surveying. Gunners were at least enthusiastic to represent pyrotechny as a geometrical art, as verses dedicated to the author John Babington suggest:

> Well hast thou then my Friend exprest thy Art,
> Since this rare Science hath the greater part
> Shar'd in thy Worke, giving it winged fire,
> To mount it up-aloft, and deck the ayre
> With splendant stars, silver and golden showers;
> These are th'effects of *Mathematick* powers[.][105]

Gunners thus viewed geometry as a means to bring order and discipline to practice and to give the appearance of discipline to superiors. As the English gunner Robert Norton explained to Charles I in the dedication of his work on gunnery: "Let Occasions be ruled with Reason, Wars managed with Discipline, Judgement and Pollicie."[106]

Gunners also looked to scholastic natural philosophy to bring order and credit to fireworks. Norton, for example, began *The Gunner* with eleven "Maximes of Naturall Philosophy necessary to be first knowne," followed by "theorems" explaining practically useful physical phenomena in Aristotelian terms.[107] The latter included accounts of how "Euery Corporall thing reposeth in its naturall place" and "Euery thing that is within the Lunar Orbe may make motion or change."[108] Others dealt with motion and resistance, the action of gunpowder, the recoil of guns, and the force of impacts: "To Strike is a matter of Mouing, as is the Time & Quantity or the distance vpon which it is made, which if the time be short . . . then it is called Swift: Therefore the shorter the Time is, the swifter and stronger is the stroake."[109] Similarly John Bate, the artisan author of *The Mysteries of Nature and Art*, which included numerous pyrotechnic recipes, began his work with "Certaine Praecognita or Principles" explaining the "causes and reasons" for effects described in the book.[110]

However, the most common allusion to the liberal arts in fireworks treatises was to the language of invention, and gunners laid great stress on the importance of the ingenious variation of fireworks as a means to impress audiences. The rhetoric of invention offered artisans a means to present their works as suited to courtly consumption, shifting their meaning away from the practical utility associated with the lowly mechanical arts toward a more pleasurable function as emblems for courtly contemplation. Tartaglia, Biringuccio, and others often presented gunnery in this way, apologizing for laying the details of a mechanical art before a noble patron, but hoping that the art might be seen as a source of pleasurable inventions. Thus, Tartaglia apologized for presenting "mechanical and common things" to Henry VIII in his *Quesiti et inventioni diverse* of 1546 but hoped that they might be taken as "inventions," or "new things according to a custome by which some use at the beginning of the yeare to present unto noble and honorable persons unripe and lower fruites, not for that any goodnesse is in them, but for daintie and new thinges which doe naturallie please men's mindes."[111] The English gunner William Bourne also apologized to the readers of his *Inventions and Devices* of 1578 for presenting "rude and base inventions" but noted: "It is possible that in the reading of these rude Inuentions, [the reader] may finde in some of them, that thing that may pleasure them."[112]

Biringuccio also made much of invention in his *Pirotechnia*, addressed to the noble patron Bernardino di Moncelesi of Salo. Biringuccio promoted military men as ingenious, as "good and lofty men of intelligence" who were "constantly discoverers of many beautiful things."[113] He explained: "There have always been in this world men of such keen intelligence that with their discourse they have been capable of infinite and various inventions that are as beneficial as they are simultaneously harmful to the human body."[114] Like Tartaglia, Biringuccio presented pyrotechnic innovations as novelties worthy of princely patronage because they "greatly astound our minds, and we remain so stupefied that when we think of them we are not able to control our faculties for some time, both from considering by what necessity or purpose these men were goaded, as well as from contemplating the profound subtlety of their discoveries, which, in truth, are such that they are awarded the greatest commendation by noble minds."[115] Such inventions were also worthy because they were capable of constant variation: "They can be used for various effects in addition to the one intended by the inventor."[116] *Varietà* was another value prized at the Renaissance court, singled out in Aristotle's *Rhetoric* as appealing to the mind's inherent enjoyment of change, and often praised as a quality of God's abundant creation.[117] Biringuccio thus chose a suitably courtly analogy, a genealogy, to describe the history of arms as a history of endless variation: "Developing then . . . various sons were born, both small and large [firearms] and of various shapes depending on the opinions or desire of the masters or of the princes who had them made."[118]

This language of invention and variety was of primary importance in subsequent treatises on fireworks, as important for pyrotechny as geometry was in articulating the art of shooting. If the gunner learned geometry to shoot a projectile in a single, repeatable trajectory, he also learned the value of ingenious invention with which to vary infinitely the trajectories and effects of festive fireworks. Variation, then, ennobled artisans. As the French gunner Jumeau wrote of his fellow fireworker Morel in 1615: "The sieur Morel . . . showed the gentility of his spirit by diverse artifices."[119] Another gunner reminded his fellows: "You must know how to make 3 or 4 [s]ortes of fireworkes at least if you will get lords wages."[120] Audiences evidently valued invention: "The last castle of fire, which bred most expectation . . . had most devices."[121] "This new invention [i.e., fireworks] allured the hearers . . . because many of them as yet had not heard of it."[122]

References to invention and ingenuity, then, abounded in both manuscripts and printed treatises dedicated to noble patrons. Authors often described the basic procedures for making a firework and then left readers to alter the procedure according to their wish. Typical was Norton, who explained in his *Gunner's Dialogue* of 1628: "Now in regard that all moueable Fire-workes, in Aire or Water, haue their motions from the force of the Rockets applied accordingly; I need not speake of them particularly, but leaue each practiser to his owne ingenious inuention."[123] Hanzelet and Thybourel wrote: "The following figure represents to you diverse sorts [of fireworks], on the model of which it is easy to invent and add others. . . . [Fireworks] are made according to the imagination, and the will of those who construct them . . . each builds them according to his fantasy."[124] Many authors followed such admonitions by giving basic recipes and then prompting readers to diversify them. Norton proposed that the structure and composition of fiery wheels "may bee . . . varied at the Worke-Masters pleasure" while grenades should be "as bigge as you will."[125] Malthus repeatedly told his readers to use sizes and quantities of materials "as you please" or "fitting for your purpose," explaining: "The industrious may adde many diversities as they shall thinke fit."[126]

## The Ingenious Gunner

Gunners thus adapted the rhetorical vocabulary of invention, imitation, ingenuity, and variety to the description and practices of their art. As in oratory, poetry, and painting, pyrotechny was now represented as entailing the creation of new visual and auditory forms that could be set together in varied compositions to produce powerful effects. No doubt, this language was critical in actually shaping the many new forms of fireworks that emerged in the sixteenth and seventeenth centuries, though, as we have seen, gunners' alchemical and practical understanding and skills also deeply informed the new kinds of pyrotechnics they invented. The liberal arts certainly shaped one dimension of pyrotechnic performance, as fireworks were during the sixteenth century increasingly framed within classical allegorical narratives, the design of which was also known as *invention,* following the conventions of poetic composition. It is testament to the shared characteristics of fireworks and other arts in the Renaissance that artists, architects, poets, and gunners could all share in the work of inventing the allegorical content of displays. In the Italian states, this work tended to be the responsibility of architects and poets. Thus, Vasari recorded in the 1560s how the design of the Florentine girandola passed from "puppeteers" to the architect Niccolò de Pericle, known as Il Tribolo, who, "with that worth and ingenuity [*virtù et ingegno*] he had applied to other things, made [a girandola] in the shape of a very beautiful octagonal . . . Temple of Peace . . . the most beautiful girandola that had ever been made up to that time."[127] The architect thus supplanted the masters Biringuccio had described as ingenious, and architects would be chiefly responsible for designing Italian displays thereafter. In Northern Europe, however, prior to the mid-seventeenth century, gunners themselves invented allegories. Morel in Paris, for example, chose classical tales such as the story of Perseus and Andromeda as the foundation for fireworks. In England, chivalric themes were more common. Thus, the five gunners responsible for the fireworks celebrating the Princess Elizabeth's wedding in London in 1613 presented their "inventions" to King James I.[128] The plans showed sea fights with ships and dolphins and pyrotechnic battles between a giant, Saint George, and the dragon, all allegorizing the elector's "clearing the world of evil enchantments" through his marriage to Elizabeth.[129] Such poetic inventions were expressly presented as a means to ennoble the gunners' art. As Morel put it: "To give more grace to my enterprise, I will execute a beautiful design which I have taken from the ancient poets."[130]

Where did gunners learn these arts of invention and poetry? How did they become proficient in the skills necessary to manipulate the worlds of court and print? Any answer is difficult because so little is known of gunners' education, training, or biography. Many gunners no doubt lacked such skills. Walton proposes that the idiosyncratic spelling and peculiar arrangement of a manuscript on gunnery and fireworks by the Elizabethan English gunner Richard Wright shows the author's unfamiliarity with printed books or even with the written word: "Wright appears to have produced a treatise imitating those he might have seen around him, although with mixed success."[131] Alternatively, Norton, the author of *The Gunner,* was the well-to-do son of the Cambridge-educated Thomas Norton, one of the English translators of Erasmus, an expert on Seneca, an official censor of printed books, and a prosperous lawyer to the Stationers' Company.[132] Robert could not have been better placed to know the subtleties of court life, humanism, and print culture. Most gunner-authors probably lay somewhere between these two cases—men such as William Bourne, literate, mathematical, and evidently able to read French and Italian, yet unschooled, serving as a town councillor in Gravesend before writing on gunnery and inventions. Others identified themselves as lacking literacy. John Babington apologized for his "rude and unpolished lines" in *Pyrotechnia*.[133] Hanzelet, alternatively, was himself a printer, publishing works in his native town of Pont-a-Mousson. Gunners also collaborated with a variety of learned professionals. Hanzelet collaborated with François Thybourel, a master surgeon, in composing their *Receuil . . . de machines militaires*. Other gunners provided fireworks for the theater, with numerous plays including fiery effects by 1600. Morel's skill in poetic composition reflected a career shared between the artillery and the staging of *ballets du court*, which often included military scenes and pyrotechnics. Another *artificier du roi*, Georges Buffequin, became the first *décorateur* of the Paris theaters in 1607.[134] Gunners also engaged directly with the nobility as employees and clients. Many sought patronage by dedicating their works to noble artillery officers, while noble residences provided another location in which gunners, scholars, and courtiers might meet and learn from one another. The poet Henry Peacham thus captured the traffic of knowledge that the noble household made possible: "When I . . . liued at the table of that Honourable Gentleman, Sir Iohn Ogle, Lord Gouernour, whither resorted many great Schollers and Captaines . . . it had beene enough to haue made a Scholler or Souldier, to haue obserued the seuerall disputations and discourses among many strangers, one while of sundry formes of battailes, sometime of Fortification, of fireworkes, History, Antiquities, Heraldrie, pronunciation of Languages, &c."[135]

Thus, a variety of avenues were open for gunners to gain access to the liberal arts and courtly idioms that they incorporated into their writings and spectacles. In so doing, they transformed fireworks according to their ingenious imaginations. Between the religious dramas of the late fourteenth century and the allegorical spectacles of the 1620s, their efforts led to a complete reconstruction of pyrotechnics, moving them from military and religious sites into new scenes of performance to exhibit princely power. Following the apocalyptic spectacle of the Girandola in Rome, gunners staged fireworks across Europe, in new arenas of civic and courtly space, where they created astonishing immersive environments of artificial meteors and marvelous motions.

It was in this context that the portraits of Suenck and Babington appeared. Gunners took pride in their achievements. "O're the face of all the

World," Captain John Smith told the gunner Robert Norton, "let thy Fames Echo sound."[136] This new confidence was reflected in a novel image of the founder of pyrotechny. The *Feuerwerkbuch* had presented Black Berthold as an alchemist, but by 1600 he appeared as a purposeful inventor, with the great power and responsibility of managing the recipe for gunpowder given to him by the devil. William Camden thus claimed that Berthold was given a four-line Latin oracle by the devil to create his new inventions, while the French military engineer Joseph Boillot portrayed him in a treatise on military fireworks of 1598 with a book of spells and a demon hovering at his shoulder as he prepared his new invention.[137] Boillot then described his own career in terms quite different than those employed by the humble alchemist-artisan: "I have dedicated the best part of my career to the research and exercise of the sciences connected to manual art, on account of the charges in which I have been employed, I have also turned my eye to reading the authors who have treated of cannons and other artificial fires, which I have amplified by many new inventions of diverse instruments, of compositions of diverse powders and artificial fires and other similar industries."[138] Boillot's words show that the appeals to liberal learning in fireworks treatises fed back into pyrotechnic practice, leading the gunner to join reading and art and to seek to create "new inventions," a skill that Boillot celebrated by writing a treatise of his own.

Alternatively, the suitably classical mechanic Archimedes was identified as the founder of pyrotechnic art. Already in 1344 the humanist Petrarch had ascribed the invention of artillery to him, to which Tartaglia, writing in the 1540s, added the invention of gunpowder. By the early seventeenth century, several authors presented Archimedes as a pyrotechnist, and Malthus was praised for his "Archimedean hand."[139] The early history of fireworks thus concluded with images of the gunner as a celebrated master and ingenious inventor. The strategy paid off. By the end of the seventeenth century, pyrotechnics was institutionalized, finding a permanent place within court society. Laboratories for making gunpowder shells and fireworks were established in many cities across Europe, at the arsenal on the banks of the Seine in Paris, in the *Zeughäuser* of Munich and Berlin, and in a great fireworks barn at Greenwich in London.[140] The appearance of these spaces, about which I shall have more to say below, together with the explosive spectacles that adorned the squares, riverbanks, towers, and other festive sites of the city, testifies to the credit pyrotechny had achieved over two centuries. As the verses dedicated to Malthus put it, gunners had succeeded in "perfecting the Pyrotecnique story."

## Conclusion

This chapter has surveyed the rise of "artificial fireworks" in Europe from the earliest military pyrotechnics and religious dramas of the 1370s to the sumptuous courtly spectacles of the early seventeenth century. Such fireworks emerged as gunners sought, like other artisans of the period, to improve their humble reputations and gain credit at court. Achieving this prompted hybridizations of artisanal crafts and liberal arts, the idiom and techniques of practical alchemy, and the rhetorical flourishes of courtly poetics and learning. The result was both a new genre of art and performance and a new literature and knowledge of pyrotechnic skills, each of which contributed to raising the status, knowledge, and confidence of the artisans who managed fireworks. In making these new fireworks, gunners also fashioned novel forms and experiences of space. Fireworks festivals joined together the battlefield, the church, the court, and the communal spaces of the city and transformed them temporarily into a remarkable, artificial nature, bringing the heavens down to earth, imitating life, and raining meteors over the heads of fascinated spectators. Such an artificial nature, and the credit it brought to gunners, did not go unnoticed by learned communities in the sixteenth and seventeenth centuries. The next chapter will consider how fireworks were soon put to use in new philosophies of nature.

CHAPTER TWO

# Philosophies of Fire: Pyrotechny as Alchemy, Magic, and Mechanics

Between the writing of the *Feuerwerkbuch* and the appearance of Suenck's portrait, fireworks had undergone a radical transformation as gunners synthesized the pyrotechnics of the battlefield and the church into dramatic new artificial fires, imitating nature's appearances and processes. The gunners' new effects charmed the courts and the princes, who dispensed increasing sums to witness fireworks, making them a regular element in the cycle of courtly festivals. By the early seventeenth century, the English court might spend between £600 and £1000 on a fireworks display, in the region of $150,000–$300,000 in today's terms. The money paid for hundreds of pyrotechnics to be prepared and set off and for scenery construction and painting by numerous laborers and artisans, overseen by the gunner.[1] Gunners secured these rewards through the creation of a practice and literature that brought together the liberal arts and pyrotechnic skills. This integration would have consequences, not only for gunners, but also for the learned themselves because, consecutively with the rising reputation and frequency of pyrotechnic displays and literature, scholarly audiences also began to take a growing interest in fireworks. The scenes of courtly festival and the world of print now became sites where gunners mingled with scholars, prompting a further integration of their respective skills and knowledges. Continuing to promote the liberal nature of their craft, gunners posed pyrotechnic knowledge as a route to philosophical understanding, while natural magicians and mechanical and experimental philosophers took up pyrotechnics as a resource for reforms of natural philosophy. Such exchanges encouraged a growing integration of artisanal and philosophical skills in the early decades of the seventeenth century, with the result that, like pyrotechny, natural philosophy gave a special place to the skills of ingenious invention and the creation of novel effects.

## Making Meteors: Natural Magic and Flying Dragons

In *Ballistics in the Seventeenth Century*, A. R. Hall wrote that military innovations were rare in the seventeenth century and meaningless for the new sciences: "Fertile inventors offered largely useless extravaganzas, and the scientist nothing."[2] Yet the coming together of artillerists and the learned in the culture of courtly fireworks had a profound effect on both communities, integrating their skills, and producing a variety of hybrid practices and ideas. Interaction hinged on new sites of exchange. The crowded scenes of performance—formal gardens, town squares, theaters, riverbanks, and fields—were all arenas where fireworks went on display, and both soldiers and scholars attended. "At night," wrote the Royal Society fellow Samuel Pepys one spring evening in 1661, "set myself to write down these three days' diary, and while I am about it, I hear the noise of the chambers, and other things of the fire-works, which are now playing upon the Thames before the King; and I wish myself with them, being sorry not to see them."[3] According to René Descartes's biographer Adrien Baillet, the gentleman philosopher should make it his business to attend courtly spectacle. Thus, Descartes witnessed the coronation of the emperor Ferdinand at Frankfurt "to the end that he might not be ignorant of what the principal actours [*sic*] of his World do represent of most pompous and glorious upon the Theatre of the Universe."[4]

One result of scholarly observations of artifice was the incoporation, beginning in the second half of the sixteenth century, of pyrotechnic effects into the performative repertoires of the learned. The first to appropriate pyrotechny in this way were the natural magicians.[5] Exotic incendiary compositions had been occasional elements in books of secrets and magic from the late Middle Ages, and, in his 1557 *De rerum varietate libri xvii*, Girolamo Cardano described an artificial bird raised by fire on a line and other pyrotechnic marvels.[6] Pyrotechnic recipes and inventions were often incorporated into magical works. The Neapolitan magician Giambattista Della Porta thus presented a whole book devoted to "artificial fires" in his *Magia naturalis*, first published in 1558 and reissued in expanded form in 1586.[7] Della Porta, also a court playwright, gave emblematic significance to fireworks, echoing the portentous message of the Roman Girandola when he wrote: "The matter is very useful and wonderful, and there is nothing in the world that more frights and terrifies the mindes of men. God is coming to judge the world by Fire."[8] He then listed a variety of pyrotechnic recipes that suggest that he had witnessed the Girandola, probably during his tenancy at the Palazzo d'Este in 1580 in the service of Luigi d'Este, since he described fireballs for "festival times" that "stream along like falling Stars."[9] He also claimed to have made these fireworks, which would "run up and down in the Air, as I often saw it at Rome, and prepar'd it."[10] Like the gunners, then, Della Porta both practiced and wrote about fireworks. However, while gunners performed fireworks on a grand scale and sought to convey knowledge to other gunners or gentleman officers, Della Porta took pyrotechnic effects as only one element in a range of wonder-filled performances. He thus presented himself as a type of "connoisseur" of diverse secrets and effects, choosing what was most rare or unusual in many arts and exhibiting them before learned or courtly audiences. This meant that the boundary between magic and craft was slender and that skills moved easily between them.[11]

Fireworks appeared in books of secrets and magic from this time on and were no doubt part of at least some magicians' performative repertoires. The Dutch engineer and magician Cornelius Drebbel displayed artificial meteors during his time in England in the service of Henry Prince of Wales. Drebbel "with divers instruments could make it rain, lighten, and thunder, at different times, as if it had come about naturally from heaven, so that people knew no differently."[12] The Jesuit director of the Collegium Romanum Athanasius Kircher and his numerous disciples also included fireworks and meteorological artifice in their compendious magical works.[13] Gaspar Schott presented pyrotechnics amid the instruments of magic in the frontispiece of his *Magia Universalis* of 1657 (fig. 9).[14]

One firework in particular proved popular among these learned practitioners. The career of the "flying dragon" shows well how fireworks moved easily between gunners, magicians, alchemists, and humanists in this period and began to figure as "experiments" in new philosophies. Dragons also offer another opportunity to consider the performative significance of fireworks and, in particular, the way in which performances worked to constitute power in the courtly festival.

Dragons, or their variants of worms, hydra, serpents, and basilisks, were a mainstay of church and courtly display from at least the fifteenth century. They appeared in many guises, including exotic exhibits in the courtly *Kunstkammern,* and as artificial figures in fireworks.[15] Festivals featured firework-packed dragons atop castles or emerging from caves; dragons figured as great monsters in battle with other creatures or, most commonly, as "flying dragons," sailing through the air exploding with sparks. This effect was achieved in two ways. Early flying dragons were probably banners or sails on a wooden frame or kite, flown in the air and containing a burning pyrotechnic mixture. Others were "runners on a line," dragon figures propelled by rockets in their interior to run back and forth along a well-soaped rope

9. Putti arrange the celestial bodies on the *theatrum mundi* above the instruments of natural magic, which include rockets, bombs, and ordnance. Frontispiece to Gaspar Schott, *P. Gasparis Schotti Magia universalis naturae et artis, sive, Recondita naturalium & artificialium rerum scientia . . . opus quadripartitum pars I, continet Optica, II. Acoustica, III. Mathematica, IV. Physica*, 4 vols. (Würzburg, 1657–59), vol. 1. University of Chicago Library, Special Collections Research Center.

10. Figure illustrating the "dragon on a line," from John Babington, *Pyrotechnia; or, A Discourse of Artificiall Fire-Works in Which the True Grounds of That Art Are Plainly and Perspicuously Laid Downe* (London, 1635), opp. 37. Edgar Fahs Smith Collection, University of Pennsylvania Library.

across a town square or between two turrets. In both cases, the figures were filled with pyrotechnics and belched flames from every orifice (fig. 10).[16]

Dragons straddled the divide of pyrotechnic, alchemical, and allegorical significance. Since the Middle Ages, a *fiery dragon* or *fire drake* was the common name for fiery exhalations or meteors seen in the sky, phenomena that appeared in the shape of dragons and portended events on earth much like comets and other extraordinary signs (fig. 11). One account thus explained: "The flying Dragon, is when a fume kindled appeareth bended, and is in the middle wrythed like the belly of a Dragon; but in the fore part for the narrownesse, it representeth the figure of the neck, from whence the sparkes are breathed . . . well knowne it is [that dragons] be wrought by the pollicie of Deuills, and inchantments of the wicked, as sundrie examples lamentable, doe make manifest at this day . . . in the yere 1532 in manye Countries, were Dragons crowned, seene flying by flocks or companies in the ayre, hauing Swines snowtes."[17]

The "flying dragon" was also a term of alchemy, signifying mercury, which "flew" from flames as a vapor, while sulfur (a dragon "without wings") did not.[18] Dragons also figured in Christian mythology and church dramas, embodying heresy, sin, or Satan himself, invariably slain by crusaders and saints to signify the Christian conquest of evil.[19] Gunners then used artifice

11. The Draco Volans, a fiery meteor, from Thomas Hill, *A Contemplation of Mysteries Contayning the Rare Effectes and Significations of Certayne Comets, and a Briefe Rehersall of Sundrie Hystoricall Examples, as Well Diuine, as Prophane* (London, 1574), fol. 25v.


to exploit the portentousness of fiery dragons to signal the power of the prince. As early as 1487, celebrations for the coronation of Elizabeth of York in London included a "Batchelor's Barge garnished and apparelled above all others [that] carried a dragon spouting flames of fire into the Thames."[20] The figure of Saint George battling artificial dragons running on a line featured in many subsequent English fireworks.[21] In Paris, Morel's display of Perseus and Andromeda in 1628 classicized this chivalrous confrontation. Dragons were included in many other fireworks.[22]

Another significant dragon appeared at celebrations on the Field of the Cloth of Gold at Guisnes in 1520, with the "appearance in the air of a great salamander or dragon, artificially constructed; it was four fathoms long, and seemed to be filled with fire which was very horrible. . . . Many were greatly frightened thereby, thinking that it must be a comet or some monster or portent, as nothing could be seen to support it."[23] The description points to a common claim made for sixteenth- and seventeenth-century fireworks. To those who understood the artifice, typically courtly audiences, fireworks were strange, curious, and pleasant, but, to those who did not, typically the "vulgar," they were fearsome apparitions, like the natural dragons and meteors that portended in the skies. Whether this is actually how audiences reacted is an open question, and certainly some images of fireworks audiences in this period varied in their representations of spectators, with no strict separation of fearful crowds from knowing elites.[24] Nevertheless, the textual division suggests an important way in which fireworks were meant to manifest power as material performances. Historians often propose that the source of spectacles' power lay in their iconography and allegory, and certainly the image of princes as dragon slayers made manifest their powers to conquer evil and Christianity's foes. But it was equally a differential access to this meaning that underwrote the force of displays. Vulgar audiences were imagined as being unable to comprehend the artifice of performances and so took artificial spectacles for real, natural, and significant events. In contrast, the higher orders—the prince and the nobility—understood artifice and took pleasure in appreciating it. Consider, for example, the contrasting reactions presented in this account of fireworks for Queen Elizabeth I in August 1572: "They in the fort shoting agayn and casting out divers fyers, terrible to those that have not bene in like experiences . . . and in dede straunge to them that understood it not; for the wildfyre falling into the ryver Aven, wold for a tyme lye still, and then agayn rise and flye abrode, casting furth many flashes and flambes, whereat the Quene's Majesty took great pleasure."[25] While the queen took pleasure in the fireworks, others who failed to understand the artifice would find them strange and terrible. Similarly, when a hundred "thunderbolts" exploded at fireworks in La Rochelle, France, in 1629, "naïve bystanders along the shore . . . were stunned and frightened and . . . thrown about pell-mell. . . . The Court alone remained still and undaunted, being well acquainted with the vanity and ostentation of these diversions."[26]

As Mario Biagioli has noted, access to the significance of emblems served the social differentiation of elites from the masses in the Renaissance court, and the same extended to fireworks performances, which, indeed, made emblems out of artillery.[27] Power was articulated through controlled access to hidden artifice, whose workings were revealed to elites in pyrotechnic treatises and the regular experience of displays but were unavailable to others. *Knowledge* of fireworks was, thus, a critical component in their political function and in the experience of displays. As Robert Laneham, witnessing another firework for Elizabeth I at Kenilworth in 1575, explained, if he did not *know* the fireworks were "in amitee . . . it woold haue made me for my part, az hardy az I am, very vengeably afeard."[28] Fireworks were "secrets," open to certain audiences but not to others, designed to cultivate "the admiration of such as know not the secrecie," or so, at least, the nobility imagined.[29] Gunners labored to provide such artifices. As the gunner William Bourne explained to his readers in 1578, artificial birds might be constructed "to flie by Arte, to flie circularly, as it shall please the inuentor, by the placing of the wheeles and springs, and such other like inuentions, which the common people would maruell at, thinking that it is done by Inchantment, and yet is done by no other meanes, but by good Artes and lawfull."[30]

Such views explain the appeal of pyrotechnics to natural magic, which also dealt in secrets and claimed to be a form of lawful or "natural" enchantment. Magical epistemology hinged exactly on carefully managed access to secret techniques, mobilized to raise the admiration of the vulgar to the advantage of the knowledgeable. Hence Della Porta's advice to the readers of his *Magia naturalis*: "If you would have your works appear more wonderful, you must not let the cause be known: for that is a wonder to us, which we see to be done, and yet know not the cause of it. For he that knows the causes of a thing done, doth not so admire the doing of it; and nothing is counted unusual and rare, but only so far forth as the causes thereof are not known."[31]

Effective magic depended on the spectacular epistemology exemplified by fireworks. Indeed, one secret that made its way from pyrotechny into magic was the flying dragon. Cardano, Della Porta, Kircher, and a number of others included directions for making dragons in their works.[32] Kircher in particular was fascinated by the dragon as a tool for astounding the vulgar on behalf of his Jesuit sponsors. The missionaries of the Society of Jesus in India, he recorded, had used the dragon to terrify "savages" (*Barbari*) into releasing incarcerated members of the holy order from prison.[33] The fathers flew dragons on strings, filled with a composition of sulfur, pitch, and wax, which illuminated the portentous message "Ira Dei" (the wrath of God) painted on each side of the dragon's body: "The savages having seen the unusually moving phantom are astonished with the utmost stupefaction, and remembering the anger of God with them and what the Fathers told them, were afraid they would suffer the punishment that had been predicted. Because of this they immediately opened the prison and let those incarcerated there go free; meanwhile the machine caught fire and ignited in flames, made a loud noise like a [thunder]clap, and stopped moving of its accord."[34] Kircher took the dragon as one of a range of pyrotechnics suitable for such religious and colonial missions. He gave recipes for "fiery rain" or "fiery fountains," burning figures of eagles, and a spinning wheel carrying statues or "the names of Jesus and Mary." Pyrotechnics should strike fear into the beholder: "I believe that using only this artifice or some marvels it would be possible to put a whole army to flight, particularly one of savages, who usually take as wonders and omens all the things whose reason they cannot understand."[35]

Jesuits commonly employed such techniques in their theater.[36] A number of accounts suggest that Europe's colonial missions also followed such procedures and, thus, that the audiences for early modern European displays of fireworks were by no means limited to Europeans.[37] In 1637, the colonial *ingenieur* Jean Bourdon of the French mission in Quebec thus performed fireworks before the native Huron for the feast of Saint Joseph. Jesuits recorded how the display of rockets, serpents, *saucissons*, and wheels was played off from a small artificial castle. They noted how, before the fireworks began, it was "said to the Savages, notably to the Hurons, that the French were more powerful than Demons, that they commanded fire; and that if they wished to burn the villages of their enemies they could soon do it."[38] The fireworks supposedly raised the astonishment of the "French, and even more of the Savages, who had never before seen anything of the kind."[39] Such techniques were not exclusive to Catholics. Protestant English colonists undertook similar displays, their reports making clear the way elites imagined the power of their artifice. The mathematician and writer on gunnery Thomas Harriot, visiting Virginia in 1585, thus described remarkable displays of Europeans' magical effects to Native Americans:

> Most thinges they sawe with vs, as Mathematicall instruments, sea compasses, the vertue of the loadstone in drawing yron, a perspectiue glasse whereby was shewed manie strange sightes, burning glasses, wildefire woorkes, gunnes, bookes, writing and reading, spring clocks that seeme to goe of themselues, and manie other thinges that wee had, were so straunge vnto them, and so farre exceeded their capacities to comprehend the reason and meanes how they should be made and done, that they thought they were rather the works

> of gods then [*sic*] of men. . . . Whereupon greater credite was giuen vnto that we spake of.[40]

Like Kircher, Harriot posed Native Americans as unable to reason out theater, which supposedly led them to think of Europeans as "gods." Thus, performances on the colonial periphery made explicit the performative politics behind pyrotechnic displays at the center.

Harriot's inclusion of "wildefire woorkes" among the more familiar instruments of magic and study is suggestive of the new directions given pyrotechnics by magicians because natural magic's spectacles were simultaneously experiments and acts of knowing nature, directing the ingenious invention associated with elaborating art to the production of new experiments to reveal nature's potential. Thus, Kircher's artificial dragon was recorded as an "experiment," while Della Porta described the procedures for constructing a flying dragon as a "mechanical experiment" and proposed that this raised the possibility of human flight.[41] Della Porta claimed that the dragon, which was made from a "quadrangle of reeds" covered in paper and linen and filled with a lantern or crackers ("Others bind a Cat or Whelp, and so they hear cries in the air"), would prompt new inventions: "Hence may an ingenious Man take occasion, to consider how to make a man flye, by huge wings bound to his elbows and breast; but he must from his childhood, by degrees, use to move them, always in a higher place."[42]

The gunners' variations of pyrotechnic recipes should here operate as variations intended to achieve a new goal, such as flight. Indeed, others followed Della Porta in varying the experiment. In 1647, Tito Livio Burattini, an Egyptologist and correspondent of Kircher's, followed the gunners seeking credit at court by touting a flying machine in the form of a dragon at the court of the Polish king Ladislas IV Vasa. Burattini's design fascinated Parisian savants, including Gilles Roberval, Marin Mersenne, and Christian Huygens. It was a "miraculous secret" of "human ingenuity," though it was never built.[43] Nevertheless, another ingenious author, Cyrano de Bergerac, also proposed the dragon as a means to fly, in this case to the moon. In his *Histoire comique . . . de la lune* (1657), Cyrano placed himself in New France, that is, Canada, the scene of the Jesuit fireworks for the Huron, where he built a flying machine for lunar voyaging. Following its first trial, however, local *ingeniuers* commandeered the machine to revive its more usual function, having the "opinion that fireworks should be fastened to it, being their rapidity, and the springs agitating its large wings, would levitate it mightily; so that none should see it, without beleeving it a fiery flying dragon."[44] To save his invention, Cyrano jumped into the machine as the fireworks were ignited, only to be transported "three quarters of the way" to the moon.[45]

Thus, the new fireworks gradually transformed into philosophical experiments, prompting their users to the contemplation of nature. Perhaps the most celebrated, though likely apocryphal, story of ingenious experiments with the dragon concerned Isaac Newton. Newton owned a copy of John Bate's 1630 book of secrets *The Mysteries of Nature and Art*, which included numerous recipes for fireworks, the flying dragon, and a fire drake, in this instance a kite to which crackers or lanterns were attached to appear "very strangely and fearefully."[46] "Kites" were closely associated with fireworks in England at this time, the word *kite* first appearing in English in John Babington's *Pyrotechnia* of 1635, which described the construction of an armillary sphere turned by rockets and flown suspended from a kite.[47] In his memoirs of Newton, William Stukeley proposed that, as a boy, Newton "invented the trick of a paper lanthorn with a candle in it, ty'd to the tail of a kite. This wonderfully affrighted all the neighboring inhabitants for some time, & caus'd not a little discourse on market days, among the country people, when over thir mugs of ale."[48] Stukeley took the episode as portentous, saying that it was an "omen of the sublimity of [Newton's] discoverys," while John Conduitt proposed that the kite construction led to natural philosophy, being the first instance of Newton's efforts to find "what body or curve would find the least resistance in a fluid."[49]

## Fireworks as Mathematical Recreations

If Newton moved from pyrotechnic experiments to natural knowledge via the dragon, this was not unusual for a seventeenth-century philosopher. When Cyrano was thrown into the air by his flying dragon, he recalled: "I had the impudence to Philosophy upon it . . . examining with my eyes and thoughts, what should be the cause."[50] Gunners and magicians posed the creation of new effects with fireworks as a practice that might prompt the production of novel "inventions" or "experiments." But they also presented the knowledge gleaned in pyrotechny as a resource for contemplating nature more generally. In books of "mathematical recreations," gunners promoted the idea that studying pyrotechny might offer resources for thinking about nature's mechanisms on the model of their own ingenious inventions. Such thinking proved fruitful for a number of seventeenth-century natural philosophers, while epistemic debates continued to take the secretive techniques of spectacle as an important point of departure.

The *Récréations mathématiques* published at Pont-a-Mousson in 1624 has often been noted for presenting a variety of mathematical problems and ingenious recipes drawn from books of secrets, magic, and the arts that proved influential for early-seventeenth-century philosophers, including Descartes, Claude Mydorge, and John Wilkins.[51] While the book is commonly attributed to a Jesuit, Jean Leurechon, or his nephew Henry Van Etten, one historian at least has argued that the real author was Jean Appier, known as Hanzelet, that is, the fireworks master and engraver of Lorraine who authored a book on military machines and fireworks with the surgeon Thybourel in 1620 in addition to a 1630 book devoted exclusively to *la pyrotechnie*.[52] If the attribution is correct, then in the *Récréations mathématiques* Hanzelet extended the gunners' efforts to promote pyrotechny as worthy of the liberal arts by presenting observations of art, not as a means to improve art for practitioners, but as a route to higher spiritual, moral, and philosophical contemplations. With this work, the audience for pyrotechny was expanded further, from practitioners and officers to gentlemen "virtuosi" who took an interest in practical matters, the same gentlemen to whom the magicians directed their collections of spectacular effects. Hanzelet wrote, he said, to "satisfie the curious, who delight themselves in these pleasant studies."[53] Furthermore, he extended exactly the spectacular epistemology typical of fireworks and natural magic into his recreational experiments: "To give a greater grace to the practise of these things, they ought to be concealed as much as they may . . . for that which doth ravish the spirits is an admirable effect, whose cause is unknowne."[54] Hanzelet pointed to exactly the differential in knowledge supposed to distinguish the noble from the vulgar in contexts of spectacle as now being vital in the performance of experiment: "Great care ought to be had that one deceive not himselfe, that would declare by way of Art to deceive another: this will make the matter contemptible to ignorant Persons, which will rather cast the fault upon the Science, than upon he that shewes it."[55]

Fireworks were among these experiments. By 1630, the *Récréations* included a whole book on fireworks and was translated into English, probably by another fireworker, Francis Malthus, in 1633. In his foreword to the English translation, Malthus explained to his readers the inclusion of fireworks: "And to this we have also added our Pyrotechnie, knowing that Beasts have for their object onely the surface of the earth; but hoping that thy spirit which followeth the motion of fire, will abandon the lower Elements, and cause thee to lift up thine eyes to soare in higher Contemplation, having so glittering Canopie to behould; and these pleasant and recreative fires ascending may cause thy affections also to ascend."[56] Here, pyrotechnic *knowledge* was to function like an emblem. Malthus echoed the proposition of Biringuccio that new inventions in artillery might offer pleasure to those who contemplated them, only now the study of pyrotechnic recipes would invoke pleasure and thought, the form of which was specifically natural philosophical. Malthus continued: "Nature having furnished us with matter, thy spirit may easily digest them, and put them finely in order, though now in disorder."[57] This move should not be surprising. Gunners had spent many decades learning to reconstruct pyrotechnics as emblematic of moral and spiritual meanings appropriate and consumable at court in the form of festive fireworks and allegories. They did so in part by integrating fireworks with geometry and natural philosophy. In the *Récréations*, Hanzelet and Malthus did the same, but now they proposed that pyrotechny led to natural and moral lessons that could be equated with the pleasure of contemplating nature. Natural knowledge, furthermore, offered distinction from the vulgar in much the same way as knowledge of artifice did in contexts of spectacle.

## The Gunpowder Theory of Meteors

While gunners sought to make pyrotechnic knowledge more philosophical, the learned made natural philosophy more pyrotechnic. Gunners made remarkable imitations of meteors and celestial apparitions with gunpowder, and, from the mid-sixteenth century, philosophers began using these spectacles as resources for explaining and describing a variety of nature's effects. Historians have long noted an enduring "gunpowder" explanation of meteors during this period that offered causes for a variety of aerial and subterranean phenomena by reference to the combustion of nitrous and sulfurous particles or exhalations analogous to the ingredients and actions of gunpowder.[58] Paracelsus, whom Smith takes as an exemplary figure in the appropriation of artisanal practices into natural philosophy, was one of the first to make this form of argument, which superseded the Aristotelian description of meteors founded on the movements and actions of hot, dry, cool, or moist exhalations or vapors.[59] Paracelsus thus proposed in his meteorological writings that a combination of the principles of niter, sulfur, and mercury could explain variously thunder, lightning, and flying dragons, by analogy to mixtures of the physical substances of saltpeter, sulfur, and mercury.[60] Over time, the analogy grew closer, until an atmospheric form of gunpowder seemed the natural explanation for meteors, as John Wallis wrote in 1696: "Thunder and Lightning are so very like the Effects of fired Gun-Powder, that we may reasonably judge them to proceed from the like

Causes. The violent Explosion of Gunpowder, attended with the Noise and Flash, is so like that of Thunder and Lightning, as if they differed only as Natural and Artificial; as if Thunder and Lightning were a kind of natural Gun-Powder, and this a kind of artificial Thunder and Lightning."[61]

The theory of an "aerial niter" took on great significance in sixteenth- and seventeenth-century natural philosophy. Philosophers including Cardano, Cornelius Drebbel, Daniel Sennert, Gassendi, and Newton all accepted some form of the gunpowder explanation of meteors. Moving from art to nature here became foundational to new methods of natural philosophy. Wallis's comparison recalls Newton's assertion in the "Rules of Reasoning" in the *Principia* that the same causes must be assigned to the same natural effects, such as "the light of our culinary fire and of the Sun."[62] Long before Wallis and Newton, however, gunners and artisans had made the equation between artificial fires and meteors. Biringuccio, for example, explained the noise of cannon fire by reference to Aristotelian meteorology, as being "born almost as thunder and lightning are generated for the same reason in the middle region of the air, from thick burning vapors."[63] Sixteenth-century engravers also visualized the correlation, picturing battles of ordnance in the skies to represent storms.[64] But Renaissance authors equally looked to *fireworks* as meteorological resources, moving from the contemplation of pyrotechnic causes to the contemplation of nature. It makes sense that, in a world where court spectacle made regular and dramatic displays of artificial meteors, explanations of natural meteors should draw on fireworks. While cannon had been in existence since the fourteenth century, gunpowder accounts of meteors emerged only much later, at just the time when gunners began imitating a great variety of meteors with pyrotechnics.[65] Thus, a display in England in 1575 showed a "blaz[e] of burning darts, flying too & fro, leamz of starz coruscant, streamz and hayl of fiery sparks, lyghtenings of wyldefier a water & lond, flyght & shot of thunderbollts: all with such continuanuns, terrour & vehemency: that the heauins thunderd, the waters soourged, the eart[h] shook."[66]

As Wallis suggests, natural and artificial meteors became tightly interlinked. If gunners imitated suns, stars, comets, rain, thunder, lightning, and snow, the authors of sixteenth-century meteorological treatises and images drew on fireworks to represent meteors. The Cambridge don and zealous Protestant William Fulke's *Meteors* of 1563 described how fiery "dancing goats" were created when two exhalations rose into the highest region of the air before igniting: "The flame appeareth to leap or dance from one part to the other, much like as bals of wild fire dance up and down in the water."[67] Fulke equated other meteors with fireworks. Hence, burning "spears" seen over London in January 1560 exploded with fiery darts "of wonderful length, like squibs that are cast into the ayre, saving that they move more swiftly than any squibs."[68] As we saw in chapter 1, the fiery meteors in Fulke's book found artificial equivalents in fireworks, among them "shooting stars," "comets, or blazing stars," and "flying dragons." Similarly, the popular almanac *Kalendar of Shepherds*, first printed at London in 1518, included "burning candles," "wildfire," "burning spears," and "firebrands" among natural meteors.

Such artificial and natural fires evidently made a deep impression on the learned. Indeed, it might be suggested that meteorology was the "paradigm science" around 1600, its phenomena marking the boundary between micro- and macrocosm, mediating between earthly and divine power, and holding great significance for diverse practices such as astrology, medicine, agriculture, navigation, and courtly spectacle. Meteors mattered, so it is perhaps unsurprising that, when the mechanical philosophy emerged in the early decades of the seventeenth century, meteors were often taken as its model applications. Once again, artisans presaged pyrotechnic explanations of meteors. As early as 1515, Leonardo da Vinci, himself a designer of courtly pageants, had asked whether the motion of wind was possible only by resistance against the air, "as is seen to be the case with the rockets driven by the fire, which is to say the percussion that the fire makes in the air which offers some resistance to it, since air is slower than fire in escaping; and if it were not so, the rocket would not move."[69] Magicians also made such claims, moving from experiments with pyrotechnics to comprehension of nature. Della Porta proposed that exhalations from gold mines generated thunder after observing the way an explosive powder made with gold produced a "terrifying" crack when it ignited: "This wonderful powder is very similar to heavenly thunder."[70]

Pyrotechny challenged Aristotle with visible mechanisms. Around 1600, such pyrotechnic explanations multiplied, taking in a great variety of meteorological and celestial phenomena. Johannes Kepler, faced with the problem of explaining the motion of comets in the interplanetary sphere, proposed that they moved "like meteors or fireworks [*ignes artificiales*]."[71] The itinerant hermeticist Robert Fludd similarly explained the divine creation of the stars by describing their ascension into heaven filled with the quintessence on the model of a high-flying squib.[72] The Roman physician Pietro Castelli understood the eruption of Vesuvius in 1632 in pyrotechnic terms: "I now come to the agent, which lights the fire in the mountain's roots. Leaving aside the little fires approximated by man, celestial bolts, as well as solar heat, I say that the bombardier that gives fire to bombards, and to fireworks, could equally ignite these underground mines. So it is, be it repeated motion and the collision of clouds in the air, and great subterranean exhalations in the interlaced caverns."[73]

Natural explanations recalled the alchemical pyrotechny of the artificer: "The World is the Alembick of nature, the air the cap of this Alembick. . . . So there ascend, salt, sulphury, nitrous, &c. vapours, which being wrapped up in clouds, put forth various effects."[74] A few years later, Descartes made meteorology the exemplar of his new mechanical philosophy and brought with it an application to nature of courtly modes of distinction through knowledge of artifice. In an undated work, "The Search for Truth," Descartes echoed the practice of gunners and magicians who revealed in writing the secrets of artificial effects in order, in part, to mitigate the wonder that those effects produced for their noble patrons, offering them distinction from the "vulgar." He planned the same in his philosophy, taking this practice as a model for natural knowledge: "After causing you to wonder at the most powerful machines, the most unusual automatons, the most impressive illusions, and the most subtle tricks that human ingenuity can devise, I shall reveal to you the secrets behind them, which are so simple and straightforward that you will no longer have reason to wonder at anything made by the hands of men. I shall then pass to the works of nature."[75]

Descartes's treatise *Météors* certainly followed this approach, extending to nature the revelations of artificial knowledge common in works on pyrotechnics (recall that Descartes was a reader of Hanzelet's *Récréations mathématiques*). The *Météors* was an important book for Descartes, intended to exemplify his philosophical method, and written while he lived near the residence of the Elector Palatine and Princess Elizabeth, whose wedding had been graced by fireworks on the Thames in 1613. The book began with a discussion of artificial meteors:

> It is our nature to have more admiration for the things above us than for those that are on our level, or below. And although some of the clouds are hardly any higher than the summits of some mountains, and often we even see some that are lower than the pinnacles of our steeples, nevertheless, because we must turn our eyes towards the sky to look at them, we fancy them to be so high that poets and painters even fashion them into God's throne, and picture Him there, using His own hands to open and close the doors of the winds, to sprinkle dew upon the flowers, and to hurl the lightning against the rocks. This leads me to hope that if I explain the nature of clouds, in such a way that we will no longer have occasion to wonder at anything that can be seen of them, or anything that descends from them, we will easily believe

that it is possible to find the causes of everything that is most admirable above the earth.[76]

Besides echoing the author of the *Récréations mathématiques* in his assertion that turning one's eyes to the wonders of the skies would lead to further knowledge, Descartes presented the claim that knowing nature's causes might, like knowledge of artificial effects, mitigate wonder. These words might be taken to suggest that he wished to banish wonder from the sciences, but this was not the case. Like gunners and magicians, he presented his revelations of causes to a select audience, whose access to knowledge might serve their distinction. Indeed, Descartes devised his own ingenious artifices for such purposes, describing in his *Optique* how to make a fountain displaying illuminated figures that "would cause great wonder in those who were ignorant of the causes."[77] Art, meanwhile, provided the model for explaining nature's effects, and, when Descartes set out to explain fiery meteors and lightning in the *Météors*, he did so by analogy to burning sulfur and gunpowder.[78] Similar artifices appeared in the works of Descartes's contemporaries and followers. Pierre Gassendi, the tutor of Cyrano de Bergerac, described the production of "fires which one sees at night running to and fro in the air" as produced "in a way similar to lightning." As did Fludd when explaining stars, Gassendi compared the generation of fiery meteors to the flight of a rocket, as pockets of igneous material ascended into high regions of the air before flaring up in burning matter and descending again.[79] Others used fireworks to consider the formation of parhelia, or multiple suns. If "Nature is much more ingenious then Art, which represents [the sun] at pleasure by artificial fires," then it seemed possible that nature gave the appearance of parhelia by igniting "one or more Clouds round and resplendent like the Sun."[80] Thus, artifice served to reveal the secrets of nature's spectacles as much as it did those of the court.

Closely related to these meteorological explanations were discussions of physiology since the bodily microcosm was necessarily linked to the heavenly macrocosm. As we saw in the previous chapter, the bodies of humans and animals were in the sixteenth and seventeenth centuries commonly represented with pyrotechnic automata, in reference to life and generation. The animated action of gunpowder prompted many to propose its ingredients as the vital agents responsible for life itself. Fire had long been associated with passionate behavior, virility, and vitality, and now this took on a pyrotechnic cause.[81] As Paracelsus claimed that an aerial niter and sulfur were responsible for thunder and lightning in the high regions of the atmosphere, he also claimed that in the lower regions the same agents were inhaled into the body to provide nutrients for the soul.[82] Paracelsus's followers went on to identify a variety of sulfurous or nitrous agents responsible for respiration, equating them with the *flamma vitalis*. Such agents explained vitality but could also explain pathological states of excess passions. Bodily convulsions and fevers were, thus, equated with microcosmic "meteors" inside the body. As the French physician Joseph Du Chesne explained: "This Niter-Sulphurus stinke is that which manifestly causeth in us fiery meteors . . . which bringeth forth innumerable passions and paines."[83] Like other proponents of nitrosulfurous meteorology, Du Chesne linked these phenomena with fireworks. While he claimed to detest the destructive forces of gunpowder, still he confessed that "it deserveth great admiration, in that it sheweth forth so great, and incredible effects, when as we being in the lower parts, it representeth thundrings and lightenings, as if they were in the aire aloft."[84]

Such pyrotechnic explanations endured into the eighteenth century. Around 1700, William Derham conceived lightning as a fiery ball that expanded and broke through a shell of air "much like that of a granado or a bomb. The flame growing bigger, it forces its passage thru its enclosing shell; and the lightning to stream out at one or more holes made throw the shell."[85] Later we shall see how these ideas provided an important context for new electrical accounts of meteors in the eighteenth century. For now it is enough to note that inventions in fireworks, as much as in ordnance, provided a significant resource for new philosophies and meteorology. As Sir Thomas Browne pointed out in his *Pseudodoxia epidemica* of 1650: "Surely a main reason why the Ancients were so imperfect in the doctrine of Meteors, was their ignorance of Gunpowder and Fire-works, which best discover the causes of many thereof."[86]

## Fireworks and the Future of Natural Philosophy

The new pyrotechnics of the sixteenth and seventeenth centuries thus proved inspirational for a number of new philosophical pursuits. Pyrotechnic practices and ingenious inventions fed into magical and alchemical experiments, where the variation of effects offered a means to apply techniques to novel ends, such as flying. Alternatively, pyrotechnic knowledge offered resources for more speculative inquiries, with the result that fireworks became models for accounts of meteors, the heavens, and the human body. Artificers and scholars contributed to these projects, relying on one another's skills for credit and innovations. With the emergence of new philosophies of nature, as with the emergence of a new courtly pyrotechny, practitioners also sought more substantial and permanent locations for their labors. Gunners had fashioned new spaces for their art in the sixteenth and seventeenth centuries, in urban sites where ingenious festivals, arsenals, and laboratories displayed their creators' powers over nature. Natural philosophers equally sought to make space for new sciences and drew up plans for scientific organizations and academies accordingly.[87] A number of projects, often presented as spectacles themselves, reflected the new scholarly interest in pyrotechnics and suggest a variety of values that philosophical communities attributed to fireworks and to the inventive skills of the gunners and artificers who made them.

Mario Biagioli reminds us that early modern philosophers sought credit for their discoveries by presenting them as emblems and incorporating them in courtly festivals. Galileo's discoveries of the moons of Jupiter thus served Medici spectacle as emblems of the ruling dynasty.[88] Proposals for the *organization* of science could similarly be presented as spectacles, while fireworks, as either military art or festive entertainment, figured in a number of early plans for philosophical institutions. The radical Dominican magician Tommaso Campanella forged influential schemes for the reform of knowledge and society amid turbulent struggles for social control in his native Naples. In *De monarchia* (1600) and *City of the Sun* (1602), he proposed a new social order founded on communal and philosophical principles, a vision that he shared with other natural magicians critical of the Spanish regime in Naples, which he had attacked by instigating a revolt in Calabria in 1599.[89] Campanella's leveling social order raised the status and role of artisanry and blended it with philosophical knowledge in spaces that invoked the arenas of courtly fireworks displays. Hence, the concentric walls of Campanella's imagined city incorporated collections of paintings depicting the arts and sciences, while a temple stood at the center, displaying an artificial vision of the heavens in its dome. Solarians elevated the cultivation of mechanical ingenuity, one wall showing "all the founders of laws and of sciences and inventors of weapons."[90] Military pyrotechny was among these prominent arts: "They also know great secrets about how to produce artificial fires for use in naval and land warfare and secrets of strategy by which they never fail to be victorious."[91] In proposals to the Spanish monarch, Campanella likewise argued for the creation of "Publick Work-houses, for the exercise of *Mechanical Arts*, to which this People is exceeding Apt; and so by this means will the Businesse of Navigation be much promoted, together with the skill of Besieging Towns, and of taking them in by the use of Artificial Fire-works. By this means the People (probably) will be taken off from their False Religion, and divided one from another; to the great Advantage of the King, and Kingdom of Spain."[92]

In a context of war, the fireworks laboratory became the scientific academy.[93] Francis Bacon, alternatively, found room for both festive and military pyrotechnics in his schemes for the advancement of science. His enthusiasm for the arts as a model of progress and experimental science has often been noted, but it is worth emphasizing the importance he placed on pyrotechnic arts in his discussions.[94] Bacon saw in the arts a model of progressive and practically useful labor that should supersede the idle speculations and submission to ancient tradition and book learning characteristic of scholasticism. Bacon took the myth of Prometheus, stealing fire from heaven, to stand for renewal of the sciences.[95] The modern inventions of printing, the compass, and gunpowder were all evidence of the value of the arts, and Bacon identified the creation of artificial meteors as exemplary of his assertion that a new philosophy would go further than that of the ancients: "Not for nothing we have opposed our modern 'There is more beyond' to the 'Thus far and no further' of antiquity. The thunderbolt is inimitable, said the ancients. In defiance of them we have proclaimed it imitable, and that not wildly but like sober men, on the evidence of our new engines."[96] Such engines then regularly featured in Baconian schemes. In the *New Atlantis* (1626), Bacon fashioned a new scientific space through a vision that Donna Coffey has likened to a courtly masque.[97] Presenting Solomon's House as the model of a scientific institution, the *New Atlantis* incorporated many artificial fires. Hence the "Engine-House . . . for all sorts of motions" included a number of effects familiar from the gunners' and magicians' works: "We represent also Ordinance and Instruments of War, and Engines of all Kinds, and likewise new Mixtures and Compositions of Gun-Powder, Wild-Fires burning in Water and Unquenchable: Also Fire-Works of all Variety, both for Pleasure, and Use. . . . We imitate also Flights of Birds; We have some Degrees of Flying in the Ayr. . . . We imitate also Motions of Living Creatures, by Images of Men, Beasts, Birds, Fishes, and Serpents; We have also a great Number of other Various Motions, strange for Equality, Finenesse, and Subtility."[98]

Notable here is Bacon's inclusion of flying mechanisms in a list of fireworks, recalling Della Porta's flying dragon, and the juxtaposition of fireworks and imitations of living creatures, recalling pyrotechny's fascination with generation and motion. Indeed, Solomon's House contained "Great and spacious Houses, where we imitate and demonstrate Meteors; As Snow, Hail, Rain, some Artificial Rains of Bodies, and not of Water, Thunders, Lightnings; Also Generations of Bodies, in Air; As Frogs, Flies and divers Others."[99] In the Engine-House, Bacon also included the water balls popular in aquatic fireworks, and he elsewhere proposed the nature of fires burning in water as a suitably curious subject for a natural history. Bacon himself explained the phenomenon by reference to the contraries familiar to gunners, namely: "a Horrible Thundring of *Fire*, and *Water*, conflicting together."[100]

For Bacon, then, both the military and the festive arts of the gunner should be part of natural philosophy. Equally important for him was the process of ingenious invention. Like Campanella, he celebrated invention in *New Atlantis*. Solomon's House contained "two long galleries" that would be filled with "all manner of the more Rare and Excellent Inventions," with statues erected to the "Principal Inventours," of ships, printing, astronomy, and "Your Monk that was the Inventour of Ordinance, and of Gunpowder."[101] In *Novum organum*, Bacon discussed experiment as an "art of invention" and recommended the study of "artificial fires" as suitable to an experimental natural philosophy, being one of several arts that "alter and prepare natural bodies and materials of things."[102] Simultaneously, he demanded that art alone was incapable of rapid progress since its discoveries occurred by accident and not through systematic experiment, of which Black Berthold's accidental discovery of gunpowder was exemplary.[103] To "anticipate accident," it was necessary to bring art and science together in order to understand the underlying forms of nature, which experiment would reveal.[104] To this notion of invention Bacon opposed the idea of invention typical of rhetoric, signifying the variation of a basic theme: "The invention of speech or argument is not properly an invention; for to invent, is to discover that we know not, and not to recover or resummon that which we already know, and the use of this invention is no other, but out of the knowledge, whereof our mind is already possessed. . . . So as, to speak truly, it is no invention."[105] He equally derided variation as a method in the arts, proposing that "small elaborations and extensions of arts" were useful but failed to produce real innovation.[106] Here Bacon underestimated the value of such practices in sixteenth-century arts since it was precisely the variation of basic forms, so often figured in works on pyrotechny, that gave rise to many of the engines presented in Solomon's House. Furthermore, gunners' notions of invention were not static. The English gunner John Babington echoed Bacon's position closely when he explained that, while his treatise *Pyrotechnia* of 1635 "may seeme to serve onely for delight and exercise, yet as by the handling of these there may bee gained knowledge in the natures and operations of the severall ingredients and their compositions; so the due consideration of the ordering of them may excite and stirre up in an ingenious minde, sundry inventions more serviceable in times of warre."[107]

Bacon thus divided two notions of invention hitherto unresolved, a split between rhetorical and operative visions of invention that would be consequential in the future.

## The Theater of Nature and Art: Furttenbach and Leibniz

Bacon sought to divorce the sciences from the rhetorical ingenuity typical of courtly arts, appropriating its productions, but discarding its reliance on the variation of existing forms as the basis of innovations. Nevertheless, in the *New Atlantis* he appealed to the highly rhetorical spaces and techniques of courtly festival to forward his schemes. The same was true for two German scholars, Joseph Furttenbach and Gottfried Wilhelm Leibniz, who planned, and even performed, their organizations as spectacle.

Joseph Furttenbach has traditionally been of interest to historians for introducing Italian Renaissance techniques of architecture and garden design to southern Germany in the middle decades of the seventeenth century.[108] But he also brought fireworks. Between 1608 and 1620, he traveled about Italy studying fortifications, architecture, and commerce.[109] A guild of fireworks masters existed in Genoa by this time, and there Furttenbach learned the art from one Hans Veldhausen, a gunner from Augsburg.[110] Returning to Germany, Furttenbach settled in Ulm, a free imperial city grown rich on commerce and the textile trade, where for several decades he operated as a merchant and city councilor overseeing building works. Furttenbach set about promoting a range of arts in Ulm, using various means. He wrote prolifically and published some two dozen books on military and civil architecture, fireworks, and artillery.[111] At his residence in Ulm Furttenbach also kept a cabinet of curiosities and a formal garden containing ingenious grottoes.[112] The *Kunstkabinett* included models of ships, pumps, and waterworks and a model of his examination firework undertaken in Genoa.[113] Evidently, all these pursuits were part of a continuum for Furttenbach, each site a display of wonders and practical knowledge intended to demonstrate his virtuosity and to instruct the people of Ulm in useful and pleasing arts.

Furttenbach also mobilized fireworks to this end. He trained his son and nephew in pyrotechny, and the three of them became members of a local guild of fireworkers, organized around master examinations requiring the performance of a *Probefeuerwerk* or examination display, for which apprentices received a certificate.[114] Furttenbach then promoted his works on the arts and sciences using fireworks. His son, Joseph Furttenbach the Younger, painted a *Probefeuerwerk* that was designed by Furttenbach himself and the Ulm merchant Johann Khoun and set off in Khoun's garden in August 1644 (plate 2).[115] The performance did not center on a castle or a temple as normal but instead showed freestanding fireworks on the ground in addition to Khoun, Furttenbach, and Furttenbach's son and nephew setting them off. This was also a proposal for the organization of the sciences. In the background, the younger Furttenbach depicted a scheme of the mechanical sciences that was printed in the same year as the frontispiece to a glossary of his father's works. The scheme laid out the order of the arts and sciences by means of a series of human figures standing on a platform. In the center stood the personification of Mechanica and below, on either side, her fourteen "sons and daughters" representing the arts and sciences, with Pyrotechny standing beside Geometry, Arithmetic, Perspective, Fortifications, and others.[116] Thus, like Galileo, Furttenbach made a spectacle of his science, but, rather than make an emblem of a single discovery (such as the moons of Jupiter), he made the very order of the arts and sciences his theme. This was an original premise, using art to promote, not religion, or a prince, but art itself.

Another enthusiast for the use of spectacle to promote the sciences was Gottfried Wilhelm Leibniz. Matthew Jones has argued that from the 1670s Leibniz followed the Jesuits and natural magicians in identifying wondrous devices and spectacular effects as powerful media for inciting rational pursuits. Since arguments by themselves were unable to move people to utilize their inherent will to do good, something was required to move their passions and stir them to rational thought. Rhetoric, the "power of persuading," might achieve this end for Leibniz, as would play, spectacle, and the wonders of natural magic. However, whereas gunners or magicians celebrated spectacle's links to contemplation, Leibniz viewed spectacle as a necessary evil, the only way to make *honnêtes gens* interested in the sciences: "It is truly . . . to make an antidote from poison."[117]

If this was his attitude, Leibniz certainly spent a great deal of time thinking about this "poison," devising, throughout his life, schemes for a "theater of nature and art" in which fireworks, optical effects, and mechanical marvels featured prominently. In 1675, soujourning in Paris, he wrote a short text entitled *Drôle de pensée* in which he described to himself an ideal academy of wonders, plays, and games for enticing the beau monde to rational pursuits. It would include, among the diverse exhibits, "artificial meteors, all sorts of optical wonders; a representation of the heavens and stars. Comets . . . fireworks . . . Mock battles; Exercises of infantry . . . Naval combats in miniature on a canal. . . . The Show could always be combined with some story or comedy. Theatre of nature and of art."[118]

Like Furttenbach, Leibniz also put art itself on display, planning stage performances of artisanal techniques. Also present were wonders such as magic lanterns, multiplying perspectives, and shadow theaters, devices often taken as seminal for Leibniz's philosophy of comprehensive wisdom and multiplying perspectives. Pyrotechnics figured among these optical shows, which included "pictures on oiled paper and burning lamps or lanterns," "figures who could walk, with a little illumination inside them," and "flying dragons of fire, etc., [which] could be made of oiled paper, illuminated."[119] Leibniz no doubt learned of such effects as a reader of books on gunnery and Jesuit and German treatises on magic and mathematical recreations such as the works of Daniel Schwenter and Georg Philipp Harsdörffer. Like Bacon, he also took the new value given to ingenious invention and variation that such works promoted as consequential for the improvement of the arts and sciences. He conceived his theater as a site productive of knowledge and particularly of new inventions: "It would open people's eyes, stimulate inventions, present beautiful sights, instruct people with an endless number of useful or ingenious novelties. All those who produce a new invention or ingenious design might come and find a medium for getting their invention known. . . . It would be a general clearing house for all inventions, and would become a museum of everything that could be imagined."[120]

Like Solomon's House or the City of the Sun, Leibniz's academy was another site for invention. Leibniz's notion of invention was quite different than Bacon's, though both sought a more systematic method of invention than those available. For Bacon invention needed to eliminate endless variations of the same store of techniques and achieve more fundamental innovation via the systematic pursuit of nature's forms. In contrast, Leibniz celebrated precisely the emblematic and courtly definition of invention: endless variation and elaboration. Drawing on the art of combinations of Lull, Della Porta, and others, Leibniz's *ars inveniendi* brilliantly proposed the resolving of all possible thoughts by varying combinations of their elemental forms.[121] As a courtier, Leibniz also busied himself constantly with material inventions, in mining, mechanics, and fireworks.[122] It was a philosophy reflected in his academy, "a museum of everything that can be imagined." The endless variety of exhibits in the *Drôle de pensée* recalls the open-ended variation of fireworks typical of sixteenth- and early-seventeenth-century pyrotechnic treatises.[123] Indeed, Leibniz well knew his local fireworks and pointed to them when he singled out the sites of greatest ingenuity. "The Germans," he wrote in 1671, "were for all time busy producing moveable works that satisfied . . . the curiosity of great men. . . . Germany, and in particular Augsburg and Nuremberg, is the mother of invention, both weight and spring clocks, the ever powerful, admirable fireworks, even of the air and water arts."[124] Leibniz also drew on fireworks to elaborate his philosophy of nature. In the *Theodicy*, he took the dragon firework running on a line as a model for the divine preestablished harmony. Attacking the suggestion that a ship's path could not be predicted, even by God, he explained: "As God orders all things at once beforehand, the accuracy of the path of this vessel would be no more strange than that of a rocket passing along a cord to a firework, since the whole regulation of things preserves a perfect harmony between themselves by means of their influence one upon the other."[125] The order of the universe in Leibniz's new science, as in many other new sciences in the seventeenth century, reflected the order of pyrotechnics.

## Conclusion

As gunners mobilized the liberal arts to transform their craft, so fireworks were put to use in new philosophies of nature. Philosophers were fascinated by the new pyrotechnics and came together with gunners and their skills in the new urban sites of festival and via the printed literature on pyrotechny. Fireworks then figured prominently in an array of efforts to create new sciences in the late sixteenth century and the seventeenth. Both gunners and scholars contributed to this process, hybridizing techniques that made pyrotechny more learned and science more pyrotechnic. Fireworks affected both philosophical practice and theory and were soon appropriated into magical and experimental repertoires, used to cultivate the mind and spirit and to imagine the mechanisms of heavenly, meteorological, and bodily phenomena. Particular fireworks, such as the fiery dragon, passed from spectacles into magic and alchemy, where new variations of their use could secure credit for ingenious practitioners like Della Porta and Burattini. Alternatively, the supposedly fearsome powers of fireworks over the vulgar might be exploited in colonial, magical, and religious settings. Ingenious artifice, creating effects and revealing their causes to select audiences, became a hallmark of natural magic. Fireworks also served more speculative studies. Gunners long worked to make gunpowder tricks into meaningful emblems, and courtly significance was extended to philosophical significance in books of mathematical recreations. In Hanzelet's and Malthus's works, knowledge of pyrotechny and other secrets not only gave elites access to the hidden mechanisms of courtly wonders but also led to contemplations of nature and the spirit. Scholars who read these works absorbed their message, using fireworks to comprehend the causes of various natural phenomena in subsequent decades, establishing a widely held gunpowder theory of meteors and the body. When philosophers fashioned new spaces for science, fireworks were again incorporated into their plans. Different schemes emphasized different elements of the spectacular and pyrotechnic in the sciences, with Bacon identifying pyrotechny as a sign of progress and an object worthy of investigation and Furttenbach and Leibniz using fireworks as a means to promote rational schemes in a theater of nature and art. Pyrotechny thus proved to be a significant and enduring element in the new sciences of the seventeenth century. How fireworks played a role in the academies that emerged from these plans is the subject of the next two chapters.

CHAPTER THREE

# A Touch of Cold Philosophy: Incendiarism and Experiment at the Royal Society

As Bacon and Leibniz took the arts of ingenious invention in new and distinct directions, so the academic cultures that emerged in the seventeenth century and the early eighteenth also diverged in their approaches to nature and, consequently, their interactions with pyrotechnics. In the next two chapters, we shall see how proponents of new sciences engaged with fireworks in two different settings, the Royal Society of London and the St. Petersburg Academy of Sciences. In each location, different political and social conditions helped shape distinctive interactions between natural philosophers and fireworks. In London, the troubled restoration of the monarchy witnessed associations of fireworks with Catholic plotting and fiery religious enthusiasm, troubles that natural philosophers sought to control through the experimental management of fiery spirits. London's markets also beckoned, prompting a peculiarly commercial interest among late Stuart philosophers and a practical taste for fireworks. In St. Petersburg, in contrast, the Academy of Sciences founded by Leibniz's successors found that natural philosophy itself was a most contentious problem among the Muscovite nobility. Fireworks could be mobilized, as Leibniz had used them in Paris, to make the sciences appeal to the tastes of skeptical elites.

Chapter 1 proposed that different styles of pyrotechnics emerged in Europe not so much along denominational lines, with "miraculous" fireworks in Catholic countries and "open" fireworks in Protestant countries, as according to the sites to which fireworks referred, courtly fireworks displaying trickery and scenery typical of church dramas, and triumphs displaying fireworks as a playful rendition of the battlefield. Historians have proposed similar divisions in styles of experimental natural philosophy as it developed in the seventeenth century. English Protestant natural philosophers thus worked hard to construct experimental traditions in opposition to Catholic beliefs and Protestant sectarians or "enthusiasts," whose claims to authority based on dogmatic beliefs were widely held as being responsible for the English Civil War. Experiments for Catholics, such as Della Porta or the Jesuit Kircher, meant creating playful or wondrous effects that, like fireworks, would astonish audiences and illustrate the miraculous and, ultimately, mysterious nature of the divine creation.[1]

Protestants viewed such performances as evocative of Catholicism in general, as nothing more than false miracles designed to force dogmatic beliefs on the credulous. They thus fostered an "antitheatrical prejudice," seeing in natural magic, as in Catholicism, *artifice, cunning,* or *craft,* terms of art that, as Rob Iliffe notes, now took on for the first time a negative meaning, as "a capacity for misrepresentation and treachery."[2] Protestant experiment was, instead, supposed to be "inartificial," making a revelation of the secrets of nature by opening up, in sober experimental performances, the hidden workings of the universe, which God had designed and intended to be open for all to see and comprehend via the medium of philosophers. For Protestants, experiment assumed and secured a world in which there were no mysteries or miracles, while its communal pursuit in academies avoided association with personal dogmatism and politically dangerous enthusiasm.[3]

Such characterizations are surely right in general but should not be seen as definitive distinctions exclusive to particular places or denominations.[4] This chapter suggests that categories were more flexible, by tracing simultaneously the history of experiment and that of pyrotechnics in post–Civil War England. In this context of Protestant experiment, one might expect a clash between the overtly miraculous and theatrical artifice of fireworks displays and the sober inquiries of experiment, but this was not the case. Rather, there existed a shifting set of interactions that changed on the basis of evolving tensions, anxieties, and solutions in relation to problems of Catholicism and enthusiasm. Without assuming an antitheatrical prejudice among English natural philosophers, we might follow E. P. Thompson, who proposed that early modern England was a site where the solution to theatricality was more theatricality of a different kind.[5] Similarly with experiment, it was often the case in the Restoration era that philosophers sought to distance experiment from artifice and spectacle and from pyrotechny in particular, which was frequently associated with Catholic plotters, political violence, and fiery passions. However, this was not always the case, and, when occasion suited, it was useful to invoke pyrotechnics in experimental philosophy, or even to practice it, to appeal to new audiences or examine novel features of nature. Experiment was another form of theater, which occasionally and increasingly during the late Stuart era could be allied with fireworks. In fact, in comparison with French and Russian, English natural philosophers would foster very close connections with the practitioners of artifice and spectacle.

## Fireworks in the Interregnum

Pyrotechnic spectacle was never the prerogative of a single community in the strife-torn context of early modern England but served a variety of opposing political and philosophical ends. Although Queen Elizabeth had been shown numerous fireworks on her visits about England, newfangled pyrotechnics were also associated with Catholics' plots. The Lopez trial of 1594 summoned various accusations against alleged Spanish conspirators, among them a plot "to burn the navy and ships with poisoned fireworks" and "to kill the Queen by means of a clergyman . . . who will destroy her by casting artificial fire, on occasion of some festival, the object being on her death to raise divisions, in which some would take the King of Castile's part, that he might assume the kingdom."[6] Guy Fawkes's Gunpowder Plot further bolstered correlations of Catholicism and incendiary plotting but met with a similarly incendiary response, the failed plot prompting an annual remembrance on the fifth of November celebrated with bonfires, effigy burning, and the ringing of church bells.[7] Pyrotechnics also served the Puritans. During the Civil War, Parliament staged anti-Catholic fireworks aimed against the king, showing "Fire-boxes like meteors, sending forth many dozen rockets out of the water, intimating the popish spirits coming from below to act their treasonous plots . . . Runners on a line, intimating the papists sending to all parts of the world, for subtle cunning and malicious plotters of mischief."[8] Royalists dismissed the shows as "squirting and squibbing fooleries."[9] *Pyrotechny* also continued to refer to alchemy, and, in mid-seventeenth-century England, the schools of Johannes Baptista van Helmont and Paracelsus, often allied with radical reform, could be denominated *pyrotechnic*. Thus, the alchemist and "philosopher of fire" George Starkey, the author of the alchemical treatise *Pyrotechny Asserted* (1658), joined Samuel Hartlib and influenced Robert Boyle to seek a "universal laboratory" and chemical utopia during the Interregnum.[10] During the 1650s, the schoolmaster and Commonwealth preacher John Webster alternatively mixed magic and Baconianism in his visions of future education, in which operative knowledge of the "sublime, and never-sufficiently praised Science of *Pyrotechny* or *Chymistry*" would free learning from the verbosity and decay of scholasticism:[11]

> That youth may not be idlely trained up in notions, speculations, and verbal disputes, but may learn to inure their hands to labour, and put their fingers to the furnaces, that the mysteries discovered by *Pyrotechny*, and the wonders brought to light by *Chymistry*, may be rendered familiar unto them: that so they may . . . be taught by manual operation, and ocular experiment, that so they may not be . . . idle speculators, but painful operators . . . indeed, true Natural *Magicians*, that walk not in the external circumference, but in the center of natures hidden secre[t]s, which can never come to pass, unless they have Laboratories as well as Libraries, and work in the fire, better than build Castles in the air.[12]

Alchemical pyrotechny thus remained salient in philosophy prior to the Restoration but was never the prerogative of a single religion or cause.

## The Restoration, Spectacular Science, and Chemical Physick

The royal cause was reasserted in 1660. At the Restoration, Charles II renewed court spectacle and pyrotechny on a grand scale. "Shew your Selfe Gloryously, to your People; Like a God" the Duke of Newcastle urged the King.[13] On Charles's return from exile, there were fireworks and pomp exceeding "the glory of what hath passed of the like kind in France."[14] To make the coronation fireworks, a Swedish engineer, Martin Beckman, was brought to London and appointed "firemaster" to the Board of Ordnance.[15] English princes had relied on foreign expertise in artifice before, with James I employing Dutch and Danish artificers in 1606. When princes needed pomp, they were prepared to supplement local skills with foreign expertise in order to attain it. The arrival of Beckman in England signaled the beginning of an increasing international traffic of pyrotechnic skills in Europe that would ultimately transform the genre.[16]

The Royal Society's activities were also spectacular in these times of renewed pomp. Meeting in rooms belonging to the mathematical professors of Gresham College in Bishopsgate, in the nascent commercial city of London, the collective of Oxford scholars and London gentlemen professed themselves heirs to Bacon and examined a variety of pyrotechnic experiments.[17] Thomas Sprat claimed in his *History of the Royal Society* in 1666 that at the Restoration "many Worthy Men . . . began now to imagine some greater thing; and to bring out experimental knowledge, from the *retreats*, in which it had long hid itself, to take part in the *Triumphs* of that universal Jubilee."[18] The virtuosi often showed fiery and gunpowder effects in the Society's early years. The Grey's Inn lawyer and virtuoso Dudley Palmer showed a powder that, "thrown into the flame of a candle, burns and sparkles with a noise like . . . loose corns of gunpowder."[19] Lord Brouncker shot cannon in the courtyard of the Society's meeting place at Gresham College to study recoil.[20] Robert Hooke blew up shells filled with high-strength gunpowder using a recipe communicated by Prince Rupert. He compared the recipe to the highly explosive *aurum fulminans* and *pulvis fulminans*, discussed by Della Porta as the cause of thunder, with a powder trier of his own design.[21] Other experiments involved candles, lamps, niter, sulfur, camphor, and various combustible oils and spirits. As in earlier-seventeenth-century pyrotechny, gunpowder linked meteors and physiology together. Failed combustion and animal deaths in Boyle's air pump suggested a common material support for life and fire. In the tradition of Paracelsian proponents of an "aerial niter," Hooke argued that saltpeter or niter might be the common cause of combustion and respiration since it burned in the air pump, in conjunction with "sulphureous bodies."[22]

Initially, Charles and the Royal Society both supported other promoters of pyrotechny. A planned Society of Chemical Physicians carried over Interregnum Helmontian utopianism into the Restoration and included among its signatories Charles's alchemist Nicolas Le Fèvre, the magician Cornelius Drebbel's son-in-law Johannes Sibertus Küffler, and the pyrotechnic alchemist Starkey.[23] Critics of these plans equated alchemical pyrotechny and enthusiasm since private alchemical labors seemed dangerously equivalent to the private inspirations of enthusiasts. Furthermore, fiery vocabularies signaled the dangers. In *Pirotechnia*, Biringuccio had celebrated the equation of fire with consuming passions and burning desires, but English critics equated fiery passions with threatening zeal and excess. Thus, William Johnson lambasted chemical society protagonists such as George Thomson as fiery enthusiasts whose doctrines were "spit out of the mouth of a Zealous Brother at a meeting, where he holds forth the Doctrine of *Vanhelmont*, as down right Gospel. . . . *Let him . . . reflect upon himself, how he hath deceived, and been deceived*. Ah! . . . How has Chymistry contributed to make him Spiritual, and his trading in the fire inflam'd his Zeal?"[24]

The Royal Society's spectacular experiments also came under attack. Accusations of fiery enthusiasm against the chemical physicians were equally directed at the Royal Society and constituted real threats as religious paranoia was stoked further a year later. The Great Fire of 1666 was widely seen as a Catholic plot to incinerate London with pyrotechnics.[25] Again, bodily passions, fire, and false religion were joined together. Fireworks allegedly belonging to Catholics were put on display in coffeehouses as evidence of plots to create further conflagrations.[26] Catholic incendiarism was linked to the

12. Frontispiece to *Pyrotechnica Loyolana, Ignatian Fire-works; or, The Fiery Jesuits Temper and Behaviour* (London, 1667). Beinecke Rare Book and Manuscript Library, Yale University.

fiery zeal of religious enthusiasm, prompting continuous analogies between papists and pyrotechnics (fig. 12). In *Pyrotechnica Loyolana,* Jesuits were literally incendiaries, instructed in making fireballs and other explosive devices to blow up London: "That, the *Jesuits* are ambitious, their *Founders* name signifies a fire-brand, quasi *ab igne natus;* and that his disposition was *Fiery*, and his profession *Military;* whereupon they affirm he came to *send Fire*. Hence *de jure* they profess the *Art* of making and casting about *Fire-balls* and *Wild-*

*fire* to burn *Houses* and *Cities*: to promote which, they have two *Colledges*, one at *Madrid*, another at *Thonon* [in France] to advance the *study* of *Artificial Fire-works*, and to subdue Protestants by *fraud* and *Arms*: they keep *stores* of *powder* in their Colledges."[27] Sermons and pamphlets multiplied condemning papists and Jesuits, "those grand *Incendiaries* in all senses."[28]

Against this perceived Catholic incendiarism, threatened communities did not shy away from artificial spectacle but fought fire with fire, asserting their own pyrotechnic theater against supposedly subversive alternatives. The royal predilection for spectacle renewed court fireworks. Another foreign firemaster, Ernest Henry de Reüs, was appointed, and both the king's and the queen's birthdays were celebrated with pyrotechnic displays in the aftermath of the Great Fire.[29] Crown institutions tried to monopolize fireworks. In 1668, the City of London imposed a ban "that no person whatsoever be henceforth permitted at any time to make, or cause to be made any sort of Fire-worke . . . within the City," with the exception of members of the Ordnance charged with royal fireworks. Repeated proclamations indicate the ban's failure.[30]

Alternatively, in 1667, in the midst of this intense suspicion of Catholic incendiaries, the Royal Society promoted numerous accounts of pyrotechny but worked hard to present itself as free from fiery spirits. Thomas Sprat's *History of the Royal Society* promoted the value of philosophy with statements against excessively inflamed passions and spectacular effects but still mobilized pyrotechny to make the point. Experimenters linked fiery dispositions with failed knowledge. Only cooled and controlled passions produced reliable knowledge, and only a philosophical community such as the Royal Society would, according to Sprat, discount the "violent and fiery" tempers of individuals in favor of more sober judgments.[31] "Confusion, unsteddiness, and the . . . animosities of divided Parties" would be stopped by "a singular sobriety of debating, slowness of consenting, and moderation of dissenting," with the result that for fellows "there was no room left, for any to attempt, to heat their own, or others minds, beyond a due temper."[32] Where "Halcyon Knowledge is breeding, all *Tempests* will cease."[33] Sprat supported the Society of Chemical Physicians by allying their opponents with incendiary religion and the chemists with temperate behavior, having "onely the discreet, and sober flame, and not the wild lightning of the others Brains."[34]

Similarly, excessively prodigious "effects" might still be studied but only to reveal their causes and undercut their drama. Here, the courtly tradition of revealing secrets to elite audiences was mobilized as religious apologetics. Experiment, unlike spectacle, was productive of knowledge and not merely the astonishment of the senses: "Things, which now seem *miraculous*, would not be so, if once we come to be fully acquainted with their *compositions*, and *operations*."[35] Sprat now claimed that experiment was an "inartificial process."[36] While experiment mobilized artifice, in the form of instruments, to reveal nature's operations, the resulting effects were not to be considered as human products. Sprat's claim, recalling Bacon, was that instruments would speak for themselves, revealing natural "matters of fact" divorced from the artifices of speech and dispute. In the context of Catholic incendiarism, the experimental philosophy would make nature, rather than artifice, a source of compulsion.[37] Sprat presented the philosophers' goal as being "to separate the knowledge of Nature, from the colours of Rhetorick, the devices of Fancy, or the delightful deceit of Fables. . . . They have attempted to free it from Artifice, and Humors, and Passions of Sects . . . not so much by any solemnity of Laws, or ostentation of Ceremonies, as by solid Practice, and examples: not by a glorious pomp of Words, but by the silent, effectual, and unanswerable Arguments of real Productions."[38]

Sprat went on to exemplify such judicious philosophy with accounts of diverse experiments on fire, combustion, and ordnance that tamed and managed fiery spirits, supposedly showing, for example, that "there is no such thing, as an Elementary Fire of the Peripatetics, nor Fiery Atoms of the Epicureans."[39] He showed how philosophers might improve the state's weapons against its enemies, incorporating accounts of gunpowder experiments and Thomas Henshaw's "Histories of Saltpeter and Gunpowder." The latter, part of the Society's "History of Trades" project, was supposed to reveal current artisanal methods of powder production, with an eye to their improvement. Philosophical pyrotechny was here clearly marked as loyal to official state pyrotechny with references to sites like the Tower and to the patronage of Prince Rupert.[40]

## Campanella Revived and the Philosophy of Explosions

Sprat's rhetorical distinction between *experiments* and the production of *artificial effects* was supposed to make experiment seem the honest mirror of nature distinct from artifice, now figured as a practice of crafty illusionism and deceit. But it was still easy to identify the Society's experiments with the pyrotechnics of popish incendiarism.[41] Hence the thrust of attacks on the Society's experimental and chymical investigations by Henry Stubbe, another opponent of that "insipid pretender to pyrotechny" the chemical physician George Thomson.[42] Among animadversions against Sprat's and Glanvill's Society apologetics of the late 1660s, it was Henshaw's histories of saltpeter and gunpowder that Stubbe singled out as especially delusive and dangerous. Evidently an expert on powder production himself, Stubbe noted the many errors in Henshaw's history and made clear the consequences: "The Narration is not only *imperfect*, but in many parts *false*, so that . . . the *History of Nature* which they propose . . . will not merit any more Credit (if so much) then [*sic*] that of *Pliny*: and these *Experimentall Philosophers* instead of *undeceiving* the age as to *inveterate Errors* will *multiply new ones*."[43]

Experiment was merely artifice and deceit. Stubbe evoked Campanella's *Monarchy of Spain* to describe the Royal Society's treacherous ends—a project to achieve a universal Catholic monarchy. Hence the title of Stubbe's *Campanella Revived; or, An Enquiry into the History of the Royal Society, Whether the Virtuosi There Do Not Pursue the Projects of Campanella for the Reducing England unto Popery*. Significantly, Stubbe pointed to Campanella's call for the erection of "Publick Work-houses for the exercise of Mechanical Arts" and the teaching of "Artificial Fire-works" as the source of his anxieties: "These are the *passages* which I think I first accommodated to the *Royal Society*, and which served me as a *Key* to expound their *History* by . . . it was a regard I had to the *Religion* and the *Education of our Youth* (which I found undermined by these *Campanella's*) which *imboldened* me never to *lay aside* my Pen. I was afraid lest our *Virtuosi* with their *trinkets* and *experiments* would serve *this Nation* such a trick as the *Pyed-Piper* at *Hammel* in the Dutchy of *Brunswic* did those Inhabitants."[44] The Royal Society's experiments were, after all, papist and pyrotechnic tricks that would lead England into oblivion. Such attacks caused real damage. In 1671, Henry Oldenburg, the Society's secretary, feared that the Society would be eclipsed.[45]

Both the court and the Royal Society reasserted their positions in the next flare-up. Anxieties surrounding Catholics multiplied again in the early 1670s. Charles's Declaration of Indulgence seemed to open the gates to papists, prompting the Test Act in March 1673. The potential heir James, Duke of York's failure to sign it betrayed his Catholicism. Antipopery exploded in the summer, and, on the Protestant holidays in November, the "mob" expressed dissent with "Pope-Burnings," processing an effigy of the pope through Cheapside, a stone's throw from Gresham College, before setting it ablaze on a bonfire: "Ye effigies of 2 divells whispering in his eares, his belly filled full of live catts who squawled most hideously."[46] Another government proclamation against fireworks was issued.[47] Royal Society fellows took note of the anti-Catholic fireworks. Robert Hooke, whose quarters were in Gresham, thus recorded the "Pope-burning" in his diary and later noted: "Saw mock Pope carryed in procession."[48]

In this incendiary location, further experiments followed at the Royal Society, which again asserted fellows' powers to keep fiery spirits subdued.

Philosophers presented experiments that disciplined wildfire. In April 1673, a month after the Test Act, Robert Boyle presented the Society with *New Experiments, Touching the Relation betwixt Flame and Air. And about Explosions*, a key text in the history of combustion and respiration sciences.[49] With the air pump, Boyle extinguished fires by exhausting the receiver. He showed fireworks under experimental control and investigated the nature of enthusiasm. Hence, an experiment "Of the Conservation of Flame under Water" examined fires that burned under water, earlier proposed by Bacon as an effect worthy of investigation. Boyle borrowed fireworks technique for the trial, ramming a powder composition into a large goose quill, the standard method for making squibs, whence "this Wild-Fire was kindled in the Air" and lowered into water, where, as any gunner knew, it burned.[50] The lesson was critical. Hooke thought that air and niter shared a common niter essential for combustion. Burning under water, Boyle argued, might show air unnecessary to combustion, but perhaps niter contained air, as "little aerial particles between the very minute solid ones."[51] He later suggested that this might be the source of "some vital substance . . . whether it be a volatile Nitre, or . . . some anonymous substance," but did not say which.[52] Avoiding hypotheses avoided accusations of enthusiasm, while evidence of a vital niter supported a gunpowder theory of physiology that placed the excessive behavior of false religionists in the realm of scientific scrutiny. Among the proponents of this physiology was the Oxford physician Thomas Willis, whose nitro-sulfurous account of the muscles and nerves used gunpowder theory to explain a variety of pathological states. Bodily convulsions, spasms, fits, and paroxysms were all explicable in terms of the explosive actions of nitrous and sulfurous particles in the blood. These were all pathologies commonly associated with enthusiastic ranting.[53] Indeed, Willis and Richard Lower had earlier attempted blood transfusions to see whether the temper of a madman might be calmed by fresh blood.[54]

In 1674, amid further pope burnings and government fears of incendiarism, there appeared a fresh assertion of this gunpowder physiology in the Oxford physician John Mayow's *Tractatus quinque medico-physici*, which constituted the most wide-ranging application of gunpowder theory yet written, arguing for a "nitro-aerial spirit" responsible for various phenomena of life, motion, meteors, and cosmology.[55] Mayow likened the production of physiological effects to a battle of the elements, familiar in pyrotechny: "The nitro-aerial spirit and sulphur are engaged in perpetual hostilities with each other, and indeed from their mutual struggle when they meet and from their diverse state when they succumb by turns all the changes of things seem to arise."[56] These changes ranged from the body to the heavens.

Mayow affirmed the nitro-aerial constitution of vitality and heat against the alternate theory of vital flames and lucid souls, which, for Mayow, "no less in Anatomy than in Religion, have always seemed to me vain and fanatical."[57] The same spirit was responsible for cosmic phenomena, modeled on the body, "since the Parts of the Air being deprived of the Nitro-aerial Spirit are raised upward, and being there impregnated afresh, return thence downward again; therefore the Air being the Blood as it were of the Macrocosm, is in a continual Circulation, and doth it self, forasmuch as in Circulating it takes in the Nitro-aerial Spirit, exercise a Kind of Respiration."[58] The whole universe might, thus, be accounted for in nitro-sulfurous terms and its explosive acts interpreted by natural philosophers.

Experimenters never referred to *pyrotechny* in these publications, perhaps now fearful of its radical associations, both with Helmontian chemistry and with Catholic incendiarism. Boyle thus distinguished experimental "chymystry" from alchemical pyrotechny, and, on the one occasion he did mention pyrotechny, he referred to it as a craft or "Arcanum," implying that the art of making fireworks was distinct from alchemical pursuits.[59] Separating the craft of fireworks from these incendiary and radical positions might make the former legitimate in philosophy. Thus, Boyle also used fireworks to undermine alchemical accounts of nature, proposing that explanations in terms of sulfur, mercury, and salt were always secondary to mechanical causes. The gunners' ingenious variations of pyrotechnics using simple ingredients offered an analogy as well as a chance for Boyle to explain away spectacle:

> This may be illustrated by what happens in artificial fire-works. For, though in most of those many differing sorts that are made, either for the use of war, or for recreation, gun-powder be a main ingredient, and divers of the phaenomena may be derived from the greater or lesser measure, wherein the compositions partake of it; . . . gunpowder itself owes its aptness to be fired and exploded to the mechanical contexture of more simple portions of matter, nitre, charcoal and sulphur; and sulphur itself, though it be by many chemists mistaken for an hypostatical principle, owes its inflammability to the convention of yet more simple and primary corpuscles.[60]

## The Cold Fire of Phosphorus

Experimentalists located at the epicenter of anti-Catholic hysteria exploited spectacular effects to assert their cosmologies but made great efforts to show how their experiments undercut hot-blooded drama and put nature in the colder light of matters of fact.[61] Perhaps not coincidentally, at the height of pope burnings and anti-Catholic hysteria over Jesuit incendiaries, phosphorus became a significant interest in the Royal Society's experimental programs.[62] Phosphors offered alternative effects to gunpowder in experiments and were intriguingly *cold*, the "icy noctiluca" being capable of emitting fumes and light without heat. There were intimate links between the white phosphorus and fireworks in continental Europe, but London experimenters played these down. Thus, Leibniz's assistant Jobst Dietrich Brandshagen promoted phosphorus in Denmark with demonstrations for the Danish court. Brandshagen smeared phosphorus on his face and on pieces of wood, a technique that King Christian V quickly related to pyrotechny:

> All the wood went up in flames like lightning. . . . The king is quite well disposed towards me and [I] attend on him almost daily, and in order to investigate the matter further he lets me learn the art of fireworks, giving me 10 thaler every month as long as I learn, but once I am done learning, I am to [serve] in the artillery, upon which I shall receive more, at present I receive as much as a fireworker who serves His Majesty, I may use as much phosphorus as I want to, as long as I report everything.[63]

When the German promoter of the white phosphorus Johann Daniel Krafft displayed the new material to Robert Boyle in London in 1677, he did so with a variety of spectacular effects, including writing the word *Domini* in twinkling phosphorus on a piece of paper at Boyle's home, which Boyle thought "a mixture of strangeness, beauty, and *frightfulness*."[64] The trick was reminiscent of magicians writing *Ira Dei* on dragons, and another of Krafft's performances involved igniting gunpowder with phosphorus. English experimenters, especially Boyle's operator Frederick Slare, performed similar feats in the Royal Society's meetings.[65] But, as Jan Golinski has shown, fellows emphasized that these displays must lead to *philosophical truths* or be rejected as mere spectacle: "If we had a mind to act Pageantries, or to spread a story of Goblins, you see how easily it might have been done, by smearing ones hands and face all over with the tincture of light, which adheres so permanently. And besides . . . the manner how it was done being concealed, the learned and ingenious might be at a loss to discover what it might be."[66]

Nature, and philosophy, should now be represented as *open*, to avoid accusations of trickery and deceit in dangerous times. *Utility* also diverted minds away from deception. Boyle listed uses for the "aerial noctiluca" in 1680, including giving light to gunners carrying powder on ships, assisting fishermen with a light under water, and the provision of time at night:

"Divers ludicrous Experiments, very pleasant and surprizing, may be made with the *Noctiluca* . . . but these Trifles, though very pretty in their kind, I purposely pass over; as also an[y] use that may be of *great*, but I fear of *mischievous*, consequence."[67] Artifice and deceit were again linked and rejected. Thus, Boyle offered icy spirits against fiery enthusiasm and restrained spectacles with the phosphorus in a time of Catholic incendiarism. The wits remembered him precisely for this:

> The middle way our Hero wisely chose,
> He had too much *Philosophy*
> An *Atheist* or *Enthusiast* to be.
>
> ....................................
>
> His Zeal, no foolish Fire that leads astray,
> That over Rocks and Precipices leads,
> Pretending pleasant Vales, and flowry Meads,
> His Zeal but trac'd, his Judgement found the way;
> His Zeal, which like the *Phosphor*, shin'd with Lambent Day,
> It warm'd, but did not burn, nor chap the Ground,
> Warm'd and enlighten'd all around.[68]

## Laboratory Effects: Royal and Experimental Fireworks after the Exclusion Crisis

Mischievous spectacles indeed continued to rage through the Exclusion Crisis of 1679–81. Shaftesbury and the future Whigs pumped up support for a bill to exclude James from succeeding Charles by cultivating the pope burnings into increasingly elaborate anti-Catholic spectacle, hiring dubious showmen to add luster to the parades. Thus, the playwright Elkanah Settle was a "lusty Fellow . . . who has an indifferent Hand at making of Crackers, Serpents, Rockets, and other Play-things, that are proper on the fifth of November; and has for such his skill received Applause and Victuals from the munificent Gentlemen about Temple-Bar."[69] The November displays were intensely controversial, prompting further bans and proclamations against fireworks and accusations of treason against Shaftesbury.[70] Only in November 1681, after Shaftesbury had been sent to the Tower, did the last pope burning take place. Those abhorring exclusion had fought Whig spectacle with their own dramas, reasserting royal power. James's progress to security in Scotland was met with loyal bonfires, liberally reported. Charles stepped up royal entries to cities, supposedly "all met with bells and bonfires" and "candles and lanthorns at every house."[71] The king rewarded pomp staged on his behalf. By the winter of 1682, opposition spectacle was routed and Tory drama triumphant.

The wake of these contests saw court pyrotechny redoubled and institutionalized by the crown. Gunners' efforts to provide spectacle for the courts now paid off. Gunners had already shifted the making of fireworks from the Tower to a "great barn" in Greenwich, the shift occurring around 1650. The barn was formally noticed as a "fireworks laboratory" by 1690, and six years later the Ordnance moved it again to Woolwich, until then the site of dockyards. From now on, the gunners' "Royal Laboratory" oversaw regular, expensive, and spectacular fireworks displays for the court. Beckman's and de Reüs's title of *firemaster* was also novel, and, in 1683, formal instructions designated ranks and responsibilities for gunners making fireworks. "Firemasters" were to instruct and oversee "fireworkers," who manufactured pyrotechnics for war and peace, assisted by "artificers" or "matrosses."[72] By 1688, now as "Comptroller of fireworks as well for Warr as Tryumph," Martin Beckman oversaw these operations.[73] Masters were enlisted to serve as fireworkers, and, once again, the crown relied on the importation of foreign artificers, in this case men hailing from the Germanies and Scandinavia. Among them were Charles's first firemaster, John Christopher Woolferman, the Danish gunner Albert Borgard, considered a key founder of the Royal Artillery, and John Henry Hopkey (Hopeke or Hopke), who became comptroller on Beckman's death.[74]

In the wake of the Exclusion Crisis, pyrotechny thus became permanent, organized, and available at all times to the crown. Both fireworks and natural philosophy were now institutionalized, and their communities would increasingly interact. The Greenwich fireworks laboratory was situated at the bottom of the hill on which Christopher Wren and Robert Hooke's Royal Greenwich Observatory was founded in 1675. Traveling artisans also served the sciences. Isaac Woolferman, probably a relation of the fireworker John Christopher Woolferman, was the first astronomer royal John Flamsteed's assistant for many years. Blackheath, also in Greenwich, served as an ordnance-testing ground, where Hooke, Lord Brouncker, and the surveyor to the Ordnance and Royal Society fellow Jonas Moore had attended cannon trials during debates over projectile motion in the previous year.[75]

The court made use of the new fireworks laboratory's skills. In 1683, the queen's birthday, two days away from Ascension Day, was the occasion for unprecedented royal fireworks. The Royal Society fellow John Evelyn attended and was impressed "with pageants of castles, forts, and other devices of girandolas serpents, the King and Queen's arms and mottoes, all represented in fire, such as had not been seen here. . . . It is said it cost £1,500."[76] James II's coronation in 1685 saw grander fireworks yet, a "Wonder-full and Stupendious" drama dispatched by Beckman from a platform of barges on the Thames. Sky and water rockets exploded around two forty-foot-high pyramids and a great artificial sun, while mortars "continually threw up in the Air, Stars, and Drops of White-Fire." It was a meteorological marvel: "Many hundred Globulous forms of Fiery matter . . . broke Perpendicularly upward . . . like a Summers Sun growing to Noon . . . those new Meteors broke into a shower of ten thousands of Stars, and with a brightness that return'd the Day."[77]

The Royal Society was also concerned with increasing performances after the Exclusion Crisis ended, seeking to expand experimental shows at its gatherings. Meeting in February 1682/3, fellows agreed to sue the exiled Shaftsbury's estate for arrears and then noted: "The Society wanted experiments at their ordinary meetings."[78] Frederick Slare and Edward Tyson were duly appointed curators and set about reintroducing arresting performances. In a more secure environment, Slare was now less concerned to separate philosophical effects from other spectacles. While he continued to exploit "cold" materials such as phosphorus and liquors, dramatic experiments now openly evoked fireworks in their heavenly and meteorological parallels. Mixing melted phosphorus and oil of vitriol, "two liquors, actually cold," produced "*sparkling* and *fiery* bodies . . . not a little surprizing" that "throw up such fiery *Balls*, like so many Stars . . . and continue to burn for some time to the great pleasure of the *Spectator*."[79] Slare showed the effect at dinners, an "experiment of a wonderful nature" making "divers suns and stars of real fire, perfectly globular, on the sides of the glas, and which there stuck like so many constellations, burning most vehemently, and resembling stars and heavenly bodies."[80] Slare promoted phosphorus as the true "aerial niter" and now proclaimed it *more* spectacular than traditional pyrotechnics. Phosphoric flashes "illustrate and resemble . . . *Lightning*, farr exceeding those made with Nitre, Gun-*powder*, or *Aurum fulminans*."[81] He posed phosphorus as a complete and superior alternative to gunpowder accounts of physiology, meteors, and cosmology. Produced from distilled urine, phosphorus was also in the body: "I am sure, that the Learned *Willis* (were he alive) would rejoice to see such a *Product* out of our *Bodies*, who was very confident of something Igneous or Flammeous, or very analogous to Fire, that did kindle and impregnate our blood."[82] Evelyn enjoyed the shows and allied Slare's effects to cosmology and generation: "It seemed to exhibit a theory of the education of light out of the chaos, and the fixing or gathering of the universal light into luminous bodies. This matter, or phosphorus, was made out of human blood and urine, elucidating the vital flame, or heat, in animal bodies. A very noble experiment!"[83] Experiments with gunpowder also multiplied at this time. One series by William Molyneux tested the effects of gunpowder and *pulvis fulminans* on the polarity of compass needles, needles having been observed to lose polarity during lightning storms. Thomas Gale presented antique manuscripts on fireworks to the Society; Denis Papin made new trials exploding gunpowder in the air pump.[84]

## Fireworks and the Glorious Revolution

Renewed fireworks at court soon suffered further controversy, however. Once again, contests hinged on alternative strategies for pyrotechny. Absolutist-style pomp, James's Catholicism, and the birth of a male heir revived fears of popery, to culminate in the Glorious Revolution. Fireworks to celebrate the Prince of Wales's birth in July 1688 thus prompted revolt. Once again, Beckman presented pyramids, sun, and ciphers. But, on news of the impending display, rumors circulated that it was part of another papist plot to burn London. Crowds attending the celebrations mocked this "little miscalculation of the court," interpreting a storm that damaged the preparations as "a judgment of God, who had felt Himself braved in the rejoicings for this imposture, for by this name the people now ordinarily characterized the birth of the prince of Wales."[85] When the display took place, Pepys and Evelyn both watched the fireworks, but this time they recorded misgivings: "They were very fine . . . but were spent too soone, for so long a preparation."[86] Once again, November saw fireworks "in spite to the Papists."[87]

Nevertheless, pyrotechny was not to be dispensed with in the ensuing revolution. Spectacle was a critical tool for legitimating the new Protestant reign of William of Orange, but it had to take account of diverse political theaters to be effective.[88] Under William, fireworks were represented as incorporating elements of both popular revels and stately magnificence, appealing to the people for support. Bonfires, dancing, and squib throwing, the elements that typified the London pope burnings and popular November holidays, were, thus, included in images of royal fireworks made by William's Dutch engraver Bernard Lens II. Royal fireworks now took place in Covent Garden, marketplace and site of popular revels (fig. 13). Whereas there were no people visible in images of James II's fireworks, Lens's mezzotints of William's displays were filled with dancers, bonfires, squib throwing, and revelers, combined with the royal insignia, crowns, and machinery

13. Bernard Lens II, *A Perfect Description of the Firework in Covent Garden That Was Performd at the Charge of the Gentry and Other Inhabitants of That Parish for the Joyfull Return of His Majestie from His Conquest in Ireland, 10 September 1690* (London, 1690).


of court spectacle. Recalling the pope burnings, effigies of Louis XIV were shown on the fires, and the popular participation was recorded in print: "While these Illuminations filled the Air with a delightful Variety of Artificial Meteors, the Streets below flam'd full of Bonfires."[89]

Despite this deference to the crowd in royally sanctioned representations, however, there was no room for unsanctioned spectacle in William's new state. The king maintained, and strengthened, the crown's monopoly on fireworks. In 1697, an act of Parliament enforced earlier prohibitions on the popular throwing of squibs, crackers, and rockets, imposing stiff fines and even hard labor in a house of correction for infringements.[90]

Philosophers also sought to secure a steady state in the era of revolution. The same period saw anti-Catholic assertions from philosophers in the form of arguments from natural theology, and these drew on pyrotechny, again presented as distinct from any alchemy, to make their effect. In 1686, Boyle stepped in with *A Free Enquiry into the Vulgarly Received Notion of Nature*, an attack on scholastic and magical attributions of agency to nature whose real target, as Jacob has argued, was the apparent papist revival under James.[91]

Boyle reworked spectacle to give proper messages of Protestant theology and philosophy. While the crowds at James's fireworks scorned the king for seeking illegitimate glory, Boyle admonished those who worshipped idols for diverting praise from the true God. False philosophers, like false kings, made a mystery of powers, attributing agency to occult tendencies and desires in things when really all actions were signs of exclusively divine agency, open to human comprehension. Against a Catholic universe of spectacular effects whose production was unknowable and miraculous, Boyle offered a theater whose effects were all intended by a designer and open for inspection. He often referred to the Strasbourg Cathedral's great automaton clock as a model of apparently autonomous motions in fact resulting from a designer, but he also considered fireworks as an equally appropriate illustration: "Some, perhaps, would add, that a Squib, or a Rocket, though an artificial Body, seems, as well as a falling Star, to move from an Internal Principle. But I shall rather observe that, on the other side, External Agents are requisite to many Motions, that are acknowledg'd to be Natural."[92] In an undated essay on spontaneous generation, possibly from the same period, a similar argument was elaborated against those who saw life generated by nature, not God:

> In the more Curious & Artificiall Fire-workes, that are sometimes made at the Births and Coronations of Princes, there are . . . wonderful Phaenomena exhibited: and you may perhaps see a Fireworke that after having a good while burn'd or smoak'd under water, will rise to the top of it, and after haveing play'd upon the surface shoot it selfe up into the Aire, and there exhibit new and surprizeing Phaenomena . . . and yet those distinct Appearances which to one that were wholly a Stranger to Artificiall Fireworkes, may seem soe many differing Creatures were all foreseen & intended by the Artist, who knew they would be produc'd, the workes on his part being order'd, as the exigency of his Designe requir'd, and the nature of the water and Fire being suppos'd. And shall we readily allow soe much foresight & contrivance to a Mechanicall artificer, and shall we scruple to allow much better Mechanismes to (the Author even of Artificers) the Omniscient God himselfe, in the production of his Great Automaton, the World?[93]

For Boyle, the world was now a pyrotechnic theater, and, as in William's postrevolutionary state, there was no room for unsanctioned agency in its spectacles. Similar arguments proliferated at this time. The details of art laid bare became models for comprehending the creator's mechanisms. John Ray thus invoked the aesthetics of variety and diverse invention in artisanry to celebrate God's skill in making an abundance of creatures in the universe: "It argues and manifests more skill by far in an Artificer to be able to frame both Clocks and Watches, and Pumps, and Mills, and Granadoes and Rockets, then he could display in making but one of those sorts of Engines; so the Almighty discovers more of his Wisdom in forming such a vast multitude of different sorts of Creatures."[94] Likewise, Thomas Burnet, whose remarkable *Theory of the Earth* (1697) recalls Castelli's descriptions of Vesuvius, identified volcanoes as the providential agents of the end of the world:

> When the fatal time draws near, all these Burning Mountains to be fill'd and replenish'd with fit materials for such a design; and when our Saviour appears in the Clouds, with an Host of Angels, that they all begin to play, as Fire-works at the Triumphal Entry of a Prince. Let *Vesuvius, Aetna, Strongyle,* and all the *Vulcanian* Islands, break out into flames; and by the Earth-quakes, which then will rage, let us suppose new Eruptions. . . . Lastly, the Lightnings of the Air, and the flaming streams of the melting Skies, will mingle and joyn with these burnings of the Earth . . . these three Causes meeting together . . . cannot but make a dreadful Scene. . . . Thus you may suppose the beginning of the General Fire.[95]

Burnet used the term *catastrophe* to describe the scenes of the earth's *revolutions*, a term that he took directly from the theater, where it signaled a change leading to the final event in a drama.[96] Once again, natural philosophers were content to draw explicitly on spectacle to comprehend nature.

## Will-with-a-Wisp: Pyrotechny and Profit

When Bacon wrote in the 1620s, practices of natural magic, alchemy, and fireworks could all be undertaken in the same "Engine-House" and were often subsumed under the broad category of *pyrotechnia*. By 1696, a distinction between fireworks and natural philosophy had been more or less constructed and institutionalized—and then relaxed. Phillips's dictionary of that year divided pyrotechny into two sorts: "Military Pyrotechny teaches the Art of making all sorts of Fire-Arms: Chymical Pyrotechny teaches the Art of managing Fire in Chymical Operations."[97] The Royal Society fellow Ephraim Chambers repeated the distinction in the 1720s, and, thereafter, it remained.[98] The division was the result of ongoing disputes over the relations of art, experiment, and alchemy and the efforts of the Royal Society to distance its labors from controversial incendiarism in years of intense religious and political strife.

With scholarly "chemistry" and artisanal "pyrotechny" secured in institutions and established in separate spheres, a traffic now began to grow between the two royal laboratories. Experiments became still more theatrical. Hence a new alliance of the terms *chemical* and *pyrotechny* in Phillips's and Chamber's definitions and less anxiety on the part of experimenters to engage in dramatic displays. In 1692, Slare revised his experiments to make flames from cold liquor. He no longer used phosphorus, which he admitted was scarce, but oil of caraway seeds and spirit of niter, a combination he called *liquid gunpowder*. Mixing this in vacuo blew the receiver off the air pump: "This Experiment will . . . not only surprize and amuse some, but please and delight others; and . . . perhaps afford some Instruction to a Philosophical Genius."[99] "Perhaps" was not the same as Beale's and Boyle's earlier condemnation of spectacle for its own sake—Slare was happy simply to entertain. Experimental and pyrotechnic accounts of meteors were also mixed together. Thus, Newton pondered Slare's explosive mixtures in query 31 of the *Opticks*, noting Slare's experiment of "liquid gunpowder," and concluding that nature would do even better: "Sulphureous Steams abound in the Bowels of the Earth and ferment with Minerals, and sometimes take Fire with a sudden Coruscation and Explosion . . . with a great shaking of the Earth, as in springing of a Mine."[100] Ascending in the air, the same sulfurous streams meeting nitrous acids would ignite, to "cause Lightening and Thunder, and fiery Meteors."[101]

As King William's revolutionary settlement was publicized in imagery that appealed to popular spectacle, so natural philosophy, properly qualified by theological messages and subdued drama, now began to be promoted in the nascent public sphere, with a growing effort to gain new audiences for the sciences through chemical and mechanical lectures. Science moved nearer to the sites of public festival. In 1710, the Royal Society left Bishopsgate for Crane Court on the Strand, close to the venues of revolutionary spectacle, Covent Garden, St. James's Square, and the Thames.[102] Like William, philosophers now appealed to the people to secure new credit for philosophy, initiating a culture of public demonstrations and lectures on medicine, natural philosophy, and mechanical arts in the coffeehouses around Gresham College and the nascent City of London centered on the Royal Exchange. Lecturers such as Francis Hauksbee, John Harris, William Whiston, and John Theophilus Desaguliers appealed to both spectacle and utility to secure new audiences for science in these performances. Appealing to a taste for spectacle, they drew "active powers," such as gravity, light, and magnetism, from matter to signal morals about the nature of God's creation, while discussions of models and mechanical principles were intended to capture the interest of merchants, traders, and investors in the City.[103] Chemical effects were also salient, identified as central to the new experimental philosophy. The chemist Peter Shaw gave lectures in London and Scarborough, proposing: "It is by means of chemistry, that Sir Isaac Newton has made a great part of his surprizing discoveries in natural philosophy." Shaw displayed the repertoire of experimental effects devised by Boyle and Slare, but now explicitly presented them as lectures on "Pyrotechny; or, Experiments Relating to Gunpowder, Explosions, and Phosphori."[104] These performances resembled the gunners' artifices since Shaw claimed that the "Use" of these effects was to imitate meteors: "Many natural Phenomena, such as Earthquakes, Thunder, Lightning, Vulcanos, the Aurora Borealis . . . &c. are imitable and explicable by chemical Experiments, particular Mixtures, and explosive Powers."[105]

As with many other aspects of courtly art, *imitation* here became an epistemic category. Della Porta had experimented with powders to comprehend thunder, while Bacon, drawing on magic and fireworks, had proposed the practical imitation of meteors as proper to a new science. In the second half of the seventeenth century, much philosophy steered away from such pursuits, investigating gunpowder, but not using it to imitate meteors. From the 1690s, however, such imitations became more common, helping establish the scientific practice of *modeling* nature. Philosophers took these artificial meteors as *explanations* of natural meteors, their degree of similitude being equivalent to their degree of accuracy.[106] The pyrotechnic and the experimental shows were by no means the same. While gunners had long drawn out morals relating to the prince with pyrotechnic imitations, experimenters' artificial meteors instead taught morals about God and nature, the distinction a product of earlier efforts to keep philosophy distant from artifice by emphasizing its piety and instructiveness. Hence, chemistry, "wherein . . . the obvious phaenomena of nature may be exactly imitated, discovers and lays before our eyes the very instruments whereby that powerful agent produces her effects," which effects "the creator hath endowed."[107] Nevertheless, the 1690s witnessed a flourishing of this meteoric modeling that was happy to allude to fireworks. Slare showed experiments of phosphorus flashing in a glass of water to make "a Parallel betwixt Lightning and a Phosphorus," and John Evelyn recorded an experiment at the Royal Society showing that gunpowder had the explosive force of volcanic eruptions.[108] Bernard Nieuwentyt proposed that subaqueous explosions could produce earthquakes by analogy to fireworks burning in water and dropped squibs in glasses of water to simulate the bubbling waters producing hurricanes.[109] Joseph Wasse investigated the effects of lightning bolts when he took "a Cohorn charg'd with three Quarter's of a Pound of very good Powder, wadded with thick Paper, and fired it against a Stone."[110]

Lecturers' experimental effects were also "useful." Sprat had argued that philosophers would cultivate useful over fanciful inventions, and Boyle promoted the utility of phosphorus as a means to divert attention from its spectacular associations. Now utility served to link philosophy to the increasingly powerful commerce of London. To appeal to audiences of traders, investors, and merchants around the Royal Exchange and the City of London, lecturers presented natural philosophy as a useful means for assessing the mechanical advantages of different machines and power sources.[111] In this context, in contrast to the sharp divide between artifice and science in the 1660s, experimenters engaged with pyrotechnics as both theatrical and useful commodities.

The steam engine became important in connecting philosophers to the establishments of the Ordnance. Vauxhall was the Ordnance Board's principal works in the seventeenth century and a worthy successor to Bacon's "Engine-House," intended "to keepe all manner of Ingenuities, rare Models and Engines which may bee useful for the Common-wealth," "to make Experiments and trials of profitable Inventions," and to provide a "place of Resort whereunto Artists and Ingeniers . . . may repaire to meet . . . and hold forth profitable Inventions.[112]

Anthony Wallace proposes that Vauxhall was a key site for the development of Thomas Savery's steam engine, a machine that was quickly made the subject of useful philosophical lectures, illustrated in John Harris's *Lexicon technicum*, and presented, via Harris's pictures, to audiences by Desaguliers. The latter helped investors decide between steam and the perpetual motion as sources of power for business projects.[113] The scenes of festive performance also continued to shape philosophical lectures. From 1713, experimenters continued to engage with fireworks, both as theatrical elements in lecturing and, when religious discontent beckoned once more, as commodities to show the usefulness of natural philosophy. Major displays of fireworks prompted these activities.

The first occurred on July 7, 1713, when the Royal Laboratory firemasters John Henry Hopkey and Albert Borgard, working to the design of the court painter Sir James Thornhill, displayed a pyrotechnic extravaganza on the Thames to celebrate the end of the War of the Spanish Succession.[114] The unpopular war cost the Whigs their royal favor and saw the Tories ascendant from 1710 until the end of Queen Anne's reign. Their rise renewed the anti-Catholic tensions of previous years, and, after the Tories brokered for peace, London Whigs promoted more pope burnings to stoke fears that the Tories encouraged popery. In November, London was once again plunged into a chaotic round of fiery agitation on the Protestant holidays: "At Nine a Clock in the Evening we set Fire to the Whore of Babylon [i.e., the pope]. . . . Honest old Brown of England was very drunk and show'd his Loyalty to the Tune of a hundred Rockets. The Mob drank the King's Health."[115] After the peace was declared, the Tory government ordered a grand firework on the Thames by Borgard and Hopkey, perhaps repeating Charles II's reassertion of spectacle to beat down Whig opposition. Certainly, the thanksgiving fireworks promoted relevant morals, showing a temple of cardinal virtues spelled out in fiery letters: "Prudence, Temperance, Fortitude, Justice, Courage, Victory, Peace, Conduct." Around them swam innumerable water rockets, fire fountains, lights, and "coll. Borgard's large and small Bees' swarms," a kind of floating bomb that sent up flights of many small rockets.[116]

The fireworks prompted both natural theological ruminations and experimental studies. Richard Steele, the proprietor of the *Guardian*, a pupil of Thomas Burnet's, and an intimate of the London lecturers, proposed that fireworks and experimental philosophy had grown close in recent years, both imitating meteors with chemistry: "Thunder is grown a common Drug among the Chymists. Lightning may be bought by the Pound. If a Man has occasion for a Lambent Flame, you have whole Sheets of it in a handful of Phosphor. . . . I am led into this Train of Thinking by the noble Fire-work that was exhibited last Night upon the *Thames*. You might see there a little Sky filled with innumerable Blazing Stars and Meteors. Nothing could be more astonishing . . . Multitudes of Stars mingled together in such an agreeable Confusion. Every Rocket ended in a Constellation."[117]

Steele thought both fireworks and chemistry's artificial meteors demonstrated "how most of the great *Phenomena*, or Appearances in nature, have been imitated by the Art of Man."[118] Steele also read these assorted artifices through the doctrine of natural theology, currently being promoted in the *Guardian*.[119] He laid out what was now a common bond of fireworks and philosophy, designating the world a *theater* whose effects were imitated by art to teach divine morals. Like Boyle, Steele took the fireworks not so much as signals of princely power as a sign of the superior spectacle of divine artifice:

> As I was lying in my Bed, and ruminating on what I had seen, I could not forbear reflecting on the Insignificancy of Human Art, when set in Comparison with the Designs of Providence. . . . I considered a Comet . . . as a Sky-Rocket discharged by the Hand that is Almighty. . . . What an amazing Thought is it to consider this stupendous Body traversing the Immensity of the Creation

> with such Rapidity, and at the same time Wheeling about in that Line which the Almighty has prescribed for it? that it should move in such an inconceivable Fury and Combustion, and at the same time with such an exact Regularity? . . . What a glorious Show are those Beings entertained with, that can look into this great Theatre of Nature, and see Myriads of such tremendous Objects wandering through those immeasurable Depths of Ether, and running their appointed Courses?[120]

No doubt, similar thoughts were entertained by another spectator of the 1713 fireworks—William Whiston, the Newtonian public lecturer and recent publisher of a spectacular paper theater of cometary paths in the *Guardian*.[121] Whiston's natural theology had recently landed him in trouble. His *New Theory of the Earth* posed natural causes behind biblical events, a comet's pass by the earth, for example, being responsible for the Flood.[122] Like Newton, Whiston shared the Arian belief, which denied the Trinity and was considered a heresy.[123] Whiston's views prompted an abusive sermon by the Tory preacher Henry Sacheverell, who condemned the natural theologians as deist idolaters, guilty of worshipping nature and not God. For his heresy Whiston lost his Cambridge chair, moved to London, and began giving public lectures at Button's coffeehouse via the assistance of Richard Steele.[124] He became a Whig cause célèbre, his supporters invoking the familiar language of fiery enthusiasm to condemn his enemies. John Harris attacked Sacheverell for the "Evil and Mischief of a Fiery Spirit" and made quite explicit the link between proper philosophy and coolheadedness: "The Heat and Fire of your Spirit and Temper makes you, as it doth all Men, ignorant and precipitant in your Judgments of things: your *fiery Zeal is without Knowledge.*"[125] Gunpowder theories explained the disturbance since the aerial niter in blood and meteors was in a stir: "There is, even now, too much of this Spirit of Thunder and Lightning . . . amongst Mankind . . . very detrimental and mischievous to the Power of Religion, very destructive to all good Government, and to the Quiet and Ease . . . of Mankind."[126] Only tolerance and debate of different viewpoints, that is, debate of the kind promoted by experimental philosophy, would secure knowledge and calm the state: "It puts the Minds of Men continually on the Enquiry . . . and promotes and increases useful Science." He continued, noting that, "when . . . Men are cool'd and come to their Senses," there would be no more "unhappy Stirs and Disorders."[127]

Whiston made efforts to show the usefulness of philosophy directly after the July fireworks, though not as a means to divert attention from religious controversy. On July 14, Whiston and his colecturer Humphrey Ditton advertised in Steele's *Guardian* a scheme inspired by the fireworks on the Thames.[128] The initial proposal, published as *A New Method for Discovering the Longitude Both at Sea and Land* the following year, turned Borgard and Hopkey's pyrotechnics into tools for improving British navigation and the global propagation of the Newtonian Arian natural theology.[129] Whiston and Ditton thus proposed solving the longitude, the long-standing and seemingly intractable problem of finding the time at sea relative to a home port to deduce an east-west bearing. Whiston thought that its discovery would help secure British commercial and territorial ambitions and make possible the extension of Arianism, or "the Propagation of our Holy Religion, *in its original Purity*, throughout the World."[130] This was to be achieved by an ingenious system of rockets or shells fired from a series of permanently anchored signal ships along British sea routes. At the peak of "Tenerife Time," chosen as the meridian, rockets would be fired to a known height—6,440 feet. To find the longitude, ships might look for these lights at midnight, gauging their bearing and distance from the signal ships (whose position was known) by a compass and by timing the difference between seeing the flash of the rocket and hearing its report or by measuring the elevation of the rocket flash. Here the rocket stood in for the pole star normally used for such measures, a brilliant application of fireworks' historical role as artificial meteors to new ends. On the Thames, Whiston had noted, "the small Stars into which the Rockets commonly resolv'd themselves . . .[were] visible no less than 20 Miles" away, leading him to seek "such large Shells as might be fir'd at a vastly greater height . . . visible for about 100 Miles."[131]

The project led to a new interaction of philosophers and pyrotechnic artisans. Whiston relied on the skills of London fireworkers, perhaps Borgard and Hopkey at Woolwich. Gunners' tacit skills were held up as guarantors of scientific accuracy. To work, firework signals needed to reach predictable heights, which the lecturers recognized in the able judgment of fireworkers: "Gunpowder may be discharged, or combustible matter set on Fire at . . . utmost height. This all that deal in Rockets, Bombs, and Mortars, do very well know. It being the business of their art to proportion the Match or Fusee to any particular time when it shall give Fire."[132] Such skills would "render the experiment more exact and infallible."[133] In August, Whiston and Ditton observed rockets and seven-inch fireballs fired from mortars on Blackheath, Greenwich, reported as visible twenty to sixty miles away.[134] But Whiston's fireworks could be even more destructive than his Arian sermonizing. On January 15, 1715, the London journals reported an explosion at a Mr. Walker's gunpowder shop in Thames Street between the Custom House and Billingsgate. The accident, which devastated several streets and killed fifty people, was put down to a boy making rockets, "as some pretend, for Mr. Whiston's Experiments."[135]

Such rumors were, perhaps, supposed to scupper Whiston's schemes. Certainly, these Newtonian fireworks had a very mixed reception. Whiston's promotions of the signaling project at Parliament in April 1714, supported by merchants and his Whig backers, led to the establishment of the Longitude Prize via a bill prepared by Steele's partner, Joseph Addison, and Lord Stanhope.[136] But Whiston's own scheme floundered, sunk as impracticable amid attacks on his enthusiasm, couched in the familiar terms of fiery spirits leading to false belief. Arianism helped inspire, and helped kill, the Whiston rocket project. Criticisms of the signaling scheme were interspersed with scoffs that Whiston was the "grand ignis-fatuis of London," "a false Fire who shakes the very Foundations of the Christian Faith."[137] Antagonists mocked Whiston's cometary account of the deluge as a pyrotechnic delusion: "There was more Wild-Fire than Water in the extravagant Fancy." They then linked the longitude scheme to Whiston's heretical Arianism: "I do not know . . . how well our Will-with-a-Wisp understands *Longitude* after all; and yet I verily believe . . . he is a profess'd *Latitudinarian* in an Ecclesiastical Sense, to all Intents and Purposes of . . . Fanaticism, against the establish'd . . . Church of England."[138] Whiston's signaling scheme dwindled soon after, presented as the epitome of a dubious *project*, the appellation now increasingly used as a term of condescension against failed business schemes. Although the scheme has retained this label ever since, it could be argued that it failed on account of Whiston's religious heresy, rather than any self-evident impracticability, and, certainly, Whiston was not the last to propose using fireworks as signals for navigation, as we shall see below.[139] Indeed, the use of fireworks as signals long outlasted the local controversies over religious incendiarism that fueled their invention. In the meantime, the English philosophers' interest in practical matters brought them into contact with the gunners and artisans responsible for London's fireworks, a new relationship epitomized in the career of Jean Desaguliers.

## Jean Theophilus Desaguliers: The Philosopher as Fireworker

In the era of emerging public science, philosophers created artificial meteors using new materials and effects, promoting a "chemical pyrotechny" whose boundaries with fireworks and alchemy were settled. Secure distinctions paved the way for increased traffic between pyrotechny and natural philosophy. In the coffeehouses, experiments became more spectacular, appealing to a wider public, while fireworks became more useful, presented in schemes like Whiston's as practical tools for the spread of Britain's empire and religion. There was much continuity with the Restoration era, however. The dispute between Whiston and Sacheverell still prompted angry accusations of incendiary enthusiasm on both sides and affected the fortunes of Whiston's longitude solution. Nevertheless, in comparison to its status in the Restoration era, the new public science was quite intimately and openly linked with fireworks.

A final case shows how much had changed since the 1660s. The career of Jean Theophilus Desaguliers as a public lecturer and Newtonian curator of experiments at the Royal Society has been well explored.[140] In coffeehouse lectures of the 1720s and 1730s, the Huguenot Desaguliers became a master at producing philosophical effects extolling the usefulness of Newtonian mechanics for judging the efficiency of machines, defining the difference between dubious showmanship and trustworthy projects using mathematics applied to models of mechanical inventions, diagrams, or live feats of strength.[141] But Desaguliers also spent at least some of his time giving fireworks displays, while his son Thomas became the chief firemaster of Woolwich's Royal Laboratory, responsible for courtly pyrotechnic performances in the eighteenth century. This side of Desaguliers's career shows how, by the early years of Hanoverian rule, English, and Protestant, natural philosophy and pyrotechny could exist comfortably together.

Desaguliers set off fireworks on at least four occasions, beginning with the thanksgiving fireworks of July 1713. Like Steele and Whiston, Desaguliers was inspired by Borgard and Hopkey's fireworks to experimental pursuits, but, unlike them, he had participated in firing the display himself, going with Colonel Samuel Horsey "out upon the Thames in a great Barge . . . where [they] play'd off some Fireworks" during the performance.[142] An accident on the barge led him to philosophize. Like Bacon and Boyle, Desaguliers was interested in aquatic fireworks. In his lectures, he explained to audiences how a water rocket had exploded under his barge, blowing a hole in it, and prompting him to wonder about the nature of subaqueous explosions. He determined "to try the Effect of the Explosion of Gunpowder under Water": "I loaded a Water-Rocket, so that it should break under the Water, and having set Fire to it, threw it into a Pond, that cover'd an Acre of Ground: And so great was the Shock, that several Persons, who stood around the Pond, felt it like a momentous Earthquake."[143] Hence another meteoric model, in an experiment that recalls Evelyn's proposals some years earlier that nitrous explosions caused earthquakes and volcanoes.[144]

Desaguliers set off fireworks again in 1718 in Hyde Park, and this time the fireworks were given for George I's birthday, "under the direction of the ingenious Monsieur Desaguliers."[145] In the same year, he translated the French academician Edme Mariotte's *Motion of Water*, which included some of the earliest speculations on the causes of rocket motion.[146] The question had been tackled before occasionally, for example, by Thomas Hobbes, who referred to an analogy between the action of fire and water: "A. What is the cause why certain squibs, though their substance be either wood or other heavy matter, made hollow and filled with gunpowder, which is also heavy; do nevertheless, when the gunpowder is kindled, fly upwards? B. The same that keeps a man that swims from sinking, though he be heavier than so much water. He keeps himself up, and goes forward, by beating back the water with his feet; and so does a squib, by beating down the air with the stream of fired gunpowder, that proceeding from its tail makes it recoil."[147]

Philippe de La Hire offered another explanation in terms of impulse in 1702.[148] Mariotte, whose hydrostatics emerged from studies on fountains with La Hire in the 1680s, also supposed that a rocket's flight was due to the impulse of the flame against the air, unless the body of the rocket was too heavy.[149] Contra Mariotte, Desaguliers argued that forces acting in every direction inside the rocket canceled one another out in a chokeless case but gave a net force upward when the case was vented at the bottom since "the Action of the Flame downwards is taken quite away, and there remains a Force . . . acting upwards."[150] Here was the first account to describe rockets in Newtonian terms of "Forces . . . equal and contrary," though Newton himself never discussed rockets.[151]

Desaguliers gave other fireworks performances during his career. In the 1730s, he served as chaplain to Frederick, Prince of Wales, for whom he gave courses on experimental philosophy.[152] He also showed fireworks at the prince's residence in Buckinghamshire and in Bristol, where in 1738 he set off pyrotechnics around King William's statue, in three acts, with the last centering on a twenty-foot-high pyramid and blazing sun.[153]

## Conclusion

Such activities were representative of careers that brought theater, spectacle, commerce, and natural philosophy together in London during the early decades of the eighteenth century, helping make fireworks a commodity for experimental philosophy's contemplation and use. In the coming decades, this would lead to the creation of an entirely novel genre of "philosophical fireworks" as performers hybridized experimental and pyrotechnic effects. But a rather tortuous relation of pyrotechny and the sciences had preceded this increasingly harmonious exchange. English experimenters had supported alchemical pyrotechny in the Interregnum and early years of the Royal Society, when foreign masters began arriving in England to renew royal fireworks. But the rising tide of Catholic incendiarism, popular insurrection, and a distaste for pomp led fellows to distance their labors from overt pyrotechnics during the later 1660s and 1670s. The shape of experiment and the changing political uses and meanings of spectacle then developed in tandem, with no simple contrast between artificial spectacle and experimental performance. Under attack for fiery Catholic enthusiasm from the likes of Stubbe, experimenters stressed the open, godly, and unspectacular nature of experimental philosophy and posed nitro-sulfurous accounts of the body, meteors, and the heavens as means to manage excess passions in religion and the state. In times of incendiary plotting, philosophical spectacle was coolheaded. But politics blew hot and cold, and, after the end of the Exclusion Crisis and the Glorious Revolution, there was more room for overt spectacle in philosophy, exploited by lecturers to promote the sciences. From the 1690s, philosophy relocated to coffeehouses and other nascent public venues, often connected to the commercial spaces of the City of London, where experiments appealed to new audiences with spectacles cast in the light of natural theology, utility, and improvement and the philosophy of active powers. Properly managed, philosophical and festive artifices could now converge, in the demonstration of artificial meteors, in the modeling of ingenious practical inventions such as the steam engine, and in experiments to find longitude prompted by the festive scenes of thanksgiving fireworks. A rising commercial context, not to mention missionary zeal, thus prompted new interactions of philosophers and fireworkers, an intimacy exemplified in Whiston's experiments and Desaguliers's career. This, however, was by no means the only trajectory by which the new science and pyrotechny became entangled in the era of early academies. In Russia, an entirely different arrangement prevailed.

CHAPTER FOUR

# Spectacular Beginnings: Fireworks in Eighteenth-Century Russia

In June 1739, the Newtonian lecturer on experimental philosophy Erasmus King traveled to Russia on a ship with three companions. One of them, Francesco Algarotti, another promoter of Newtonianism, explained how "Mr. King . . . intends to exhibit a course of experimental philosophy, in presence of the Empress. Imagine now what quantity of machines we are provided with, to demonstrate to the Russias the weight of the air, the centrifugal force, the law of motion, electricity, and all the other philosophical discoveries. What, however, undoubtedly excels them, is our ample provisions of lemons and exquisite wines; and, above all, our French cook."[1] Algarotti's preference for the luxuries of court life over experimental demonstrations was a taste that, as we shall see, would be of much significance in Russia. Also on board was Thomas Desaguliers, the nineteen-year-old son of Jean Theophilus Desaguliers, who had recently been introduced to the Royal Society and was being sent to Russia by his father, Algarotti supposed, "to learn the practice of navigation."[2] Evidently, Thomas did not take to the art of navigation, returning to London by January 1740 to enroll in the Royal Artillery as a cadet. Soon, he became the chief firemaster of London's Woolwich Arsenal, where he would preside, like his father, over numerous royal fireworks displays.[3] Thomas Desaguliers perhaps learned to appreciate fireworks attending the celebration of the marriage of the Russian empress Anna Ivanovna's niece to Prince Anton Ulrich of Brunswick. It was in order to attend these ceremonies in an official capacity that the third man traveling with King, Lord Baltimore, was sent to Russia.

The interactions of science, government, and fireworks that they may have witnessed at Prince Anton Ulrich's wedding were expressions of a very different culture than that in London. The fireworks were ignited from a platform on the river Neva on July 9. The scene was an elaborate "invention," or allegorical design, presenting the name of God in shining rays, the monograms of the wedded couple entwined, and the female figures of Germania and Russia beneath mottoes announcing the empress's divine provision for and maternal care over the happy union. The figure of Venus, probably representing the empress herself, rose in a shell amid playing swans and water nymphs while fireworks were set off all around.[4] It was perhaps enough to entice the young Desaguliers away from navigation.

The designer of these allegorical inventions was not a gunner but a mathematician, Christian Goldbach, a correspondent of Leibniz's, the author of a treatise on infinite series, and a leading member of the St. Petersburg Academy of Sciences.[5] Founded in 1725 by Tsar Peter the Great, the academy was a showcase of foreign talent in the sciences, housing some thirty German, Swiss, and French scholars whose task it was to bring European science and civilization to the Muscovites.[6] By the time King, Desaguliers, and Baltimore arrived, designing fireworks was a regular part of the academy's activities, and Goldbach shared the task of inventing allegories for court fireworks and illuminations with several academicians. Together with St. Petersburg's fireworkers, they succeeded in making Russia the site of some of the most spectacular pyrotechnics in the eighteenth century, spectacles that led the *Encyclopédie* to declare: "The Muscovites are superior to the rest of Europe in the combination of figures, motions, the contrasts of artificial fire."[7]

This chapter explores the history of fireworks in Russia between the reigns of Tsar Peter I (1696–1725) and his daughter Elizabeth Petrovna (1740–60).[8] During this period, Russia underwent a transformation, as Peter imported Western European culture as a means to build the state and diminish the power of the Orthodox Church and the traditional Muscovite nobility. Fireworks were a part of this process of reform, which gave European military and later court culture a central role in the new state. Peter's reforms also shaped the development of the sciences in Russia, where no significant tradition of natural philosophy existed prior to the early eighteenth century. When Peter decided to found an academy to bring the sciences to Russia, he gave it a role teaching, not just the content of science, but also the manners and conduct of Europeans, making education in "civility" a key part of the institution's work. However, after the Academy of Sciences opened in 1725, its mostly German academicians soon found that the sciences failed to interest the Russian nobility, and they turned to alternative strategies to secure the success of their institution. Fireworks then came to play an important role in the academy, unprecedented anywhere else, as a powerful vehicle for promoting the sciences, securing patronage, and educating Russian audiences. And, if ingenious mechanical invention served experimental philosophers in England, so the rhetorical side of invention served Russia's academicians. For more than sixty years, the professors of St. Petersburg composed allegorical designs, or inventions, for Russian court fireworks and illuminations. This enterprise was not without difficulties and led academicians to compete with Russia's artillerist fireworkers and among themselves in order to promote their pyrotechnic designs. But the history of Russia's academic pyrotechny demonstrates how fireworks could be of great significance for the fate of the sciences in the eighteenth century and how the relations of natural philosophy and pyrotechny varied significantly in different regions. While commerce and public culture promoted links between the sciences and pyrotechny in England, in Russia such connections were forged by the court; and, whereas engaging with fireworks brought London's scholarly and pyrotechnic communities together, St. Petersburg's academicians, despite their close involvement in designing displays, remained distant from Russian fireworkers and made efforts to diminish the value of their skillful contributions to displays. At the end of the chapter, a short foray into contemporary events in absolutist France further brings into relief some of the distinctive features of the Russian case.

## Peter the Great: Prince and Pyrotechnist

Fireworks, like the sciences, were a foreign import in Russia. Tsar Ivan III Vasil'evich first introduced fireworks to Russia in the 1470s, hiring foreigners to manage ordnance in the realm, while gunners in the emperor Charles V's service entertained Tsar Ivan IV with fireworks in Moscow, prompting their use by Russians in wars against the Tartars and at the siege of Kazan.[9] Like the Catholic Church, the Russian Orthodox Church used gunpowder effects to add drama to liturgical plays during the sixteenth and seventeenth centuries. Between 1543 and 1648, the Furnace Play (*peshchnoe deistvo*) told the story of the prophet Elijah who rained fire on the enemies of Israel and ascended to heaven in a fiery chariot. Performed in Orthodox churches on the Sunday before Christmas, the drama showed three Israelites rescued from a furnace adorned with fireworks and children rescued by an automaton angel.[10] Religious drama reflected the church's dominant position in Russian culture in the seventeenth century. But this was a position diminished by schism and the growing efforts of the Russian tsars to assert the power of the state during the seventeenth century. When the Romanovs adapted motifs of church ritual and cosmology to build a more secular court culture, they began to exhibit fireworks as royal, rather than religious, spectacle, following the pattern set by princes across Europe. Tsar Aleksei Mikhailovich thus ordered fireworks for the Shrovetide festival of Maslenitsa in 1672, which culminated in the burning of an effigy of Chuchilo, that is, the devil.[11] A few years later, fireworks shown by a Dutch envoy in Russia astonished Muscovites, who imagined that they had seen a "fiery snake," perhaps akin to the dragons displayed elsewhere.[12] Fireworks thereafter became a regular occurrence at court, taking place annually in Moscow at Maslenitsa.[13] Russians emphasized the pleasure, rather than the artifice, of fireworks, calling them "fiery amusements" (*ognennye potekhi*) or "fiery diversions" (*ognennye zabavy*).

Peter I ascended the throne in 1696 and set about a major series of reforms to reduce the power of the Orthodox Church and modernize the Russian military and state. Fireworks were among the practices he reformed. Peter was fascinated by fireworks and eliminated the religious associations of pyrotechnics in favor of grand princely displays to accompany the new year—itself a novelty since he replaced the Orthodox calendar with the Julian. Russia's triumphs in ongoing battles with the Ottoman Turks and the Great Northern War with Sweden were also celebrated with fireworks. Hence, a vast display held near Moscow in 1697 celebrated the capture of Azov from the Turks. The fireworks were recorded in a fine engraved print, a new medium for Russia, by the Dutch artisan Adrian Schoonebeck and centered on a great two-headed eagle, representing Russia, blasting the Turkish crescent moon with a rocket before a triumphal column bearing the Latin motto "Victoria." The figure of Neptune, symbolizing Russian power over the seas, could be seen together with transparent paintings of Russia's newly extended army and navy battling the Turks. All around, mortars projected fireballs over martial trophies and illuminated pyramids, watched by a male audience depicted by Schoonebeck drinking, dancing, and pointing in wonder at the pyrotechnics (fig. 14).[14] Such dynamic scenes became typical of Petrine fireworks. Another display of January 1, 1710, showed the Swedish lion conquered by the Russian eagle, each represented by automaton figures running on lines, amid allegorical transparent paintings illuminated from behind.[15]

Imported foreigners and Russian gunners of the tsar's personal Preobrazhenskii regiment of artillery assisted Peter in making these displays, and Dutch engravers captured the events in elaborate engravings handed out to audiences to record and clarify the fireworks' meanings.[16] As was done in London, Peter established permanent locations for gunners to make fireworks, first in Moscow, and then in his new capital, St. Petersburg, in the bastions of the city's Saint Peter and Paul Fortress.[17] From here, fireworks were taken to the nearby Troitskii Square or to barges on the river Neva to

14. Fireworks for the taking of Azov, February 12, 1697, at Krasnoe Selo, near Moscow. Adrian Schoonebeck (engraver), *Ignium artificialium, jussu invictissimi Petri Alexii Fil: Magni Russorum Caesaris, etc, etc, etc, post Ianaim Urbem, ipsius armis feliciter, expugnatam, praeparatorum et accensorum, die 12 feb. 1697, prope Moscuam* (Amsterdam, 1698–99). Rijksmuseum, Amsterdam.

be performed by new ranks of artillerist "fireworkers" (*feierverkery*). However, it was the tsar himself who took the most active role in devising and setting off these fireworks. Peter recorded and translated numerous recipes for pyrotechnic effects and even devised his own, including color-tinted fires and a "white candle composition" made from "twenty-seven pounds saltpeter, fifteen pounds sulfur, eight pounds ground gunpowder, twenty-two lots rosin [*kanifol'*]."[18] In 1710, displays showed "beautiful light blue and green fires, invented by the tsar himself."[19] On Peter's orders, German books on artillery and pyrotechnics were translated into Russian, and Peter, like the seventeenth-century gunners, also designed the allegories for his displays, using Cesare Ripa's *Iconologia* and Dutch books of emblems and devices whose translation into Russian he also oversaw.[20]

Peter's involvement in fireworks was not simply for pleasure but part of an effort to teach foreign civility to Russians by example. Contemporaries noted how Peter was enthusiastic that his subjects participate in these fireworks. "The tsar, being the captain of the fireworkers, constructed this firework himself. Standing among the company, he explained the meaning of each allegory surrounding them while it was burning."[21] This enthusiasm for personal didacticism was not a characteristic unique to Peter but part of a Russian monarchical tradition. The tsar, in Orthodoxy, was *obraz,* the image or icon of divinity whose exemplary morality and conduct were to be imitated by the Orthodox. But, rather than act as a spiritual role model, Peter instead behaved as a skilled technician and statesman, busily practicing useful arts, hoping that his audience, the Russian people, would imitate him. Thus, he presented himself variously as an artisan, a practical shipbuilder, a soldier, a sailor, a fortifications expert, and a fireworker. He viewed fireworks in practical terms. When asked about the source of his interest in pyrotechnics, the tsar allegedly explained: "Had it cost more, it would give me little concern. My object is to accustom my subjects, in this agreeable manner, to the more serious fire of musketry and cannon. Experience has taught me that the more we are used to this sportive fire, the less we fear to brave the other, notwithstanding its danger."[22]

Just as Leibniz planned to teach the arts and sciences through dramas or "comedies," Peter thought that fireworks might serve the same end. Indeed, there were German precedents here in addition to Russian. German "prince-practitioners" identified virtue and power with displays of learning, taste, and collections.[23] By the early seventeenth century, this applied as much to fireworks as to courtly cabinets of curiosities or patronage of the arts, another sign of the gunners' successful integration of court and pyrotechny. At Dresden, a city much admired by Peter, in the early seventeenth century, the Saxon prince Johann Georg II liked to devote himself to the design of fireworks for family and state occasions.[24] Elector Augustus of Saxony built a vast collection of artisanal tools and curiosities at his *Residenz* in the city, later visited by Peter.[25]

## Founding the Russian Academy of Sciences

Leibniz was also an important figure for Peter, serving the tsar as an adviser, meeting him on several occasions, and offering numerous proposals for the reform of Russia's government and education.[26] Leibniz also suggested designs for an academy of arts and sciences, to bring prestige to Russia, to serve the state by advising on technical matters, and to act as a bridge between Europe and the Far East. Peter's ambassadors already traded with China, and their cargoes occasionally included Chinese fireworks, another interest of Leibniz's.[27] Leibniz's plans continued to reflect his earlier schemes, promoting science through the pleasures of courtly entertainments. As late as 1714, projects for an academy included grottoes, a menagerie, botanical gardens, mineral cabinets, art galleries, and "en un mot theatres de la Nature et de l'Art."[28] Peter also saw a key educative role for the academy. Since there was no scientific tradition in Russia, any scientific institution would have to serve as a gymnasium, university, and academy and be staffed by foreigners who would teach Russian students to disseminate the new learning across the empire. Peter was aware that the introduction of Western science (in Russian *nauka*, with the same broad meaning as the German *Wissenschaft*) might provoke the hostility of elements in the Orthodox Church and the Muscovite nobility, but, while the project remained under his patronage, it would be secure.

In the event, Peter's academy was not a theater of arts, and there were no fireworks. Plans changed in the interval between Leibniz's death in 1716 and the opening of the St. Petersburg Imperial Academy of Sciences in late 1725. The final project drawn up for the academy centered its activities on the sciences, not the arts.[29] The institution would employ an elite of professors to provide scientific advice to the government and to serve as emblems of Russia's newly civilized status in Europe. It would continue Peter's policy of teaching by example but offer models of civil, rather than artisanal, conduct for Muscovites to emulate. Thus, as Michael Gordin has noted, if the Royal Society drew on local gentlemanly conventions to forge scientific culture, in Russia the academy was charged with using scientific culture to teach the same social conventions to locals. The academician should be another *obraz*, demonstrating in his conduct the polite etiquette requisite for making the secular knowledge necessary to the success of the Petrine state.[30] Science was considered the exemplar of the making of secular knowledge, and so the professors were not artisans, or technicians, or even, with the exception of the physicist Georg Wilhelm Krafft, experimenters, but scholars of mathematics, mathematical physics, law, and history chosen by Leibniz's disciple Christian Wolff.[31] These men were highly educated professors and did not lower themselves to the work of the hand, the only practical activities in the academy being carried out by a book illustrator, a mechanic, and a printer, all imported since training in the necessary skills was not available locally.

The academy itself was also an exemplary venue. In contrast to the concerns of the fellows of the Royal Society in its early years, St. Petersburg's academic organizers had no concern to downplay spectacle, which suited their close relationship to the Russian imperial court and the academy's role of publicizing etiquette and Russia's new scientific status. Unlike the Royal

Society's meager meeting rooms in London, the St. Petersburg academy encompassed a grand purpose-built baroque edifice and requisitioned palace, situated opposite the imperial Winter Palace on the banks of the river Neva. These buildings housed a library, offices, and the substantial Kunstkamera, which was based on Peter the Great's own collections.[32] There was also a press, and the academy soon dominated secular publishing in Russia.[33] The buildings were designed for show, the Kunstkamera being open to the public, while the court and imperial family were invited annually to attend scientific lectures in soirées or "assemblies" in the academy's buildings. Staged with much fanfare, these lectures joined education in science with a demonstration of philosophical etiquette, consisting as they did of a conversation between two professors, a lecture and reply, on topics such as the history of mathematics, geography, and the heliocentric theory of Copernicus, then still largely unknown in Russia. Lectures were printed and circulated using the academy's press.[34]

Russia's academy was, thus, no gentlemen's club but a state institution, organized and managed from above, with little place for demonstrations of the arts or experimental science. Nevertheless, as in England, in Russia changing local circumstances could quickly alter the tenor and direction of academic activities. The most salient force in this regard was not religion, the Orthodox Church having been much weakened during Peter's reign, but dynastic politics. Although the academy was an institution, it had no formal charter, and its fate was largely tied up with the sovereign and patronage from the high nobility.[35] This was obvious from the beginning. In 1725, Peter I died, leaving the academy without its principal defender even before it opened. Only after reassurances that the academy would be supported by Peter's widow, now Empress Catherine, did its foreign professors agree to come to St. Petersburg to begin work. Without Peter's patronage, the academy, like Russia itself, would be subjected to many years of instability amid turbulent courtly politics.

Despite initial good fortune, the academy's scientific and manners projects both faltered. Prince Alexander Menshikov, Peter's favorite and a supporter of Petrine reforms and Western learning, became the de facto ruler of Russia during Catherine's short reign (1725–27).[36] Under Menshikov and Catherine, the academy flourished, initiating its annual assemblies, praising the empress with verses and orations, and presenting Western science to an audience of nobles, government officials, and military officers favorable to Petrine reforms and under whose authority the academy remained open and funded. Lectures on the sciences began to prove troublesome, however.

In March 1728, the Russian translation of a lecture by Daniel Bernoulli and Joseph Delisle on Copernican heliocentrism was abandoned for fear of objections from the Orthodox Synod.[37] Simultaneously, nobles proved apathetic when it came to the sciences and learning. Hence verses describing the nobility's reaction to lectures on Copernicus:

> Why plot the courses of the stars, and why
> Sit up all night for some spot in the sky?
> Why give up sleep for curiosity
> To learn if it's the sun that moves or we?[38]

To the nobility in Russia, science was *irrelevant*. One foreign visitor noted the problem: "The Russian nobility have no wish to learn philosophical reflections."[39] Another explained: "A great many consider themselves fully educated if they have learned to read and write in their own language . . . great book learning does not interest such a person."[40] This may have been one reason why the academy's public assemblies were short-lived, the last being given in the spring of 1732, with no more for a decade.[41] To make matters worse, dynastic politics again intervened. In 1727, on the death of Catherine, the infant Peter Alekseevich, the grandson of Peter the Great, was pronounced Tsar Peter II, and soon after Menshikov was swept from power and exiled by the reactionary Dolgorukii family, who disdained Petrine reforms.[42] The court now moved back to the traditional capital of Moscow, where it remained until 1732. Menshikov's fall from grace left the academy without a powerful patron, with a diminishing number of students in its gymnasium and university, and mounting debts and internal rivalries. Some ministers considered closing it.[43] In 1731, the academy's secretary, Johann Daniel Schumacher, a former librarian to Peter the Great, predicted its "most complete collapse."[44]

## Patronage and Invention: The Academy and B. C. Münnich

The potential audience for science in Russia was small, and, unlike their English counterparts, Russia's academicians had no community of commercial traders and merchants to appeal to. If they did not interest the court and nobility in science, there was no alternative audience to turn to. Academicians did, in fact, relate to the few government ministers who organized engineering and mining projects for the state, but for significant patronage they needed to appeal to the nobles and, above all, the court. In England, engaging with mechanical inventions like the steam engine helped forge links to commercial audiences. In Russia, the Academy of Sciences also found invention critical to forging links with the court. However, Russia's academicians turned to the rhetorical notion of invention to achieve this.

This came about through the academy's efforts to find patronage. To resolve their mounting problems, academic administrators decided that it would be useful to find a powerful patron, or "protector," at court. The courtier Johann Albrecht Korff explained to Schumacher: "Everywhere where there is an academy of sciences, it is understood that they have some strong personality at the court in the capacity of a protector. This does not necessarily have to be a learned man, so long as he takes care of the academy. Just think how valuable it will be for the academy to elect for this purpose a man who is powerful in our court."[45] One such man was Burchard Christoph Münnich, a military engineer of Saxon origin engaged in Russian service by Peter I in 1721 to build the Lake Ladoga canal.[46] A friend of Peter's and rival of Menshikov's, Münnich became the governor of St. Petersburg after Menshikov's fall, from which position he sought to ingratiate himself with the new tsar, Peter II, and his protectors.

To do this, Münnich turned to pyrotechnics and the expertise of the Russian artillery and the Academy of Sciences.[47] When Peter II was crowned emperor in Moscow in early 1728, Münnich sought to display his loyalty to the new tsar by staging an elaborate illumination in front of his residence in St. Petersburg. Such acts were not unprecedented in European courts, where nobles competed for princely favor with impressive fireworks and illuminations. To design the allegory for his display, Münnich turned to the new Academy of Sciences, supposing that its foreign members were expert in these matters. Schumacher, who was less sure, chose the professor of jurisprudence Johann Simon Beckenstein to design the display, and it was performed on March 3 in front of Münnich's residence. Engravings and descriptions were printed at the academy and circulated, naming Münnich as patron of the display.[48] For this performance, Münnich soon had reason to be pleased. In May 1729, he was promoted to head the Chancellery of Artillery and Fortifications, Russia's equivalent of England's Board of Ordnance, located in St. Petersburg's new arsenal, situated upriver from the Winter Palace on the Fontanka Canal.[49] He thanked the academy with a dinner, drinking "a great goblet of wine to toast us."[50]

The academy's search for patronage and Münnich's desire for advantage at court thus came together in a collaboration on fireworks that initiated an increasingly regular exchange between the academicians and Münnich.

Münnich served as a protector of academic interests, in exchange for which academicians provided regular allegorical inventions of fireworks and illuminations. Münnich was not alone in these efforts. The academy's professor of astronomy and the only French member, Joseph Delisle, also invented illuminations to display his power in Russia and abroad. As an unofficial representative of the French crown in Russia, Delisle organized celebrations of the birth of the French dauphin at the observatory in the Kunstkamera's tower in 1730. Delisle publicized his celebrations in the Parisian *Mercure de France*, describing how the Kunstkamera and observatory were illuminated with pyramids of light and a "great number of terrines," that is, illuminating lamps. Transparent paintings with French and Russian mottoes also graced the celebrations, while guests were ushered from their boats by trumpet players to attend a dinner, concert, and ball that lasted until eight in the morning.[51]

Then came another dynastic crisis. When Peter II died in 1730, the reactionary nobility replaced him with a Courland princess from the court of Mitau, Anna Ivanovna. Expecting that she would be compliant to their wishes, the nobility soon discovered her to be a powerful autocrat in her own right, dominating the nobles, and returning the court to St. Petersburg.[52] There, Anna began constructing a vast permanent court rivaling Versailles. The Italian architect Bartolomeo Francesco Rastrelli was commissioned to embellish St. Petersburg with a new Winter Palace and a vast *manège*. French, Italian, and German master artisans filled them with tapestries, portraits, jewelry, silverware, and imported Chinese porcelain. A court theater and opera house were founded, the first in Russia, and at the Winter Palace Anna built up a more formal court, creating numerous new positions for attendants and pages. Ranks were sewn into new fashions and uniforms, as Peter's ubiquitous German smock coats were replaced by expensive silks, gold embroidery, furs, and brocades. Courtiers attended a constant round of engagements and entertainments, balls, masquerades, and dances. The obligation to engage in such costly style helped maintain the nobility under Empress Anna's control.[53]

The staging of fireworks and illuminations was also a key element in this courtly competition. Already in 1730, the English ambassador present in Moscow for Anna's coronation noted: "Great preparations are making for her Majestys Coronation, which its said will be the 16th Instant. The Duke of Liria is erecting before his House a very fine Triumphal Arch, which is to be illuminated for three days after the Coronation. His Grace doth his utmost, to distinguish his Joy, above all the other Ministers, which is no small mortification to Count Wratislaus."[54] Anna's new reign put a premium on all such displays, a fact that Münnich and the academicians were also quick to exploit, as preparations for the arrival of Anna in St. Petersburg in 1732 make apparent. To welcome the empress, Münnich ordered the construction of an elaborate "fireworks theater" on the Neva, positioned opposite the imperial Winter Palace on the southeast corner of Vasilevskii Island, close to the buildings of the Academy of Sciences. The only purpose-built venue for fireworks and illuminations in early-eighteenth-century Europe, and testimony to the importance Münnich attributed to pyrotechny, the fireworks theater was a six-hundred-foot-long wooden jetty raised on piles to two stories in height, projecting out over the river. Münnich planned it to be capable of carrying some thirty thousand illuminated lamps.[55]

To execute the illuminations for Anna's arrival, Münnich brought a number of experienced artillerists from Moscow.[56] To design them, he turned again to Schumacher and the academy. By now it was evident to Schumacher that designing fireworks offered the academy an opportunity to earn favor at court. Consequently, to meet Münnich's request, he hired a new adjunct in applied arts expressly for the purpose of inventing pyrotechnic allegories. His choice was the former Dresden court poet Gottlob Friedrich Wilhelm Juncker.[57] As the academy's historian Gerhard Friedrich Müller recalled:

> The court arrived from Moscow. . . . Mr. Schumacher wished to be distinguished by splendid illuminations in front of the academy building, library, and *Kunstkamera*. For this, it was necessary to find someone to devise the imagery and describe it, partly in prose, partly in verse. Mr. Juncker did everything he could manage, and . . . according to contemporary taste . . . his inventions were found to be as good as they could be. . . . Mr. Schumacher wished them assiduously and perfectly copied into a grand book in royal folio, in which splendid illustrations would be drawn. It was proposed that the book would be given to the empress, and from this he expected advantages for the academy.[58]

The illuminations were clearly considered as a means to foster patronage, but, in this instance, they served Münnich better than they did the academy. After Anna's arrival, Münnich was made the chief or *general-feld'tseikhmeister* of the Russian artillery. The academy was less fortunate since, for unknown reasons, the book prepared by Juncker was never given to Anna, and "a rich garment, sown with thick silver galloon, ordered by Mr. Schumacher for Mr. Juncker, was in vain."[59]

## The Process of Invention: Academic Pyrotechny in the 1730s

From this time on, the academy regularly invented fireworks and illuminations for Münnich and, after Münnich departed on a military campaign in 1734, taking Juncker with him as his official historiographer, for his successors.[60] Münnich quickly institutionalized his relationship with the academy, securing access to its members' skills, and setting in place a procedure whereby it regularly supplied allegorical inventions to the Chancellery of Artillery and Fortifications. Anna was serious enough about fireworks to enshrine them in law, an act typical of Russia's centralized state and Anna's determination to build a splendid court. A decree of 1733 thus ordered that fireworks take place every year on the empress's birthday, name day, and coronation day and for the new year.[61] Fireworks also celebrated triumphs, peace celebrations, and imperial weddings and births. On appointed days, crowds gathered on the shores of the Neva to view the fireworks or illuminations, the empress and court watching from the balconies and windows of the Winter Palace. Displays went off to the sound of trumpets, kettledrums, and cannon salutes.[62]

Following the process of invention that lay behind these performances brings out how important they were considered in the academy and how they were intended to serve its various goals. Schumacher and the academic presidents devoted much energy to ensuring that allegorical inventions were of the highest quality. In 1735, following Juncker's departure, Schumacher replaced him with the Swabian musician and poet Jacob Stählin, a member of Johann Christoph Gottsched's circle in Leipzig and a friend of Johann Sebastian Bach's. Stählin, who was named "professor of eloquence and rhetoric" in St. Petersburg, claimed to have learned pyrotechnic composition in Zittau from an otherwise unknown Italian named Montallegro and drew the academy's attention after composing fireworks for the king of Poland-Saxony, Augustus III. Designing scores of fireworks for the Russian court, in addition to operas and ballets, Stählin became by the 1760s the most prominent figure in the academy after Leonard Euler, though today he is largely forgotten.[63] His prominence was justified since his inventions, which he claimed were based on the emblems of Filippo Picinelli's *Mundus symbolicus*, helped secure the value of the academy at the Russian court through much of the eighteenth century.[64] The mathematicians Christian Goldbach and Vasilii Adudorov also composed fireworks and illuminations during Anna's reign.[65]

To further ensure quality, allegorical inventions were subjected to close scrutiny. Schumacher might commission two academicians to invent a pyrotechnic allegory and then choose the one he deemed most suitable. One or more compositions were then forwarded to Münnich or the empress herself for approval.[66] Judgments hinged on the *decorum* of the invention, usually presented as a written narrative describing the scenery and verses signaling the allegory's symbolic contents and meaning. In his *Great Art of Artillery*, Siemienowicz explained the notion of decorum as a choice of imagery and narrative fitting to an occasion and the rank of the inventor's patron. Inventions should ideally follow suitable classical, biblical, or historical themes, presenting at a wedding, for example, the pyrotechnic or illuminated figures of Juno, Venus, Diana, or Cupid with garlands of flowers, fruit trees bound with festoons, and palaces, castles, or fountains following the orders of architecture.[67] Peace was to be celebrated with figures of doves, olive branches, and female personifications of Peace and Felicity, holding a caduceus and cornucopia, while triumph demanded a triumphal arch and military standards and paraphernalia.[68] Decorations should always be appropriate to the occasion but could still be arranged in ingenious ways, according to "the Fancy and Discretion of the Persons who are led by their *Genius* . . . to cultivate our *Art*."[69] Russian academicians probably followed Siemienowicz's widely known work, which also recommended "oeconomy" as part of invention, that is, the making of drawings and models in wood, wax, or pasteboard showing the distribution of fireworks and scenery for artisans to construct.[70] Such activities are in evidence in Russia, and some drawings remain of Russian designs, while Stählin made a model of at least one of his inventions.[71]

An important consequence of this arrangement was that gunners, traditionally central in the inventive process, were now left out of it, being expected only to build the approved scenery and to set off or "execute" displays. Here gunners suffered from their own success since the growth of fireworks at court led patrons like Münnich to divide the labor of pyrotechny and seek out specialists in the different skills entailed. Thus, after an allegorical invention was approved, it passed on to the Chancellery of Artillery and Fortifications at the St. Petersburg Arsenal. The arsenal was the site, in the early 1730s, of a number of artillery reforms presided over by Münnich that further institutionalized pyrotechny. Münnich thus constructed a fireworks laboratory at the arsenal and gave artillery ranks a clear definition, with *feierverkery* serving under an *oberfeierverker* answering to Münnich.[72] The laboratory soon became a busy scene, described as "a large square room, one side of which measured thirty arshins [seventy feet]. There were a great number of people in this [room], all kinds of masters, some composing plans, others filling fountains [fireworks] . . . in addition, a few persons from the bombardiers of the Preobrazhenskii regiment worked under instruction [and] carpenters, turners, metalworkers were located in the corners of this room."[73]

These artisanal skills were indispensable to the academy's pyrotechnic projects, fireworks providing the spectacle that glorified academic allegories. According to the *Encyclopédie,* Russian fireworkers were also purveyors of ingenious inventions for the court during the 1730s and 1740s, creating excellent "combinations of figures, motions" and "contrasts of artificial fire." No manuals on how Russians made fireworks were in circulation, so the *Encyclopédie*'s judgment may have been based on observations of fireworks prints or on oral reports. Indeed, one French author heard reports of Russian fireworks, and his printed description was probably the source of the *Encyclopédie*'s comments.[74] He wrote:

> In the fêtes at celebrations in Muscovy all sorts of figures in fire are represented, such as tableaux of designs outlined with continuous fiery traces; a person who has been in St. Petersburg told me he had seen gigantic representations in these celebrations, which endured in fire for the space of one or two hours; they elevate a facade . . . more than one hundred feet tall, which is prepared without any decoration, inasmuch as during the day one sees only the appearance of a black wall, and then at night, when it is ignited, the fire suddenly spreads to all the traces of artifice, which burn with a clear fire.[75]

Stählin also kept a personal record of how fireworks and illuminations were set off at this time, describing these great boards as "plans" (*plany*) that were painted *à la detrémpe* to depict allegorical scenery such as temples, pyramids, gardens, or balustrades. These were illuminated in "trace work" with small fireworks or lamps outlining the decorations.[76] The technique is also evident in colored gouache paintings of displays produced by the artillery in this period in which the black plans and traced decorations are visible (plate 3).[77]

Although their opinions are scarce, artillerists noted how well this work was received. Mikhail Vasil'evich Danilov, a graduate of St. Petersburg's drawing and artillery schools who joined the arsenal's fireworks laboratory in 1743, thus recorded his achievements on illuminating a palace at Vseviatskoe Selo near Moscow for the empress: "For such an event the whole house was illuminated with lamps. I carried this out for the first time alone, without my commanders, and earned myself praise for this, which for a young man brings the desire to receive more."[78] "Praise" here was important since it implied potential promotion to a higher rank. Social rank (*chin*) was of great importance in Russia, formalized in the "Table of Ranks" for civil service, military, and noble communities since the time of Peter the Great. To officers like Danilov promotion in the Table of Ranks, rather than financial remuneration, was the highest reward.[79]

Audiences were also thoroughly impressed by the artillerists' displays. The Russian spectator of another performance noted: "The illumination really was a worthy vision, and I rubbed my eyes as I watched and admired it."[80] In fact, judging from foreign accounts of Russian fireworks, the artillerists' skills often raised more admiration among audiences than did the academicians' allegories. Thus, the Swedish ambassador Karl Berk recorded how the "allegorical figures are all painted, but spectators pay still more attention to the bombs, bouquets, and rockets that explode not one after the other but in their hundreds."[81] Pyrotechnics could also signal the progress that the academy was supposed to represent, as the English ambassador observed on seeing fireworks in 1730: "The ceremony concluded with a fire work, perhaps not to be out done in any part of the world, to such a superlative degree of polite luxury are they arrived in so few years."[82]

This was not the response academicians required of Russian fireworks if they were to bring glory to the academy rather than the artillery. Hence, another aspect of their inventive endeavors served to direct attention back toward their allegories. Academicians already denied artillerists an inventive capacity by insisting on a division between execution and invention in pyrotechny, implying that artillerists only fired off what academicians thought up. To secure this division, the academy exploited its dominance of publishing. Most fireworks performances were, thus, accompanied by the publication, via the academy's press, of printed descriptions (*opisaniia*) of displays, usually written in Russian and German, and bound together with impressive engravings of allegorical scenery.[83] Scores of these *opisaniia* were printed during Anna's reign, making them one of the academy's most prolific endeavors.[84] Both images and texts directed attention away from pyrotechnic ingenuity. Schoonebeck's images of Petrine fireworks showed bustling groups of courtiers and artillerists watching and setting off displays. In contrast, engravings of the 1730s produced at the academy showed only the scenic arrangements of fireworks, with no sign of the fireworkers who set them off, and only highly conventional representations of the pyrotechnics (compare fig. 14 with fig. 15 on page 121). Similarly, while the accompanying texts occasionally listed the types of fireworks seen in a display, most referred only in a general manner to the "pleasure fires" being used.[85] The bulk of the texts focused on describing and explaining the significance of the allegorical symbols in a display. This fulfilled a key goal of the academy, to teach the Russians Western manners, for which a knowledge of classical iconography was considered essential. Descriptions thus went to some length to explain classical imagery, while Stählin recalled Leibniz in noting the educative role of spectacles: "Theater is similar in all respects to a school, where people learn virtue just the same as in philosophical meetings."[86] *Opisaniia* thus focused audiences' attention on academicians' contributions to fireworks at the expense of artisan fireworkers. The academy relied on fireworkers' skills, and praised them when necessary, but always made sure that the emphasis was on its members' own inventive allegories, seeking to ensure their visibility to the court. Indeed, even the academy itself was subsumed into allegorical inventions. Juncker thus presented the Kunstkamera as if its sole purpose was to serve as a display for the court, suggesting: "This . . . structure [the fireworks theater] is positioned . . . on the bank of the Neva on an arrangement of a thousand piles that has been so capably constructed that it seems that the right side and observatory of the *Kunstkamera* and the bell tower of the Peter and Paul Fortress, when they too are illuminated, have been built for no other purpose than to be observed from this place."[87] The academy itself became the focus of spectacle. Later, in 1749, the academy ordered its own "fireworks theater" constructed over the water in front of the Kunstkamera, while its buildings were used on at least one occasion for a courtly masquerade.[88]

## Love for the Sciences: Empress Anna and Fireworks as Praise

Perhaps the most important use of allegorical inventions in the 1730s was their ability to shape the image of the sovereign and her rule. If ingenious mechanical inventions in England transformed the direction of English natural philosophy in the eighteenth century, so the allegorical inventions of the academicians performed a similar function in Russia since they helped secure assistance from the empress that ensured the academy's survival. Fireworks and illuminations educated Russian audiences, not only in iconography, but also in ideology, making manifest a form of state and government that Münnich and the academicians deemed fitting to Anna's reign. Anna approved these images, but she was also beholden to them. Throughout her reign and her successor's, academicians produced many allegorical inventions for the empress, not only fireworks, but also odes, dedications, and verses, to the point where one foreign visitor noted: "The publications of the academicians are quite few if one does not count the odes and verses that they produce for any sort of opportune event and that are flattering in the extreme."[89] This was not groveling on the part of academicians but a form of political interaction since laudatory verses and inventions constituted what Margaret McGowan terms *praise literature*.[90] Since courtiers could not criticize princes directly, they praised them using descriptions of virtue that they were obliged to live up to. Thus, Erasmus, whose work was influential in early-eighteenth-century Russia, noted: "By having the image of virtue put before them, bad princes might be made better."[91] Allegorical inventions were a gift to the empress, who was expected to live up to the image by which she was represented.

Inventions then could be used to guide and publicize policy, and Münnich and the academicians exploited this. Münnich, who was also a minister, promoted the continuance of Petrine reforms, the expansion of the military, and a form of German cameralism or "police," all of which were presented as attributes of the empress's rule and Russia's prosperity in fireworks and illuminations during the 1730s. Thus, Juncker's invention for illuminations celebrating Anna Ivanovna's birthday in 1733 made a performance of the "well-ordered police state," presenting "'Good Order,' in the form of a woman having a crown on her head and wavy hair, signifying, not only that good order is the greatest beauty there is in any state, but also that all those things in which it is found bring respect. In her right hand, she carries a clock and in her left a circle and scale, which shows that good order in any state is achieved through precise observation and the division of time and deeds."[92]

Alternatively, academicians incorporated images and texts in their inventions that demonstrated the sovereign's "love for the sciences [*liubov' k naukam*]" or "love for virtue, the sciences and arts, and the art of war."[93] Figures of mathematical instruments might appear in displays, while martial and academic values were combined in the figure of Minerva, the goddess patronizing the military arts and sciences, a personification that academicians would give Russian monarchs for the remainder of the century (fig. 15).[94] The printed description of the academy's fireworks for Anna's coronation day in 1734, for example, carried an engraved portrait of the empress as Minerva, above "kettledrums and other military equipment."[95] The accompanying text celebrated her love of science by describing an illuminated "Temple of Wisdom" based on the Temple of Solomon, with seven pillars, and fronted by statues symbolizing the virtues.[96] These images, which recall Furttenbach's displays of the sons and daughters of Mechanica in Ulm, tied the rationality of the sciences to the power of the monarch and displayed

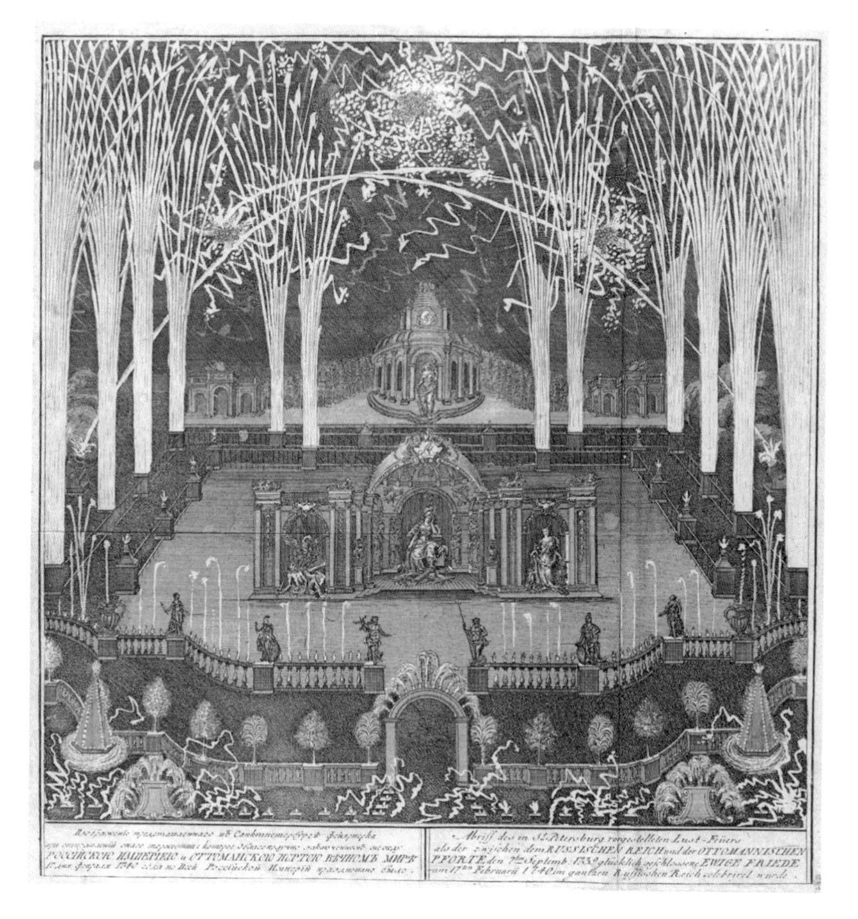

15. Fireworks for the peace between Russia and the Ottoman Empire, St. Petersburg, February 17, 1740. Empress Anna appears as Minerva, goddess of war and protector of the arts and sciences. *Izobrazhenie predstavlennago v Sanktpeterburge feierverka pri otpravlenii onago torzhestva; kotoroe o blagopoluchno zakliuchennom mezhdu rossiiskoiu imperieiu i ottomanskoiu portoiu vechnom mire, 17 dnia fevralia 1740 goda po Vsei Rossiiskoi Imperii prazdnovano bylo* (St. Petersburg, 1740). Research Library, the Getty Research Institute, Los Angeles, California.

the connection for audiences since "edifying reason [*nravouchitel'nyi razum*], which resides in the person of Her Imperial Majesty," was shown seated inside the temple.[97]

Through images such as these, fireworks were both a prescription for and an appeal to Russians, and, above all, to Anna, to support and nourish the sciences and military arts. These sentiments were echoed in academicians' odes and verses, such as those of the Russian chemist Mikhail Vasil'evich Lomonosov that proclaimed: "Rejoice now sciences! Minerva has ascended the throne. She . . . with her generosity and zeal will build a beautiful paradise for us."[98] In shaping the image of the sovereign and the state through allusions to love for the sciences and martial values, academicians both affirmed the sovereign's power and pleaded with the sovereign and courtly audiences to conduct themselves according to such values. Thus, pyrotechny—and this might be said of the academy's civilizing efforts in general—did not simply explain or disseminate knowledge and manners but demanded reciprocation, operating a gift relation whereby the celebratory images presented in fireworks demanded real support from the sovereign and elites.[99]

Such an approach is exemplified in a firework staged for the birthday of Anna Ivanovna in January 1735 (fig. 16). A dominant theme in academic fireworks and illuminations was cosmological imagery. As elsewhere in Europe, princely power could be figured through the imagery of artificial stars and meteors, which captured in artifice the portentous power of the skies and heavens and redirected it to celebrate earthly powers. Many fireworks were explicitly astrological in their iconography. In Russia, princes had long been enamored of astrology, and Anna employed the academic professor of physics Georg Wolfgang Krafft to cast her horoscopes. This was a problem for Krafft. Other academicians ridiculed him over astrology, and he avoided mentioning it in his autobiography.[100] But astrology was an imperial taste that could not be ignored by the advocates of the sciences, and Krafft's horoscopes were always made "on the appointed day, fulfilled to sustain imperial favor for the Academy."[101] Furthermore, academicians could use Anna's astrological interests to forward their own rational programs. Cosmological imagery was exploited to teach astronomical lessons and to present the consumption of astronomy as a courtly virtue. In January 1735, academicians celebrated Anna's birthday with a firework consisting of a central armillary sphere "in which the sun stands in the sign of the zodiac under which the birth of Her Imperial Majesty took place."[102] Above this was another sun, atop a temple of virtue, representing the empress. On each side of the central armillary sphere were "two other armillary spheres, in one of which the sun is seen according to the system of Tycho, and in the other according to the system of Copernicus, the two main views by which the physicists of our times present the world and all its bodies."[103]

This is a salient example of the use of courtly fireworks to promote the sciences. Like William Whiston in London, Russia's academicians directed the traditional iconography of court spectacle to novel ends. However, while Whiston mobilized stars to solve commercial and navigational problems, the academicians used fireworks to forward praise of the empress and cosmo-

16. Fireworks for the birthday of Anna Ivanovna, January 28, 1735. Ottomar Elliger III (engraver), *Plan Illuminatsii i feierverka kotorye Genvaria 28 dnia 1735 goda v vysokii den' Rozhdeniia eia imperatorskago velichestva Samoderzhitsy Vserossiiskoi v Sanktpeterburge predstavleny byli* (St. Petersburg, 1735). Sächsische Landesbibliothek, Staats- und Universitätsbibliothek, Dresden/Deutsch Fotothek.

logical theory. Few Russians knew of such ideas, though some had learned of Copernicus through Christian Huygens's *Kosmotheoros*, translated into Russian in 1717. As Gordin argues, the *Kosmotheoros* had provided Russians with a model of civic conduct and the "natural philosophical form of life" through exemplars of polite disputation over the Copernican theory.[104] The

1735 firework made a dazzling spectacle of such themes, again prescribing models of conduct through celestial representations. The display neatly linked contemplation of Tycho's and Copernicus's systems with adoration of the sovereign. As in the *Kosmotheoros*, no dogmatic assertion was made concerning which system was correct, but modes of conduct *were* asserted. Anna was linked to the cosmological source of all order and action, made into a Sun Queen in the manner of Louis XIV. Beneath the Tychonic system was the motto "Profert Magnalia Curso" (You bring in your course great things) and beneath the Copernican "Stans Omnia Movet" (Your constancy sets everything in motion). Between these world systems, "on either side of the pedestal recline the figures of wonder and pleasure, which present themselves, not only when one looks at the sun, but also in discussion of the great qualities of Her Imperial Majesty."[105] Courtiers thus learned that to discourse on the celestial motions was to give praise to the sovereign.[106]

These fireworks marked polite astronomy and imperial praise as connected elements of proper courtly conduct. Anna Ivanovna reciprocated. Several weeks later, Anna summoned Krafft and the astronomer Joseph Delisle to court to present her with astronomical and physical demonstrations, the first time any experiments were displayed to a courtly audience following the demise of the public assemblies and Bernoulli and Delisle's earlier controversial lecture on cosmology. Following Krafft's demonstration of a burning lens: "Her majesty deigned to observe, amongst other things, Saturn with its ring and satellites through a Newtonian telescope, which was seven feet in length. Her imperial majesty declared her most gracious pleasure and decreed that physical as well as astronomical instruments should be acquired for the continuance of such observations at court."[107]

This was a striking contrast to the apathetic response of nobles to lectures on astronomy just five years earlier. Now, the fireworks of 1735 drew Anna to astronomy and led her to enact the role of patroness and astronomer-princess. Indeed, the academy's "Copernican" firework is a good illustration of the manner in which pyrotechnics could achieve in Russia what more typical forms of scientific mediation could not since lectures failed to appeal to the nobility. It also highlights the value of allegorical inventions in Russia, just as prolific and original as inventions in commercially focused England, but geared to the court rather than mining or manufactures. Not that the Russians made only allegories and the English only machines. Russian fireworkers showed remarkable pyrotechnics, while the English could be rhetorical when occasion called. Thus, Desaguliers's long poem *The Newtonian System the Best System of Government*, dedicated to Princess Caroline of Anspach in 1728, included the same mix of courtly praise and cosmology as the Russians' fireworks: "Attraction now in all the realm is seen / To bless the Reign of George and Caroline."[108] The verses served to foster princely patronage and helped earn Desaguliers his position as chaplain to the Prince of Wales, for whom he performed fireworks.

## The Consequences of Academic Pyrotechny

Academic inventions paid off. The academy received substantial funding at the end of Anna's reign, and its existence was secured. In fact, fireworks were not the only form of invention undertaken by the academy to appeal to the court in Anna's reign. Schumacher understood the value of such projects and hired a growing number of artisans at the academy to produce commodities for the court. Their work included the production of prints and engravings, portraits, and cityscapes, the preparation of a coronation album for the empress, and the hiring of artisans to make ivory turnings, snuffboxes, medals, and architectural designs for the court. There was even an academic stone-cutting mill founded to make jewelry near the imperial palace of Peterhof.[109] These technicians constituted a major component of academic personnel by 1740, when 140 artisans worked in the academy, compared to 13 academicians and 15 adjuncts.[110] Fireworks and illuminations were also expensive, costing 339 rubles in 1732, 664 a year later, and 2,525 in 1735, the year of the Copernican display. Since the academy's budget was set at 24,912 rubles per annum and an average professor's annual income was about 1,000 rubles, this was a major expenditure.[111] Such proliferation and expense met resistance from academicians, though complaints did not criticize the idea that professors should compose fireworks.[112] Rather, they focused on the costs of maintaining so many artisans, and, once again, the professors showed their prejudice against practitioners of the mechanical arts. Complaints began as the number of artisans grew. On February 6, 1733, professors sent a decree to the Russian Senate criticizing the proliferation of arts in the academy: "Such diverse activities are not united in any other Academy in the world."[113] Anna's reluctant astrologer Krafft, the only academician to work with his hands as professor of experimental physics, did not sign. However, when only a few months later the academy's newly appointed president, the influential Courland courtier Hermann Keiserling, questioned academicians on the value of the arts, most tempered their criticism, saying only that the expense of the arts was detrimental to the academy, though arts were useful to the sciences and to the state. After Keiserling petitioned the empress with this report, Anna boosted academic funds, and the complaints declined.[114]

## France: Fireworks in an Alternative Absolutist Culture

Before concluding this chapter, it is worth taking a detour to France, where fireworks were institutionalised alongside the sciences in another absolutist culture during the reign of Louis XIV. The French context helps clarify what was distinctive about the relationship of fireworks and the sciences in Russia, despite France's and Russia's shared absolutist context. In Paris, as in St. Petersburg, fireworks served to glorify the prince, and, in France, as in Russia, a great deal of money and effort was expended on displays. The French, like the Russians, hired experts to design fireworks in place of gunner-*artificiers*; however, they chose local and foreign artists and architects rather than scientific academicians to perform this work. While Peter and Münnich borrowed from Dutch and German pyrotechny, the authorities in Paris turned to Italians, or French artists trained in Italy, to take over the business of inventing displays. Already in 1649, the Bureau de la ville, which regularly organized Parisian fireworks, rejected plans for fireworks offered by the king's latest "ingénieur des feux d'artifices," Thomas Caresme, in favor of designs by the Italian-trained engraver Jean Valdor, who proposed firing the display around an Italian-styled classical temple.[115] In the same year, the Italian engraver Stefano della Bella produced an elaborately illustrated manuscript fireworks book in Paris, similar to earlier Italian manuscripts that used bold illustrations to depict pyrotechnic art (plate 4).[116] Fireworks were also institutionalized in Paris. From the early 1660s, the Administration de l'argenterie, menus-plaisirs et affaires de la chambre du roi oversaw displays for the court, while fireworks in the city of Paris were organized by the *prévôt des marchands* and his *corps des echevins*.[117] As in London, laws prohibited anyone besides the royal gunners and the nobility from setting off fireworks.[118] During the reign of Louis XIV, these bodies turned for fireworks designs to Italian and French painters, architects, stage designers, and writers, men such as the Modena-born architect Gaspare Vigarani, the painter and academician Charles Le Brun, and the Jesuit man of letters Claude-François Menestrier. The artillerists Maximilièn Titon and Denis Caresme had to be content with setting fireworks off.[119]

As in Russia, then, the French court created a hierarchical division of labor in which the task of designing displays was taken away from gunners, passed to more learned experts, and insitutionalized. Furthermore, the labors of Parisian artificers, like those of their Russian counterparts, were erased from representations of courtly fireworks. As in St. Petersburg, elaborate engravings and poetic court literature regularly described the sovereign's spectacles for posterity but made no mention of the labors of artificers in fireworks displays. Like the Russian academicians, Parisian men of letters distracted attention from artisanal labor in competition for praise from the sovereign. However, in Paris, praise of the sovereign demanded that *all* labors be erased from view since the king's ruling image identified him as a miracle worker, under whom the laws of nature and the elements bowed, and whose power was the cause for all events in the state, including festivals.[120] Thus, the Jesuit Menestrier explained the meaning of fireworks in the reign of Louis XIV in the short treatise *Advis necessaires pour la conduite des feux d'artifices*.[121] Menestrier ignored the art of making fireworks altogether, in favor of a discussion of the emblems, symbols, and decorum suitable to pyrotechnic allegories. He also explained that it was not men who made festivals but nature: "Four great workers of marvels of nature labor incessantly for the glory of our incomparable Monarch."[122] These were the elements, air, earth, fire, and water, and of these fire, "since it holds the highest rank in the order of the world," had "the advantage over all the others in these public festivities": "There is nothing more bountiful than it [fire], it communicates its qualities to all that approach it, & makes continual profusion of its light. There is nothing more bustling, it is in continual restlessness, it attaches itself to all bodies, it works on all sorts of matter, & transforms the substance of everything it penetrates, its workings create miracles of art and nature. It changes sand into crystal, poisons into remedies, flowers into essence, & earth into gold."[123] In Menestrier's fetishistic argument, the element fire bent under the will of the king to perform spectacles in his honor. Engravings and paintings of the Sun King's fireworks displays reiterated this image, removing any sign of human labors from the scenes depicted (plate 5).[124]

As the elements worked only at the command of the prince, the prince also took on the attributes of a great, even supernatural artificer. In descriptions of fireworks at Versailles by the court poet André Félibien, Louis XIV appeared as a kind of alchemist, under whom the elements transformed themselves into gold: "All that was seen in this great stretch of more than three hundred fathoms was no longer either fire, air or water. These elements were so mixed together that they became unrecognizable and there appeared a new one of a most extraordinary nature. It seemed to be a composite of a thousand sparks of fire, which as a thick dust or, rather, as an infinity of atoms of gold shone in the middle of a greater light."[125] In France, as Roger Chartier observes: "All the sovereign's acts were prodigies that bent time to his law and nature to his will."[126] The same ruling image would become prominent in Russia, but not until the reign of Catherine II.[127] Before then, the image of the ruler as Minerva dominated Russian displays, which moreover celebrated a range of themes, not just the power of the monarch.

Another difference between Russia and France lay in the role that the sciences played in these pyrotechnic festivals. While the Russian Academy of Sciences oversaw pyrotechnic designs, in the French system of academies established in Paris by Jean-Baptiste Colbert during the 1660s the Academy of Sciences operated independently of the Academy of Arts, and it was the Academy of Arts that provided expertise for producing spectacle. French savants also had much less trouble interesting French noble society in the sciences than their Russian counterparts had interesting the Russian nobility. Descartes had set a trend for noble interest in new sciences, and, after the turmoil of the Fronde declined, lecturers and savants such as Jacques Rohault, Nicolas Lemery, and Pierre Polinière set out to publicize the sciences among Parisian polite society. When the Paris Academy of Sciences was founded in 1666, it stood at the apex of a number of salons and informal academies interested in the sciences and so did not have to work as did the St. Petersburg academy to secure its place in society.[128] For aristocratic audiences, philosophical experiments proved as popular as other spectacles, as accounts of the chemical lectures of Nicolas Lemery suggest. While Russian nobles ignored scientific lectures, the best of Paris attended Lemery's: "His Laboratory was less a Room than a Cave, & almost magical lair, illuminated solely by the glow of the Furnaces; nevertheless the affluence of the crowd there was so great, that there was scarcely enough room for his operations. The most famous names were entered on the roster of his audience, Rohault, Bernier, Auzout, Regis, Tournefort. Even Ladies swept up by the fad had the audacity to come and show themselves in these very learned gatherings."[129]

Natural philosophy was a "fad," though this did not preclude Lemery from including fireworks and other elements appealing to the nobility in his performances and publications. Around 1700, he modeled "an Etna or Vesuvius." He buried fifty pounds of fermenting paste of sulfur, water, and iron filings in the ground, and the fermenting mixture cracked it open, seeming to support recent suggestions that pyrites, of iron and sulfur, provided the fuel for volcanoes and earthquakes.[130] Lemery's chemical lectures were published in 1675 as the *Cours de chymie*, which "sold like a work of Galantry or Satire."[131] Lemery also published *Le nouveau recueil de curiositez . . . de la nature & de l'art*, which included recipes appealing to the nobility, on beauty, jewelry and pearls, perfume, and hunting and fishing. Later, he included sections on artificial fireworks, an inextinguishable wildfire, and a "Fire that burns on armour."[132]

In Paris, then, experiment, spectacle, and fireworks all took up a place within fashionable society, where their techniques were often interchangeable. The physician Pierre Polinière began the eighteenth century with lectures on *pyrotechnic*, the term he used for chemistry, showing color changes and phosphorus effects indebted to Slare, an imitation of thunder, and other artificial meteors. "Everything," Polinière told his audience, "belongs to pyrotechnic: Rain, hail, tempests, lightning, congealing, distilling, fermenting, and dissolving are all pyrotechnic operations."[133] As in London, meteors might be brought to order by laws of nature mobilizing pyrotechnic resources.

And here was the key contribution of the *sciences* to royal pyrotechnics in Louis XIV's France. Peter Dear has noted the distinctive concern of the Paris academy under Louis XIV to discover the regular laws that characterized nature. London's Protestant experimenters, who believed that the age of miracles had passed, did not assume nature to be entirely regular and gave much attention to exceptional and singular phenomena as being of natural interest. But Catholic French academicians, for whom the age of miracles continued, regarded unusual phenomena as potential miracles and sought to demarcate them by identifying rigid laws of nature.[134] Dear reads this contrast in religious terms, relating to divine miracles, but it could equally apply to court culture and its miraculous spectacles. Thus, while the Paris Academy of Sciences made no direct contribution to fireworks, its image of a regular, lawlike nature helped make the miraculous actions of the prince stand out. Such miracles were then recounted in festival literature and demonstrated in performances organized by the other academies in Colbert's system. The arrangement recalls the thoughts of Francis Bacon that he presented in a play addressed to a king, the "Prince of Purpoole," and wrote before England revolted against absolutism: "When all other miracles and wonders cease by reason that you shall have discovered their natural causes, yourself shall be left the only miracle and wonder of the world."[135]

## Conclusion

Russia and France both constituted absolutist cultures in which centralized academies managed grand public spectacles for the court. Yet their pyrotechnic cultures, and the relations of fireworks and the sciences that these entailed, were quite different. In Paris, the Academy of Sciences enjoyed substantial support from the crown. Academicians played no direct role in festivals, nor did they need to. Experimental science was à la mode, while other academies furnished expertise, often Italian in inspiration, to make fireworks for the city and the court. These fireworks erased the labor of both their inventors and their executors in the name of the king's ruling image of a miracle-working prince. In Russia, by contrast, the academy's situation was more precarious and insecure, prompting academicians to cultivate a role in the production of courtly spectacle that they used to gain patronage and publicity among noble audiences largely apathetic to the sciences. Here, too, artisanal labor was erased in accounts of fireworks, but, in Russia, academicans went to great lengths to make their own contributions visible. Furthermore, in academic allegories, the Russian empress appeared, not as a miracle worker, but as a benevolent patroness, encouraging the arts and sciences.

Just as French and Russian absolutist cultures gave rise to different relations of art and science, so Russia's academicians also differed from English natural philosophers in their relations with artificers. Russia did not yet witness a Desaguliers, moving from fireworks displays to experiments, nor was there any interest in creating more practical or profitable uses for fireworks outside spectacle.

Nevertheless, skill in *invention* was just as important in courtly Russia as it was in industrializing England. If London's gentlemen virtuosi appealed to utility and theology to promote experimental science in the wake of England's religious turmoil, St. Petersburg's academicians used skills of praise and poetry to appeal to the court. Making science work in Russia demanded such appeals. As the Russian empress expanded the court, the Academy of Sciences sought to make itself an indispensable source of expertise in poetic inventions and allegories, though it had to divert attention from other forms of invention to achieve this. Russia's experience, then, saw, amid struggles to control creditable spectacle, a competition between philosophers and artisans. Simultaneously, experimental science was negligible, unimpressive to the tastes of Anna's court. Fashionable elites flocked to witness Lemery's bustling laboratory in Paris, but St. Petersburg's experimental space was risible. When Erasmus King, Thomas Desaguliers, and Francesco Algarotti arrived in St. Petersburg in 1739, Algarotti went to visit the Academy of Sciences. He was surprised at what he saw: "The experimental physics laboratory is so poor and sad an assemblage as to make anyone coming from Italy or France laugh, let alone anyone from Holland, or even more so, from England."[136] Only in the 1740s did experiment begin to flourish at the academy, after Anna's reign was over and yet another coup brought Elizabeth Petrovna, the daughter of Peter the Great, to the Russian throne. Dynastic change brought epistemic change, and public experiments revived under Elizabeth, a noted Francophile. Academicians renewed their assemblies, the first on a suitably spectacular topic, the multicolored light and sound machine known as the *ocular harpsichord* of the French Jesuit Louis-Bertrand Castel.[137] After this time, public lectures became more common in St. Petersburg.[138] But allegorical invention remained critical to securing patronage, and the academy's pyrotechnic displays continued. A new reign, however, saw further struggles over spectacle, as we shall see in chapter 5.

CHAPTER FIVE

# Traveling Italians: Pyrotechnic *Macchine* in Paris, London, and St. Petersburg

Consider one account of the splendid Chinea festivals in Rome during the 1720s: "In the publick Piazza . . . was a Machine, built in very handsome Architecture, rais'd on an arch of Rock-work, with several large Figures, for the Fire-works; the four principal figures representing the four Quarters of the World. These, with others at a distance, which they call *Girandole*, whirling in a thousand Varieties before the Eye, and so numerous a Chorus of admirable Musick filling the Ear, gave a surprisingly magnificent entertainment."[1] The Chinea was one of the highlights of pyrotechnic spectacle in eighteenth-century Rome. Staged simultaneously with the Girandola, on June 29, the annual ritual entailed the giving of a tribute, traditionally a white horse (*chinea*), to the pope by the constable of Naples. From 1722, on the evenings prior to and following the constable's annual procession to St. Peter's, spectacular fireworks organized by the Neapolitan embassy in Rome lit up the public squares. To design these displays, the constable turned to the best architects in the city, men such as Niccolò Salvi, Alessandro Specchi, Gabriele Valvassori, and Nicola Michetti.[2] Since the days of Il Tribolo, architects had been central in Italian pyrotechnic displays, acting as the inventors of elaborate temporary edifices for fireworks. Architects exploited these opportunities to build fantastic imaginary structures, full-scale temples in three dimensions, constructed with wood and iron frames, and hung with trompe l'oeil painted cloth and papier-mâché and stucco decorations prepared by a small army of carpenters, turners, painters, and sculptors. These *macchine*, or "machines," which became typical of displays across the Italian states, were the most elaborate fireworks decorations of the eighteenth century, their popularity shifting the focus of fireworks spectacles from pyrotechnics to fantastic architectural experiments (fig. 17).[3]

17. The Chinea procession travels from the Piazza Farnese, on the left, to Saint Peter's Basilica, passing the Castel Sant'Angelo, and a great *macchina* representing a temple of Minerva and a burning altar, designed by Louis Le Lorrain in 1746. Attrib. Claude Gallimard, *Solemnis Equitationis ad Vaticanum Incessus* (Rome [?], ca. 1746). Research Library, the Getty Research Institute, Los Angeles, California.

In the eighteenth century, Italian artisans took this style of fireworks and spread out across Europe and Russia showing displays. Their performances, with an emphasis on architectural and mechanical ingenuity, prompted as much interest as the artificial meteors of the gunners had in the previous century. Up to now, we have been concerned mostly with the process by which those gunners, together with natural philosophers, succeeded in making their art a permanent and fixed element of local, national cultures. Arsenals, laboratories, and scientific academies all emerged in the seventeenth century as sites for pursuing natural philosophy and pyrotechny, a testament to the success of these communities in allying their skills with the courts. By articulating their labors in terms of valuable skills and depending on local conditions and politics, the practitioners of fireworks and natural philosophy gradually succeeded in securing permanent places for their talents. At the same time, however, an equally important component of this early modern pyrotechnic and philosophical culture was *travel*, and it is to traveling skills that this chapter is devoted.

We have already noted the travels of German academicians to Russia, Swedish fireworkers to England, and French Jesuits to India and New France. In making these voyages, artificers and scholars helped establish the traditions that, for the sake of brevity, I identify as "French," or "Russian," or "English," but that were really hybrid cultures forged from the skills of people from many places. This was especially the case in the eighteenth century, when a variety of practitioners of fireworks transformed European pyrotechnic culture by traveling from city to city and showing remarkable performances in fire. Many of these men, and probably several women, hailed from Italy, and it could easily be said that fireworks in Europe after the 1740s were made in emulation of, or in reaction to, Italian pyrotechnics, though Russia and China also became significant points of reference, as we shall see below. Following these Italians' travels thus tells us about the ways in which European pyrotechnic traditions were transformed in the eighteenth century. But, more important, they offer another useful means to consider the regionality of relations between artisans and philosophers because we can compare the different ways in which local communities responded to much the same performances and performers, as the latter traveled about different cities showing their unique skills.

What follows, then, is a voyage through Paris, London, and St. Petersburg that assesses these reactions and compares them to build up further a geography of early modern art and science. The particular pyrotechnic skills of an architect and stage designer named Giovanni Niccolò Servandoni and the Ruggieri family of artificers of Bologna are described, together with their activities in Paris showing fireworks during the 1730s and 1740s. Subsequent sections examine different reactions to their skills in Paris, London, and St. Petersburg. In Paris, a flourish of new pyrotechnic treatises followed the Italians' shows. Publishing books became a characteristically Parisian means to consider relations of arts and sciences. The English preferred practice, and, when the Italian spectacles arrived in London, they prompted hostility from the public and efforts to turn them to alternative, useful, and experimental ends. Finally, the Italians arrived in Russia, where the court and the Academy of Sciences continued to place great value on pyrotechnics. In Russia, Italian *macchine* and fireworks prompted competition from the Russian artillery and the academy, which attempted to outdo the Italians in pyrotechnic innovations. In the course of their travels, Italian fireworks raised critical questions concerning the openness and secrecy of pyrotechnic knowledge, the relative worth of theory versus practice, and the proper hierarchy of the savant and the artisan in pyrotechnic productions. The diverse ways in which spectators and authors responded to these questions testifies to the impact that traveling Italian pyrotechnists had across Europe in the middle decades of the eighteenth century and to ongoing variations in the relations of art and science.

## Traveling Artisans in Italy and France: Servandoni and Ruggieri

As Italians worked at the court of Louis XIV, in the eighteenth century French artists frequently traveled to Rome to observe its arts and antiquities, multiplying ties between the French and the Roman artistic communities.[4] The traffic affected pyrotechny as much as other arts. From the late 1730s, French masters in Rome designed the Chinea *macchine*, among them the painters Pierre-Ignace Parrocel and Charles-François Hutin and the architects Louis-Joseph Le Lorrain and Ennemond Petitot. There were also strong links with Spain, creating a remarkable Catholic network of artisanal skills. The Neapolitan embassy in Rome employed Spanish architects to design Chinea structures, while Roman architects traveled to Spanish Naples for commissions.[5] Italian architects and artificers also continued to travel to France, staging theatrical shows and fireworks displays that soon earned them reputations and demand in other European cities and courts. The most prominent of these traveling Italians in the first half of the eighteenth century were an architect, Giovanni Niccolò Servandoni, and a family of artisans, the Ruggieri brothers, who from the 1730s dazzled Paris with their pyrotechnics.

Servandoni was the son of an Italian mother and a French coach driver traveling the road between Lyon and Florence, territories that the younger Servandoni crossed frequently during his career.[6] In Florence, he learned the new *scena per angelo* perspective stage-design techniques of Francesco Galli-Bibiena. Moving on to Rome in 1715, he studied perspective and drawing with the architectural engraver Giuseppe Ignazio Rossi before setting out for Paris and the opportunity to exhibit his talents. By 1724, Servandoni had become established as the director of stage design at the Paris Opéra, one of a series of official royal theaters, along with the Comédie-française and the Comédie-italienne.[7] Despite their subsidies from the crown, which were small, the theaters needed to secure audiences to maintain income, and innovations were welcome. Servandoni obliged, bringing to Paris the innovative *per angelo* techniques of Panini and Galli-Bibiena. Scenes of Egyptian ruins, the Temple of Minerva, and a gem-studded Palace of the Sun soon earned Servandoni a glowing reputation and the role of principal painter and designer to the Opéra.[8]

Fireworks and theater had been entangled since the earliest church performances of the fourteenth century, a connection that continued to be strong in Italy while gunners in Northern Europe developed military locations for pyrotechnics. Servandoni blended theatrical and pyrotechnic innovations. In 1730, he introduced vast *macchine* in the Italian style to the French court and the city of Paris that would become models for European fireworks thereafter. On January 21, on the order of the Spanish ambassadors of Louis XV's grandson Philip V, Servandoni celebrated the birth of the dauphin, Louis, with a colossal *macchina* on the Seine. The intended performance, divided into three acts, was supposed to be spectacular. Following a battle of sea monsters on the river, a great pyrotechnic sun, forty-nine feet wide, would ascend over two rugged Pyrenees mountains, before a rainbow carrying the goddess Iris traversed the mountains to show how the birth of the dauphin united France and Spain (fig. 18).[9]

In the event, the display did not go quite according to plan, at least according to an English report on the festival: "The Rising sun was omitted for Want of Time; and the Rainbow, when made, was too unwieldy and unmanageable; and therefore could not appear; nor the Goddess Iris . . . these should have been describ'd as designed only, and not executed."[10] In Paris, accounts printed in the *Mercure de France* and engravings of the display by Gabriel Pierre Martin Dumont le Romain made the performance appear perfect, with the sun and the rainbow in full glory. The fireworks included fixed and turning suns, serpents, grenades, balloons, rockets, and *jets de feu*, which sent up great fountains of fire in imitation of *jets d'eau*.[11]

18. A machine on the Seine designed by Giovanni Niccolò Servandoni. Dumont le Romain (etcher), *Plan et vue du feu d'artifice tiré sur la riviere le 21 Janvier 1730* (Paris, 1730). Research Library, the Getty Research Institute, Los Angeles, California.

Although these were not the first *macchine* Paris had witnessed, nothing on this scale had been attempted before, and, after the performance, Servandoni was rewarded with election to the Académie de peinture et de sculpture. He then turned to permanent architectural inventions, designing the west facade of the church of St. Sulpice in Paris, a design that helped usher the neoclassical style into French taste.[12] Another neoclassical marvel came in 1739 when Servandoni was again called on to design fireworks for the marriage of Louis XV's daughter Louise Elizabeth to Philip, the infante of Spain. On August 29, he presented the grandest firework yet to take place on the Seine, assisted by the royal architect Jacques Gabriel and the *artificiers du roi* Testard, the Dodemant family, and a Prussian artillerist named Elrich. The display, which cost thirty-five thousand livres, was recorded in a monumental festival book by the architect Jean-François Blondel, also a champion of Servandoni's St. Sulpice designs. Blondel detailed Servandoni's inventions, which included an octagonal floating orchestra pavillion and an illuminated Temple of Hymen, in a series of engravings, foregrounding the architecture above everything else (plate 6).[13] The fireworks were nevertheless spectacular, presenting cannonades, flights of honorary rockets, fixed and turning suns, and an abundance of fire imitating water, in pyrotechnic "cascades," "fontaines d'artifice," and "jets de feu" whose sparks Blondel showed streaming above the Pont Neuf.[14]

Servandoni thus initiated a new era in fireworks and theatrical performances in Paris, aggrandizing elements already present in French fireworks to a new, and vast, scale. In these performances, neoclassical architecture and perspective scenery dominated displays, but French *artificiers* also offered ingenious fireworks to accompany them. New pyrotechnics emphasized the play of fire and water, in fiery imitations of *jets d'eau*, fountains, and cascades. Audiences also contributed to the displays, experiencing them no longer as curious and strange marvels but more like opera or theater productions. By this time, the crowds of urban spectators assembled at displays were no longer represented as unable to comprehend the artifice, while elites distinguished themselves less through exclusive knowledge of fireworks than through *nonchalance*, a dispassionate etiquette literally "without heat." Custom permitted theatergoing audiences of the time to arrive late and depart early and to appear to ignore the play, conversing over the top of it.[15] Engravings of fireworks in France during the 1730s suggest that aristocratic audiences engaged in similarly nonchalant conversation before the explosive effects of fireworks, an attitude equally reflected in works on natural philosophy. Spectators at fireworks for the dauphin at Meudon in 1735 thus resemble the relaxed interlocutors on the frontispiece of Fontenelle's *Conversations on the Plurality of the Worlds* engraved a few years earlier (figs. 19–20). Everyone was impressed by the spectacles of nature and art, but too sophisticated to show it.

Servandoni was not the only Italian making a reputation in Paris for his pyrotechny at this time. Fireworks at the theater were also the specialty of the Ruggieri family, who hailed from Bologna, long a center for theatrical innovations and fireworks. The family consisted of five brothers—Gaetano, Petronio, Antonio, Francesco, and Pietro, the eldest—who arrived in Paris in 1743 to produce fireworks at the Comédie-italienne.[16] The Comédie was home to a troupe of Italian actors and had existed since 1660, with a brief intermission, staging French dramas and commedia dell'arte.[17] Like the other theaters, the Comédie-italienne competed for revenue and in the 1730s began adding novel spectacles or *petite pièces* to repertoires to attract audiences.[18] Already in 1729, these referred to fireworks, the Comédie's playwrights Dominique and Romagnesy staging "une petite Piece nouvelle" entitled "Le feu d'artifice; ou, La piece sans dénoüement." The play centered on the enduring association of fireworks and the passion of love. As the Vaudeville put it:

19. Charles Nicolas Cochin le fils (draftsman and etcher), *Dessein de l'illumination et du feu d'artifice a monseigneur le dauphin à Meudon le 3e septembre 1735* (Paris, 1735). Research Library, the Getty Research Institute, Los Angeles, California.

> L'Amour est un Artificier
> Qui mieux que moi sçait son métier;
> Qu'il fasse des yeux d'une Belle
> Partir une seule étincelle:
> Pan, pan, pan,
> La poudre prend;
> Tout est en feu dans un instant.[19]

The Ruggieri's arrival brought real fireworks to the Comédie, with performances that the brothers named *spectacles pyrriques* filled with technical tricks that perplexed their audiences. These centered on fixed and rotating fireworks, set on iron axles on the stage of the Comédie, showing "various transformations of different figures of fire . . . that succeed either by horizontal, vertical, or inclined rotations or by the interposition of figures varied by their arrangements or their shapes, such as suns, girandoles, pyramids,

20. Bernard Picart, plate illustrating Fontenelle's *Entretiens sur la pluralité des mondes*. From Bernard Le Bovier de Fontenelle, *Oeuvres diverses de M. de Fontenelle, de l'académie françoise. Nouvelle édition, augmentée et enrichie de figures gravées par Bernard Picart*, 3 vols. (The Hague, 1728–29), vol. 1. © Roger-Viollet/The Image Works.

cradles, fountain jets or cascades, wheels, globes, crosses, polygons, and pointed stars."[20] An ingenious opening in the roof of the Comédie allowed the smoke from these fireworks to be carried away. The most intriguing part of the performance, however, involved the communication of fire from a spinning to a fixed wheel, whereby one ignited the other without any apparent intervention from the artificer. This was a feat considered impossible by French audiences and artificers, and, in 1743, it earned the Ruggieri the approbation of the king.[21] Soon after, the Ruggieri were appointed *artificiers* to the city of Paris and then *artificiers* to the king, for whom they began performing royal displays on a magnificent scale at Versailles. Servandoni and the Ruggieri thus came to dominate pyrotechnic inventions and displays in Paris during the 1730s and 1740s. The Italians had arrived.

## Revealing Secrets: Perrinet d'Orval and the Ruggieri

The *artificiers* of Paris did not take kindly to the arrival of foreigners and their pyrotechnic innovations. To fire Servandoni's 1739 display, the Saxon artillerist Elrich was brought to Paris because, according to the authorities, "they are very experienced in fireworks in Saxony. They make fires much superior to ours."[22] Parisian artificers disagreed, and,

> as the fireworks were performed by different artificers and among others by the Saxon captain, they made a mess of the execution out of envy for one another. . . . The king was displeased with the lack of respect of these people. This is why the following day M. Prévot des Marchands put in execution the *lettres de cachet* that had been sent him for the [French artificers] named Dodemant, father and son, and Testard. They were brought on the king's orders to the Hôtel de Ville, where they were [kept] for a whole month, and the city did not give them their payment until a year afterward.[23]

Artificers who disputed pyrotechnic divisions of labor and the authorities' choices were, thus, locked up, though the artificers did not give up. Two years later, the king issued a *brevet,* or letter of privilege, giving the same French artificers—Testard, Dodemant, Charles-Nicolas Guérin, and Guérin's son Charles-Denis Guérin—the right to stage fireworks for twelve years on the feast of Saint Louis.[24] The artificers performed only one display, but the episode suggests their enduring efforts to compete with foreign artificers.[25] Indeed, other disputes flared after the Ruggieri arrived. Etienne-Michel Dodemant and Pietro Ruggieri staged parallel displays before Paris's Hôtel de Ville on a number of occasions in the 1740s. In 1749, Horace Walpole recorded how "there were forty killed and near three hundred wounded by a dispute between the French and Italians in the management, who quarrelling for precedence in lighting the fires, both lighted at once and blew up the whole."[26]

The appearance of the Italians in Paris thus led to an aggressive response from angry artisans, and other fights would occur later. A more enduring form of response to the Italians was in the publication of books, with a wave of new treatises on fireworks following in the wake of Servandoni and Ruggieri's performances. Between 1741 and 1757, eight new treatises or editions of treatises on fireworks appeared in Paris, all inspired by the public taste for the Italians' spectacles.[27] Gunners had long used print as a medium to promote their art, rendering pyrotechnic secrets open to noble audiences. Now, in the works of another artificer, Jean Charles Perrinet d'Orval, the techniques of the Ruggieri prompted discussions of the role of openness and secrecy in pyrotechny and criticisms of the declining place of ingenious invention among artificers. Perrinet was the first person to publish on fireworks in France following the arrival of Servandoni and the Ruggieri. His *Essay sur les feux d'artifice pour le spectacle et pour la guerre* appeared in Paris in 1745, no doubt inspired by the interest that Servandoni's and Ruggieri's performances were raising at this time.[28] Little is known about Perrinet save that he was born around 1707 and died around 1780, was a native of Sancerre, and worked as an artisan *artificier,* or fireworks maker, with an atelier near the Port Louis in Paris.[29] A number of such artificers had premises on the outskirts of Paris by the 1740s, after an ordinance of 1706 banned the making or sale of pyrotechnics within the city limits.[30] Later in life, Perrinet made military fireworks for the state and may have spent time in the army or artillery because his book dealt with both festive and military fireworks and resembled the gunners' manuals of the seventeenth century. Certainly, learning the craft of fireworks was now part of artillery training, which was further institutionalized during the 1720s with the foundation of artillery schools at La Fère, Grenoble, and elsewhere.[31] Bernard Forest de Bélidor's *Bombardier françois,* intended as a textbook for the new schools, included fireworks recipes, while manuscript books of instruction at the schools show that artillerists underwent a formal education in pyrotechny.[32] In line with the French division of labor between execution and invention, this instruction gave little room for artificers to exercise their ingenuity, teaching, for example: "Different compositions employed by order in the turning sun of nine cartridges by Sieur Colbert artificer, it is necessary not to change anything I say because the fires will be distorted and lose their beauty, don't do anything."[33]

If Perrinet's training was in the military, this would make sense of his response to the Italians because, while he praised their exquisite skills in fireworks, he also sought to appropriate those skills by representing them as a "secret" that only he could reveal. Like a number of nonmilitary artificers in the eighteenth century, the Ruggieri family kept notebook accounts of their techniques, but these were informal and never published.[34] Such a practice did not fit the gunners' tradition of describing skills in print, which suited Perrinet as a means to reveal the secrets of his rivals. Above all, Perrinet was impressed with the Ruggieri's effect of making moving pieces ignite fixed ones: "The difficulty that [the Ruggieri] have found the secret of surmounting consists of making fire communicate from a moving to a fixed thing, by means of which they may carry it successively and at once to all the parts of their artifice, to make it play and produce effects that would have seemed impossible before this was discovered. The mechanism seemed so ingenious to me that I left it to myself to reproduce it, charmed by the idea of rendering it public if I should succeed."[35]

Perrinet followed the tradition of gunners' revelations of secrets, only his audience consisted of the public, rather than the nobility. Perrinet went on to describe his method, which he claimed allowed both himself and others to give similar spectacles. This was no simple task since the Ruggieri had evidently constructed a remarkable piece of machinery, one whose complexity suggests why audiences found their fireworks so impressive (fig. 21).[36] Perrinet's version was ingenious, consisting of an iron axle carrying three "suns," the first of which would be "fixed," firing out a shower of sparks in imitation of a sun, before communicating fire to a second "soleil tournant," which spun on the axle. When this stopped rotating, it would ignite a third sun, again fixed, with the potential to add further pieces if necessary. The machine concluded with jets rotating on the ends of six long axles to form a great star, making the whole piece some six to eight feet in diameter. The communication worked by using a "match" or fast-burning fuse that passed from the last burning jet of the first sun to the first jet of the second, via grooves in the cylindrical axle, and through the long axles to ignite the final star.

Thus, the secret of Italian pyrotechnics was revealed. However, Perrinet was also critical of French artificers and men of letters. He began the *Essay* by emphasizing the need to articulate technique accurately and by attacking the lack of inventiveness in his fellow artificers. Of seventeenth-century treatises on fireworks he wrote:

> In general these authors have no more than a vague theory of their subject, and it is hard to believe they could really put into execution the many things

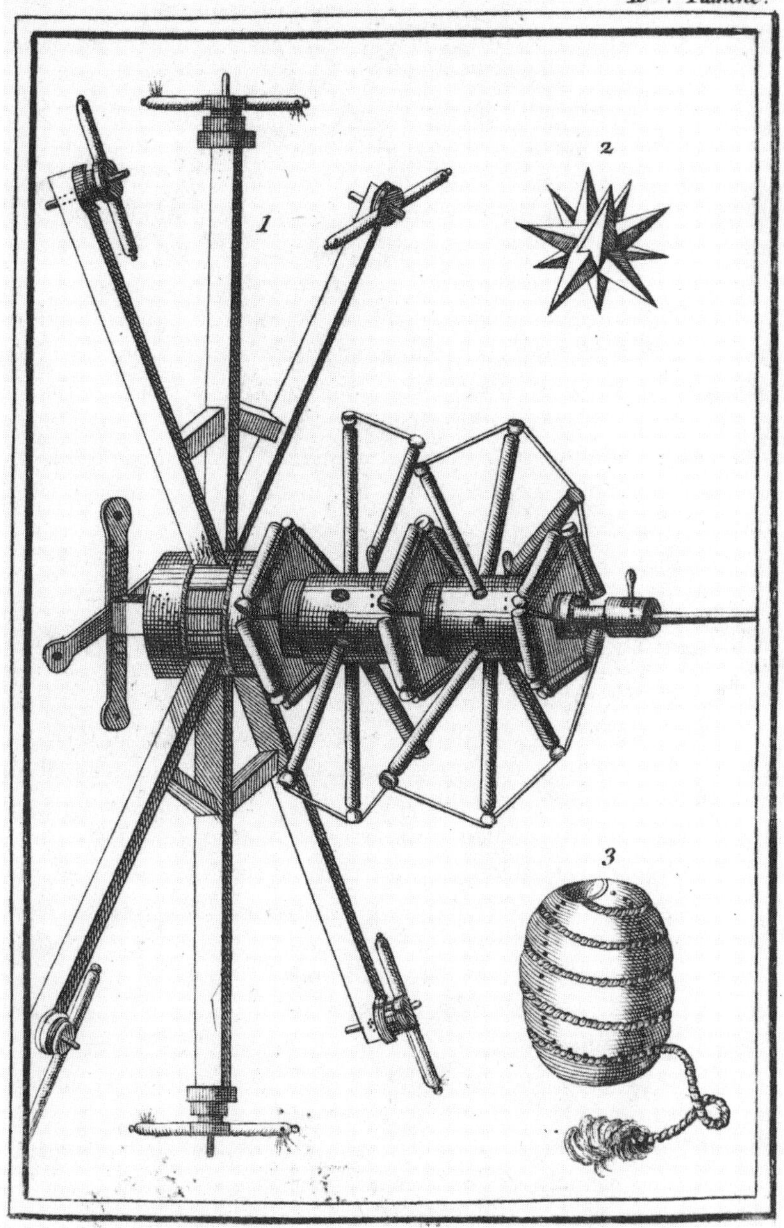


21. Perrinet d'Orval's imitation of the Ruggieri's "Spectacle pyrrique," showing "the machine all set up and each piece in particular, of which I shall give an explanation." Plate 10 of Jean-Charles Perrinet d'Orval, *Essay sur les feux d'artifice pour le spectacle et pour la guerre* (Paris, 1745). Courtesy of the Hagley Museum and Library, Wilmington, Delaware.

> whose success they announce as infallible. But, if the artificers are those who demand the most rigorous clarifications, one may also reproach them on account of their uniform manner of work and their lack of industry in imagining new things: many among them know only how to make like machines [*machinalement*] what they are shown; less artists than artisans they are in no state to form the least reasoning over that which daily occupies them.[37]

Utilizing knowledge without experience or practice without innovation based on knowledge was barren. Both knowledge and experience should be united in the artificer. Perrinet thus lamented a lack of ingenuity among gunners in evidence in the division of labor being promoted in the artillery schools. To Perrinet, gunners had become the tools for showing the inventions of others, and they now needed to reassert their independent imagination. Perrinet's imitations of the Ruggieri's technique were, then, designed to do just that, combining theoretical insights and practice to reproduce the ingenious methods of others: "I have reunited in these *Essays* that which experience has shown me to be good in the speculations of authors and in the practice of artificers."[38]

## Frézier: The Man of Letters as a Manager of State Spectacle

The artisan Perrinet thus followed a long tradition of pyrotechnic authorship promoting a combination of practical experience and ingenious invention as the ideal for artificers. In contrast, the learned engineer Amédée-François Frézier followed the works of the emblematist Menestrier in seeking to secure the status of men of letters as the exclusive managers of state spectacle, presiding over both architects like Servandoni and artificers such as Perrinet or the Ruggieri. Frézier, like Perrinet, chose to make his arguments about fireworks in a learned publication. Frézier was a well-known, if notorious, figure among men of letters in Paris and London.[39] Following infantry service in Italy, he was educated in astronomy and mathematics under Pierre Varignon and Philippe de La Hire at the prestigious Collège Mazarin. He then joined the *corps du génie* before being chosen to undertake an overseas voyage from 1712 to 1714 to survey Spanish fortifications on the coast of South America.[40] Subsequent disputes with Edmund Halley and Paris academicians ensured that Frézier remained an outsider to the scientific establishment.[41] But he continued managing fortifications construction in Saint-Dominique, Philippsbourg, Landau, and Brest while publishing treatises on architecture, stonecutting, and pyrotechny. The latter first appeared in 1706, as a *Traité des feux d'artifice pour le spectacle*, mostly derived from

Siemienowicz and other seventeenth-century works. But, in 1742, aware of growing public interest after Servandoni's fireworks, and concerned by the appearance of a pirated version of the *Traité*, Frézier announced a new expanded edition of his work, which finally appeared, much augmented, in 1747.[42]

Like Menestrier earlier, Frézier devoted much of his *Traité* to speculations on the history of fireworks and to the decoration of fireworks theaters and allegorical inventions. In contrast to the gunners, Freziér ignored military pyrotechny altogether, and, like Menestrier, he sought to give the man of letters a principal role in pyrotechny. In line with popular enthusiasm for Servandoni, he presented the Italians, and the ancient Romans, as authoritative in pyrotechnics. After identifying ancient Rome as the originator of fire spectacles, he pointed to Biringuccio's account of Florentine and Sienese pyrotechnics and the Castel Sant'Angelo's girandola to show how pyrotechny was "already magnificent in Italy" in the sixteenth century.[43] According to Frézier, the subsequent history of fireworks was a story about their staging, all fireworks being imitations of those in Rome and performed on theaters and machines whose architecture the third and last section of his treatise examined. Here, unsurprisingly, he chose Servandoni's designs as the model for a number of ideal pyrotechnic theaters.[44]

Because he was an architect and enthusiastic antiquarian himself, it is little wonder that Frézier sought to make the Romans and Servandoni's fireworks theaters exemplary. However, while he praised the role of *macchine* and architecture in spectacle, he still viewed them as subservient to his own skills, arguing that, ultimately, the most important element of producing fireworks was the invention or poetic composition that guided the display and decorations as a whole. Fireworks should be the work of "architects, painters, and sculptors, but particularly those men of letters who know how to present through agreeable ideas subjects that give occasion to rejoicings."[45] Like St. Petersburg's academicians, Frézier sought to focus attention on the intellectual contribution to fireworks, above that of artificers or architects. Fireworks might "surprize the eyes, but connoisseurs and men of letters wish more that their spirit find satisfaction in being . . . presented with agreeable ideas, taken from poetry and history and applied appropriately. . . . At the rejoicings made in Paris and at court in 1739 one saw decorations of a grandeur unlike anything we have seen up to now; . . . however, it seemed that in some of them the designs of architects had more of a part in their invention than men of letters."[46]

Beneath the men of letters and architects in this hierarchy came the artificers, whom Frézier, in contrast to Perrinet d'Orval, was sure had no ingenuity at all. Their contribution, according to his account, should be no more than that of carrying out the intellectual's ideas, fireworks being marked only "by an ingenious invention of the subject of the decorations and a prudent execution of the artifices."[47] Frézier was also sure that all artisans were secretive, in contrast to Perrinet's claim that only the Italians were guilty of this conceit. He thus expressed contempt for artisans who refused to reveal their methods to him when he first set about writing the *Traité*: "I immediately searched for a living voice to instruct me; but having found only a master jealous of his secret, I was forced to have recourse to books."[48]

Frézier went on to criticize both the Ruggieri's and Perrinet's labors. Despite his enthusiasm for Rome and the skills of Servandoni, he revealed the secrets of the Ruggieri by reprinting Perrinet's description of their techniques from the *Essay* "to enable the reader to imitate all that we have seen executed on the Italian stage."[49]

Italians arriving in Paris thus prompted deep debates over the proper nature of pyrotechny in the 1740s, raising issues such as the openness and secrecy of knowledge, the correct balance of theory and practice, and the proper hierarchy of pyrotechnic labors. While fighting was a common reaction of the *artificiers* to foreigners, Perrinet and Frézier wrote novel books on fireworks to articulate their response. Both men no doubt sought to profit from the Italians' successes with these works, but their representations of pyrotechny contrasted significantly, Perrinet seeking to reveal the secrets of competing artisans, and Frézier attempting to secure a hierarchy of letters over architecture and artifice.

## England: The Grand Whim for Posterity to Laugh At

In the meantime, Italian fireworks became the dominant fashion, soon spreading abroad. In 1749, Servandoni, Gaetano Ruggieri, and another Bolognese artificer named Giuseppe Sarti traveled to London to stage the grandest firework seen in England up to that time. Celebrating the end of the War of the Austrian Succession and the Peace of Aix-la-Chapelle, the performance presented Londoners with England's first grand pyrotechnic *macchina*, designed by Servandoni in the style of the temples erected in Paris and Rome. The fireworks included the Ruggieri's trademark communication of fire between fixed and moving pieces, and there was music by Georg Friedrich Handel—his well-known "Music for the Royal Fireworks."[50] But, if such delights found favor among audiences in Paris, it was a different story in London, where many criticized the Italian fireworks as a waste of money in the aftermath of war. Such skepticism was reflected in philosophical responses to the fireworks, which, following the practices of Desaguliers and Whiston, sought to turn the Green Park pyrotechnics to practical ends, encouraging further interactions between artificers and philosophers.

Italian fireworks were a real novelty in London, where there had been no grand display since the Peace of Utrecht in 1713. After the demise of Robert Walpole's Whig administration and a prolonged period of peace, the subsequent rise of more Tory and court interests in government provided the context for a revival of pomp.[51] However, the Green Park firework followed an age of English distaste for foreign extravagance, in favor of more commercial, instructive, and godly entertainments. Remnants of religious contests over pyrotechny remained. Voltaire, in London in 1726, recorded the Quakers' laments when fireworks were set off to celebrate victories.[52] Public attitudes toward the consumption of splendor were equally fraught. Addison and Steele, the promoters of Whiston's utilitarian pyrotechnics, scoffed at efforts to plant Italian-style opera in London, lambasting it as contrived, sensual, and ridiculous.[53]

Alternatively, there was growing enthusiasm among the commercial classes for instructive public lectures on natural philosophy and an expanding market for philosophical instruments and books. Commercial entertainments also flourished, the age of Walpole witnessing the rise of theaters, public houses, and pleasure gardens, where small-scale fireworks helped attract audiences and secure revenues.[54] Pleasure gardens such as Mulberry, Cuper's, Ranelagh, and Marylebone opened across the city, offering visitors promenades, refreshments, and music in leafy surroundings.[55] Jonathan Tyers founded gardens at Vauxhall, formerly the site of the Ordnance Board works where Savery may have invented his steam engine. Here, as elsewhere, illuminations, and from about 1740 fireworks, enticed spectators on summer evenings, with audiences ranging from those who could afford the ticket price (usually one to five shillings) to the king, whose birthday was celebrated with fireworks at Marylebone as early as 1718.[56] Spectators enthused over Vauxhall's illuminations: "Glittering Lamps in Order planted, Strike the Eye with sweet Surprise; *Adam* was not more inchanted, When he saw the first Sun rise."[57] Not all marveled at the displays, however, and commercial sensibilities could be turned against the new venues, through traditional allusions to enthusiasm and deceit. Thus, the mock apocryphal *Book of Entertainments* damned pleasure gardens and fireworks in 1742, invoking the enthusiastic falsehoods of William Whiston to condemn "Mother Cuper" of Cuper's Gardens for duping the public:

36. And she said, when there appeareth a Comet in the sky, do not the people go forth at Midnight? do they not gape and stare, and are not they greatly alarmed?

37. And do not the old men go forth, and the Prophets prophesy? Yea, doth not Whis[to]n the Prophet prophesy exceedingly, albeit it cometh not to pass?

38. Thus they are alarm'd, both small and great! Come now therefore let us make unto ourselves Comets of Gun-powder, and Comets of Salt-petre, and it shall be, that while they gape and stare, I will pick their Pockets.[58]

Similar scepticism, and above all a concern over the financial viability of fireworks, greeted the Italians' fireworks in 1749. The court organized the display, with artillery officers of the Royal Laboratory at Woolwich called on to assist the Italians, led by Laboratory Comptroller Charles Frederick, working for the Duke of Montagu, master general of the Ordnance. Frederick was assisted by Thomas Desaguliers, the son of Prince Frederick's chaplain, Jean Theophilus Desaguliers, and now, several years after his voyage to St. Petersburg, chief firemaster at Woolwich.[59]

The English fireworkers evidently shared their French counterparts' fascination with Rome and their disdain for foreigners. Frederick was a passionate antiquarian and a leading fellow of the Society of Antiquaries. In the 1730s, he traveled to Rome and Constantinople in search of ancient buildings and became known for drawings of architectural antiquities published after his return. Recalling Frézier, for Frederick the Green Park display was an opportunity to imitate and surpass the ancients. As Horace Walpole noted on a visit to the preparations in the park: "I wish you could see him making squibs of his papillotes, and bronzed over with a patina of gunpowder, and talking himself still hoarser on the superiority that his firework will have over the Roman naumachia."[60] Making this conquest of Rome succeed proved extremely difficult, however, because the public often insisted on viewing the approaching fireworks only on financial terms. Many exploited Londoners' growing anticipation of the fireworks by selling cheap prints and souvenirs depicting Servandoni's machine, while the London journals attracted readers with articles on pyrotechny (plate 7).[61]

Financial interests also lay behind more critical views. The peace declared in October was lambasted as capitulation to England's enemies, France and Spain, and immediately the lack of commercial advantage this entailed was contrasted with the extravagance of celebrations. As Walpole wrote, the peace "does not give the least joy; the stocks do not rise, and the merchants are unsatisfied . . . but the government is to give a magnificent firework."[62]

22. *The Contrast* (London, 1749). © The Trustees of the British Museum.

Like Addison and Steele, Samuel Johnson asserted the unsuitability of Italian pomp for English audiences: "The first reflection that naturally arises is upon the inequality of the effect to the cause."[63] More important, paying for fireworks with money that could be spent on the national debt or alms for returning soldiers was financially irresponsible and morally inept: "There are some who think not only reason, but humanity offended, by such a trifling profusion."[64] Satirical prints made a similar point. In *The Contrast, 1749*, beneath the heading "No Money with Fire-Works," a portly Dutchman representing "Money with Commerce" scorned an Englishman biting his nails, his pockets empty—"Myn Heer—you have been at war—what have you got?" (fig. 22).

These were remarkable debates on court pyrotechnics, unparalleled anywhere else in Europe at the time, and did not bode well for the Italians and their assistants from the Royal Laboratory. Montagu and Frederick did their best to manage the situation. To secure appropriate significance for their work, the Board of Ordnance used tactics similar to St. Petersburg's academicians, issuing authorized descriptions, and promising plans and elevations of the machine to counter "inaccurate and faulty" popular prints.[65] The resulting *Description of the Machine*, published in April, described Servandoni's *macchina* as a vast Doric temple, 114 feet high and 410 feet long, flanked by wings and two pavilions, all "adorned with Frets, Gilding, Lustres, Artificial

Flowers, Inscriptions, Statues, Allegorical Pictures, &c."[66] Unlike the Russian academicians, the Board of Ordnance did not leave artificers out of the description, and in the remainder of the pamphlet Ruggieri and Sarti outlined the order of the fireworks—a grand salute of cannon, followed by flights of honorary rockets and twelve acts of changing pieces, including forty-foot-high *jets de feu* and fires communicated from moving to fixed pieces, the Ruggieri's speciality. A huge girandola burst of six thousand rockets would conclude the entertainments, rehearsing scenes from Servandoni's earlier Paris displays.[67] Everything was supposed to glorify the king and the peace, though one motto seems to have taken public criticisms into account, showing that the "happy Re-establishment of Commerce under the auspices of the best of Kings, is the Joy of the British People and Senate."[68]

This was the only place where the performer's knowledge of the display was to be exhibited in print, the description serving as both an advertisement of the artificers' names and skills and a way to counter unruly criticisms surrounding the display. It was not unusual for such descriptions to be issued on the Continent to fix the meanings of pyrotechnic scenery. London's Board of Ordnance, however, proved incapable of handling problems from then on. The promised plans and elevations never materialized.[69] Worse, crown inexperience in Italian pyrotechny left a dearth of skills when it came to staging the event. Frederick and Desaguliers may have known what they were doing, but otherwise Woolwich Laboratory needed assistance. So, as John Brewer has shown, royal pyrotechny relied on commercial entrepreneurs. When the Vauxhall operator Jonathan Tyers offered to provide skilled assistants and equipment to illuminate the temple, Montagu and Frederick accepted. In return, Tyers won the right to stage a rehearsal of the music for the royal fireworks, recently composed by Handel, whose *Atalanta* had earlier been accompanied by fireworks at Cuper's.[70] Handel and Tyers's rehearsal at Vauxhall, a few days before the fireworks, went off brilliantly.[71]

The official performance of the Green Park fireworks was famously less successful. It began with Handel's music at six in the evening, after which the king and Montagu reviewed the machine, now standing brightly painted in the Green Park. At eight thirty the fireworks began, but many failed to light in the damp weather. Then disaster struck, as one of the temple's pavilions caught fire and burned to the ground. The gunners continued as best they could, showing the illuminated sun and girandola, before the final illumination at eleven o'clock.[72] Then the English and Italian fireworkers began brawling. Servandoni accused the English of incompetence and drew his sword to duel with Frederick. Servandoni was arrested but later had to be freed, departing London with three thousand pounds, much to English disgust.[73] Soon after, Servandoni's passionate actions prompted a typically English attack that portrayed him as a Catholic enthusiast, full of excess and pomposity. *The Green Park Folly* imagined the Virtues standing on Servandoni's *macchina*, regretting their inapplicability to the character of this "hungry strolling elf."[74] Another pamphlet represented Servandoni as a Paris surgeon, selling spare French body parts for Englishmen injured at the fireworks, evoking fears of foreign intrusion.[75] Servandoni's "magnificent temple" was now cast as "Cracker Castle" or the "New Folly-Castle."[76]

The Ruggieri also suffered since audiences condemned the fireworks as much as the poor organization. Some in the audience were impressed, notably by the girandola: "One Explosion particularly which they say was of Six Thousand was beyond all Imagination, & excepting to poor Mrs. Talbot who was frightened out of all her Wits. . . they were no less beautiful."[77] But most thought the performance dismal. Walpole recorded: "The rockets, and whatever was thrown up into the air, succeeded mighty well; but the wheels, and all that was to compose the principal part, were pitiful and ill-conducted, with no changes of colored fires and shapes: the illumination was mean, and lighted so slowly that scarce any body had patience to wait the finishing; and then, what contributed to the awkwardness of the whole, was the right pavilion catching fire, and being burnt down in the middle of the show."[78] Soon the satires and criticisms redoubled. Comparisons to Rome were made, but not in the way Frederick intended: "All the ancient free states fell by the force of luxury; and the people were converted into slaves, upon the credit of those flaming principles that now . . . operate . . . on the minds of Britons."[79] Less cataclysmic critics depicted the machine on fire as "the Grand Whim for posterity to laugh at."[80]

Evidently some English audiences for fireworks still had little time for Italian machines and pyrotechnics. When the dandy Captain Robert Jones published a new treatise on fireworks in 1765, he did so in order that English artificers would not have to rely on French and Italians for their spectacles.[81] Yet Gaetano Ruggieri remained in England, given a post in Woolwich's Royal Laboratory, where he served as an artificer and trained several apprentices before his death in 1776.[82] In addition, despite the cries of the merchants, the fireworks proved profitable for at least some communities. Prints sold well, judging by how many remain, and the pleasure gardens undoubtedly benefited from both the credit and the fiasco of the display. When tied to commercial labors, fireworks could be successful. Fireworks proliferated in the pleasure gardens after 1749 and grew fashionable. Cuper's Gardens staged miniature machines based on the Green Park temple, while Cuper's artificer and proprietor, Samuel Clanfield, began advertising fireworks for sale at the "Royal Fireworks, in Hester Lane, West Smithfield," reproducing Servandoni's Green Park temple on his trade card.[83] Thus, in 1749, it was not so much fireworks in general that the English objected to as *this* firework, its timing, and its expense. In the pleasure gardens, fireworks flourished both before and after the "Grand Whim."

## The Philosophers' Response: Useful Fireworks

The philosophers' reaction to the fireworks of Sarti, Ruggieri, and Servandoni closely reflected and contributed to the public criticisms of the "Green Park Folly" as a waste of resources in the aftermath of war. While Perrinet and Frézier offered competing visions of pyrotechny in the wake of Italian displays, fellows of the Royal Society chose a more practical route, reflecting the tradition of experimenting with fireworks that had grown up at the beginning of the century.

Fellows of the Royal Society may have contributed directly to the Green Park fireworks. In 1748, the Union fire insurance office offered a reward to anyone supplying information on unauthorized persons "making, giving, selling or vending any squibs, serpents or other such fireworks" in London, after a significant explosion destroyed a "squib or serpent seller's" premises in Blackfriars.[84] Royal Society fellows were well connected with the insurance offices.[85] In April of the same year Stephen Hales, the Royal Society experimentalist and Desaguliers's collaborator, made experiments to secure buildings from fire by layering earth on floors and roofs to slow the progress of the flames.[86] When Hales's results were published in 1749, Cromwell Mortimer added observations claiming that he had looked over a room inside the Green Park *macchina* two days after the display had been given. Mortimer "was greatly pleased to see the Doctor's Scheme confirmed by the Practise of the Engineers upon that Occasion" since the interior deal-wood floor was covered with sifted gravel, an inch deep, while the walls were painted with powdered lime, size, and water.[87] Mortimer failed to mention that much of the rest of the machine had burned down and whether Hales's methods had been used to protect it.[88]

Hales's experiments may have brought him into contact with the Green Park fireworkers. This was certainly the case for another experimenter who used the display for philosophical pursuits. In November 1748, an article appeared in the *Gentleman's Magazine* proposing a "geometrical use" for the fireworks.[89] Its author was probably the mathematician Benjamin Robins, who made the improvement of artillery his specialty. In 1742, Robins's *New Principles of Gunnery* had laid out improved Newtonian calculations of projectile motion. The *New Principles* applied Newton's second law to measures of the projectile velocity of musket balls at various ranges to obtain the force of air resistance acting on the balls. This qualified Galileo's ballistics theory, which assumed projectiles traveling in a vacuum, and led Robins to devise new methods for computing the motion of projectiles accounting for the resistance of the air. Practical range tables based on these results and many other ballistic investigations in the 1740s earned Robins the Royal Society's Copley Medal in 1747 and, later, a preeminent reputation among artillerists.[90]

If Robins was the author of the article in the *Gentleman's Magazine*, he shared the Green Park fireworkers' reliance on a mixture of state, commercial, and public resources. The article proposed a use for fireworks fitting to England's commercial and maritime interests and so typical of English philosophers' approach to pyrotechnics. Chapter 3 noted that William Whiston's signaling ideas for solving the longitude were by no means a wild project doomed to failure, and, indeed, schemes to make signals with rockets continued thereafter. As recently as 1745, one William Gee wrote to the Duke of Newcastle with proposals for anti-Jacobite rocket signals in Scotland: "I think by placing Signals at proper distances, the whole Country would in one Hour, be upon their Guard, or as the Signal was made, march directly to their assistance."[91]

Now Robins took up the challenge again, proposing rockets as signals: "Rockets . . . besides the beauty of their appearance, are, or may be, of very great use in geography, navigation, military affairs, and many other arts . . . for all kinds of instantaneous discourse between different stations."[92] Knowing the height to which rockets of a given size ascended offered a means to determine the distance between an observer and the point at which the rocket ascended, but for this one needed to know the height that such rockets reached. Whiston had apparently made such experiments, but these went unmentioned in Robins's accounts. The author instead appealed to the public and invited readers of the *Gentleman's Magazine* to take the Green Park fireworks as an opportunity to make the needed observations. Anyone within fifteen to fifty miles of the park should check the sky to see how far away rockets could be observed, and more local observers should measure the angle of the rockets at their highest altitude. Comparing published results in the magazine would reveal the answers. Soon, other correspondents to the magazine recommended the experiment, refining methods to make the observations.[93]

Robins's plans, however, came to little. After the fireworks, only one letter appeared in the *Gentleman's Magazine*, from a Welsh reader in Carmarthen claiming to have seen flashes in the sky on the night of the display but without any measures of height.[94] The *Gentleman's Magazine* thus turned out to be an ineffective forum for public knowledgemaking. But Robins had also organized his own observations, using different methods than those recommended for the public. While magazine correspondents were asked to locate the position of rockets against the fixed stars or nearby buildings to determine their angle of elevation, Robins had a friend use an instrument, probably a form of quadrant taking in ten degrees of altitude, to observe the fireworks. The expert witness was, thus, distinguished by his instruments and also by his venue for publishing results—not the *Gentleman's Magazine*, but the more prestigious *Philosophical Transactions*.[95] According to Robins, Ruggieri's single rising rockets and those of the concluding girandola reached heights of up to six hundred yards, which suggested that rockets could be visible at distances up to fifty miles.[96]

That Robins chose the Green Park firework as a chance to see the highest ascents of rockets indicates his high expectations of the skills of Italian pyrotechnists, an assumption that he shared with many members of the London public. However, in the event, he expressed a dismay that echoed the public contempt for the fireworkers in the aftermath of the failed spectacle. In the *Philosophical Transactions*, he complained that the "imperfect manner" of fireworkers' rocketmaking had led the rockets to move irregularly in their ascent, limiting the heights attained: "They may be made to reach much greater distances."[97] Consequently, a second round of observations followed, this time to ascertain how high rockets could reach if their construction were "improved." So now, like Whiston earlier, Robins set out to make better rockets capable of fulfilling his philosophical ambitions.

Once again, experimental pursuits led philosophers and artisans to interact. Initially, Robins turned to a "Person . . . employ'd in the Royal Laboratory" to make the necessary rockets, but, when his fellow philosopher John Ellicott reported the results, he thought them unimpressive.[98] Trials faltered through "the Negligence of the Engineer," and Woolwich's supposedly improved rockets went no higher than those at the Green Park show. So experimental fireworks suffered from the same problems as those of the court, which needed Tyers's pleasure garden skills to manage the Green Park display. Unable to secure appropriate expertise from Woolwich, Robins also turned to private entrepreneurs—a Jewish merchant in London's Devonshire Square named Samuel Da Costa, "of an extraordinary Genius in Mechanics," and a Mr. Banks, "many Years practic'd making Rockets" but otherwise unknown.[99] Da Costa's rockets then trumped the Royal Laboratory's, rising to 1,254 yards, "double to any of those fired in the Green Park."[100] On the basis of these trials, Ellicott finally reported that the best rockets to be used in any future observations should be between 2.5 and 3.5 inches in diameter, not the highest in ascent, but offering the best balance of production cost weighed against height attained.[101]

Italian fireworks thus led to quite different enterprises in England than in France. As in the past, philosophical relations to the art of pyrotechny reflected local conditions and differed in different sites. In Paris, Italian *macchine* proved popular with the public and prompted a significant literature exploiting their fashionable status. Servandoni and Ruggieri's pyrotechny then provided a spur to debates in print over secrets, artisanry, and the hierarchy of pyrotechnic labors. Artificers, in contrast, resented foreign interlopers and used their fists to show their displeasure. In England, the same resentment and fighting occurred among artificers, but English philosophers sought practical and commercial benefits from fireworks, echoing public sentiment, which viewed fireworks from an increasingly utilitarian and economic perspective.

Indeed, the practical use of fireworks grew significantly in the wake of Robins's experiments, contributing to enlightened and imperial sciences. Signal rockets became common in schemes for land and marine surveying, astronomical observations, and communication over long distances. In 1769, for example, Robins's collaborator John Canton used rockets to make signals during his observations of the transit of Venus.[102] In 1784, Francis Wollaston determined the longitude of observation points at Chislehurst in Kent using rocket signals, from which he sought to measure the relative positions and magnitudes of stars.[103] Robins's methods were soon after recommended to British imperial surveyors in India and used in Ordnance surveys.[104]

## Competition and Chemical Innovation: Responses to Italian Fireworks in Russia

Another place where Robins's experiments were considered useful was in Russia. In the early 1770s, they were replicated by General Fyodor Orlov in St. Petersburg before the English admiral Charles Knowles. Orlov, however, did not turn to private fireworkmakers for his equipment but relied on the expert skills of the Russian artillery. Using two *fusées d'honneur* of 2 puds (about 70 pounds) and two of 4 puds (about 140 pounds), fixed in frames and fired one after another, an artillery officer measured the height to which they ascended—the first to 200 sazhens (1,400 feet) and the second to 300 sazhens (1,820 feet), about half the height managed by Da Costa.[105] That everything remained in the hands of the artillery for these experiments was indicative of Russia's reaction to foreign pyrotechnics, including those of the Italians. Fireworks and the sciences remained exclusively state institutions well into the eighteenth century, with skills cultivated in the academy and the artillery rather than pleasure gardens or public lectures. When Italian fireworks arrived in Russia, they thus faced competition from skilled Russian artillerists, who sought to outdo Italian novelties with innovations of their own. Problems arose for these fireworkers, however, not from the Italians, but from their own superiors.

The occasion for these events was the appearance in Russia of Giuseppe Sarti, another Bolognese pyrotechnist and an associate of Gaetano Ruggieri's, responsible for the Green Park fireworks in London. Nothing more is known of Sarti save that he probably traveled to Russia in 1754 and remained until 1760.[106] He first appeared in Moscow, presenting fireworks to the imperial family at the Moscow opera, before moving on to St. Petersburg to show further displays. For his efforts, Sarti was rewarded, like Ruggieri, with a position in the Russian artillery, but his skills prompted competition.

As in France, Russia's fireworks had already begun to draw on Italian precedents before the arrival of Sarti. After the death of Anna Ivanovna in 1740, yet another coup brought Elizabeth Petrovna, the daughter of Peter the Great, to the Russian throne. Elizabeth exhibited a taste for French, Italian, and traditional Russian culture that the learned sought to accommodate. The professors of the St. Petersburg Academy of Sciences continued to design fireworks and illuminations for the empress. In the early 1750s, the academy's only Russian professor, the chemist Mikhail Vasil'evich Lomonosov, joined Jacob Stählin in the invention of allegorical fireworks. A student of Christian Wolff's, Lomonosov earned praise for his odes and verses and for his promotion of Russians in the academy.[107] Simultaneously, Stählin cultivated an "academy of arts" within the Academy of Sciences whose members included an Italian painter, Giovanni Valeriani, his Bolognese apprentice, Antonio Peresinotti, and the Veronese draftsman and scenographer Francesco Gradizzi.[108] Stählin was also an aficionado of the music and stage scenery of Italian opera, which had begun to be shown in Russia during the 1730s.[109] This led him to recommend changes in the staging of court fireworks. From April 1742 he combined the usual three plans that made up Russian displays into a single allegorical panel, painted in perspective like a stage set, and a year later he introduced moving scenery, or "figures moving in fire by means of mechanical traction, as seen in the operas."[110]

French and Italian tastes were also cultivated by prominent courtiers, particularly by Ivan Ivanovich Shuvalov, Lomonosov's patron and the empress's favorite. Another promoter of opera, Shuvalov was equally enthusiastic over Italian and French pyrotechnics, which he encouraged vigorously. As Stählin claimed: "Thanks to Count Shuvalov fireworks . . . became so fashionable at court that not one festival passed without them."[111]

This new taste for French and Italian spectacle no doubt lay behind Sarti's arrival in Russia, perhaps at Shuvalov's invitation. His first performance was given in the Opera House at Moscow, where, like the Ruggieri in the Comédie-italienne, Sarti "showed a firework of his art . . . after a tragedy, to the great pleasure of all the spectators; it consisted of various changing figures, burning one after the other with great order and accuracy, the figures made up of rockets of white fire, with moving wheels and fountains."[112] Here was the same mixture of theater and fireworks that typified the Ruggieri's Paris performances and evidently the same technique of communicating fire from moving to fixed pieces that the Ruggieri pioneered. Sarti also showed Italian *macchine*. Following the Moscow display, Sarti gave a firework celebrating the new year of 1756 in St. Petersburg, with a vast *macchina* representing the "Temple of the Russian Empire" on a great gallery surmounted by statues and surrounded by a circular balustrade. The allegorical invention, by Stählin, continued earlier cosmic themes in Russian pyrotechny, showing a youth before the temple above whom rose "the moon with other planets in a happy state."[113] Natural and poetic references were woven together in the allegory. Beside the temple stood personifications of Love of the Fatherland and Love of Mankind, holding a lodestone to his chest.[114] The scene was recorded in a print reminiscent of images of the Chinea, engraved by Francesco Gradizzi, whom Stählin had recently brought to the drawing department of the Academy of Sciences (fig. 23).

Shuvalov, Stählin, and the court were, no doubt, impressed with Sarti. But, just as English and French artificers resisted the Italians, Russian fireworkers from the St. Petersburg Arsenal laboratory were determined to upstage him. When Sarti arrived, two Russian artillery officers were in charge of performing fireworks, the *ober-feierverker* Matvei Martynov and the *shtykiunker* Mikhail Vasil'evich Danilov, whom we met in chapter 4. Both men reacted sharply to Sarti, who arrived "contrary to our expectation," according

23. Fireworks for the new year 1756 in St. Petersburg, executed by Guiseppe Sarti. Francesco Gradizzi (draftsman), Vasilii Petrovich Sokolov (engraver), *Izobrazhenie feierverka predstavlennago v Sankt-Peterburge v Novyi 1756 God vvecheru na Lugu pred Zimnim eia Imperatorskago velichestva Domom* (St. Petersburg, ca. 1756). Research Library, the Getty Research Institute, Los Angeles, California.

to Danilov's memoirs. On seeing Sarti's Moscow displays, the Russians, like Perrinet d'Orval, resolved to imitate his techniques. However, Danilov explained,

> I must confess that we had not a little difficulty in making fires and sparks similar to his, which by their great size appeared excellent; and on hearing praise for this Sarti from many courtiers, we resolved not only to make a firework similar to Sarti's, but also to show something still better of another kind. Martynov asked whether it was possible for us to show a firework in the same opera house. We applied all our strength to the invention of all sorts of rarities for presentation to the spectators; however, we would not have been victorious if an unexpected event had not led me to try to make a trial of green fire, which had never been found in all the world. . . . I took Venetian *iar*, dissolved it in alcohol, soaked gun cotton with it, set it alight, and saw that it burned with a very green flame. I continued to make many trials with this and succeeded in making it burn having made figures from it.[115]

Unable to replicate Sarti's secrets, the Russian artificer Danilov presented an alternative wonder to compete with the Italian interloper. This was a significant wonder because Russian green fire would later become the model for all colored fires in pyrotechny, as we shall see below. Indeed, nineteenth-century writers suggested that the search for a green fire had been the "desideratum" of earlier pyrotechny.[116] Little evidence of this remains, but Danilov was certainly not the first European to propose methods for coloring fireworks green. Many sixteenth- and seventeenth-century treatises included lists of ingredients for tinting flames green by using verdigris (copper acetate) or sal-ammoniac (ammonium chloride).[117] In 1739, the Italian Raimondo de Sangro, prince of San Severo, claimed to have made a green fire, and, in 1743, a similar claim was put forward for Count Friedrich August Rutowski, the illegitimate son of Augustus the Strong of Saxony.[118] The Russians had also made green fire before, with fireworks in St. Petersburg in 1710 showing "beautiful light blue and green fires, invented by the tsar himself," and another display in 1737 included "palm trees made with white and green fire."[119] Probably these colors were only pale, and later writers suggested that the intense fires of niter and sulfur in gunpowder destroyed any color produced by additives in these displays.[120] It is not clear what Danilov's method added to these compositions, but it must have surpassed available recipes.

Neither did Danilov explain why he chose to make his fire *green*. The color was evidently difficult to make and so worthy of credit, while green was resonant with symbolic meanings in European culture. In Russia, green was important because Russian courtly allegory hinged on the imagery of a "happy garden state," in which greenery—palm trees, plants, flowers, and parterres—stood in for the perfect situation to which the sovereign had brought the Russian Empire.[121] Perhaps more significant than the color, however, was Danilov's choice to make a *chemical*, rather than a mechanical, innovation. It might seem obvious that a pyrotechnist would make chemical innovations, and gunners had been inventing novel fireworks effects since the sixteenth century. But recent technical innovations in fireworks had all focused on machinery rather than compositions. Hence, the Ruggieri's communicating fires depended on clever mechanisms, Stählin's innovations involved "moving machinery," and Servandoni made his reputation with *macchine*. That Danilov chose to produce a novel composition may have been related to his training as a painter of heraldry for the artillery. "Venetian *iar*" was a form of verdigris, used until the nineteenth century to make the most brilliant green pigments for painting.[122] A Russian painter's manual of 1768 explained how to make green paint using verdigris and also recommended using sal-ammoniac or a solution of copper in strong spirits.[123] In making green fire, Danilov transferred the skills of the painter to pyrotechnics. It is notable that Hanzelet, who had proposed using verdigris for green fire in the seventeenth century, also worked as an artist.

The subsequent fate of Danilov and Martynov's green fire testifies to the particular premium that the Russian court continued to place on fireworks. On hearing of his invention, Danilov explained, the commanding officer of the artillery, General Matvei Andreevich Tolstoy, "who considered his knowledge of artillery and fireworks above everyone, not wishing to hear that anyone had come near his knowledge, ordered Martynov and myself to come to him and show our green fire."[124] The artificers demonstrated their invention and greatly impressed Tolstoy, who thought that "the secret of the green fire was no less amazing than that of Columbus's egg."[125]

Tolstoy exploited his status to secure credit for the fireworkers' skills. As we saw in chapter 4, social rank (*chin*) was of great importance in Russia, formalized in the Table of Ranks since the time of Peter the Great. While financial incentives motivated pyrotechnic innovations in England and France, Russians preferred promotion in the Table of Ranks as the highest reward.[126] Rank also affected the process of invention, and, in this case, it mattered more in defining authorship than who had actually made the invention, with no protection for lowly artillerists' ingenious innovations.[127] Seeing an opportunity to gain praise for his family, Tolstoy ordered Danilov and Martynov to collaborate with his nephew Ievlev in demonstrating the green fire to the empress. The unfortunate Martynov and Danilov were, therefore, joined by Ievlev when they first showed the green fire to a noble audience at the opera house, and after the successful display all three were offered promotions.[128] Tolstoy, however, was still unsatisfied since Martynov's promotion to major of artillery would effectively sideline his other nephew. He therefore "tried to persuade Martynov to take his reward in money instead of rank."[129] When Danilov and Martynov refused to cooperate, Tolstoy ensured that neither received any reward. Sarti, meanwhile, was promoted to the post of *ober-feierwerker* at the St. Petersburg Arsenal laboratory and received a gift of a thousand rubles, "despite having seen that locals could match his productions."[130]

## Artisans No Innovators: Academic Chemistry and Green Fire

Just as the city authorities in Paris had punished French *artificiers* for fighting with the Italians, so the projects of Russian fireworkers were frustrated by noble officers. Their plans were also contested by another institution with interests in pyrotechny, the Academy of Sciences. Here, the composition of fireworks and illuminations continued to be a vital element of academic labors, earning the institution patronage and prestige. However, by the early 1750s, the rewards that pyrotechnic inventions might bring to academicians prompted significant competition between rival professors. Between 1751 and 1755, Lomonosov thus took over Stählin's traditional role as designer of fireworks and illuminations allegories, probably thanks to the intervention of his powerful patron, Shuvalov.[131] Stählin resisted and eventually returned to his former role after Lomonosov abandoned pyrotechnic design to pursue new experiments making colored-glass mosaics.[132]

Lomonosov evidently took an interest in both the works of Sarti and the Russian artificers' green fire, both of which appeared during his competitions with Stählin. As a chemist, with a laboratory in the Academy of Sciences, Lomonosov may have identified practical innovations in pyrotechny as a means to distinguish his inventions above those of Stählin, and certainly his illuminations designs included references to novel fireworks such as "spiraling rockets, invented by myself and demonstrated in tests."[133] Rivals' skills could also be undercut, and, in July 1755, a translation of Perrinet d'Orval's revelations of the secret of Ruggieri's *spectacles pyrriques* appeared in the academy's Russian journal, perhaps at Lomonosov's instigation.[134] Lomonosov also tried to replicate Danilov and Martynov's green fire. His notebook for December 1755 recorded: "Today, under my direction, Klementev the laboratory assistant is investigating how to make high-reaching green stars for the fireworks."[135] However, no further experiments on fireworks by Lomonosov are known. His efforts must have failed. But this did not stop him from attempting to appropriate the fireworkers' invention. Since the 1730s, Russian academicians had directed attention away from artisanal skills and toward their own contributions to fireworks. Lomonosov now did the same, by arguing that artisans could not produce legitimate chemical innovations without an understanding of chemical science.

The occasion for this claim was one of the Academy of Sciences' renewed public assemblies, held before members of the imperial family and government, in July 1756, where Lomonosov presented an oration on the origins of light.[136] By this time, Martynov and Danilov's green fire had been appearing regularly in courtly fireworks for more than a year, displayed on decorations of royal monograms, laurel crowns, and palm trees.[137] Sarti also continued to perform fireworks, at the palace of Ivan Shuvalov, and for the imperial court in St. Petersburg.[138]

Lomonosov used his public oration to undercut his rivals, a tactic that he had employed before. While Stählin was busy building up a cohort of fine artists at the academy in 1751, Lomonosov had given an oration on the use of chemistry in which he claimed: "Painting . . . depends entirely upon chemistry. Should we take away the means for preparing pigments, we would be deprived of the pleasures of portraiture."[139] Lomonosov used the same tactics in his oration on the origin of light. His scientific lecture was itself a court spectacle, linking the pursuit of knowledge, like the Copernican fireworks of the 1730s, with celestial bodies, praise of the sovereign, and the iconography of Russia as a happy garden state. Lomonosov thus addressed the Grand Dukes Peter and Paul, sons of the empress Elizabeth: "Bloom thou amidst the abundant spread of our garden of the whole Russian Empire, renewed . . . by . . . our most gracious Sovereign. Grow thou in the radiance of a sun without beginning."[140] Light was a fitting subject for study, one that turns the heart to joy and "enriches your eyes and hearts and does not leave you in darkness."[141]

The bulk of Lomonosov's oration outlined his ideas on the origins of color, explained by a corpuscular theory in which different mixtures and motions of ether particles produced different colors. Lomonosov relied on his experience experimenting with fireworks and glass mosaics to arrive at this theory, in which the three primary colors corresponded with the three chemical principles salt, mercury, and sulfur.[142] Changing recipes produced different colors. Thus, a yellow-tinted firework might be made using antimony, "a body rich in mercury," which certified that mercury was the ether particle giving all yellow bodies their color. In addition: "A flame of green color, though shown by many burning bodies, comes most of all from copper . . . when this melts, the whole flame becomes green when fresh cold charcoal is thrown on it . . . the heat of the flame is decreased by the cold charcoal, the acid material of the hot copper loses its rotary motion force while the inflammable and mercurious materials [sulfur and mercury] are heated enough by the weak heat for motion. Thus without motion of the red ether [produced by salt], the yellow and blue present green to the sense of vision."[143]

Lomonosov thus provided a chemical explanation for the production of green flames using copper and charcoal, though he did not explain how to make a green firework, in which these chemical constituents would have to burn with *gunpowder* to produce the color—the problem that makers of green fire always faced and that Lomonosov himself seems to have failed to resolve. Nevertheless, he was quite specific about the importance of such a chemical theory of light. At the conclusion of the oration, he declared that, since nature functioned chemically to produce color, knowledge of chemistry was fundamental to those wishing to produce color in art: "Those who, because they turn their praise to chemical practice alone, do not venture to raise their heads above soot and ashes in order to seek the reason and nature of primary particles composing bodies . . . are to be considered vain and sophistical. For knowledge of the primary particles is as necessary in physics as the primary particles themselves are necessary for composing sensible bodies."[144]

In particular, Lomonosov sought to distinguish himself from those who made "simple discoveries without any supporting evidence and without enough work to justify them."[145] Following the performances of green fire by uneducated pyrotechnists, it is quite possible that Lomonosov's criticisms were aimed at Danilov and Martynov, who had recently produced a remarkable discovery with no chemical education. Lomonosov, unable to replicate their green fire, then asserted the necessity of theoretical knowledge in chemical innovations. Only the trained chemist could legitimately offer novel inventions in the realm of color.

In the context of the competition over green fire, Lomonosov claimed that theory *must* be included in practice. Such claims diverted attention away from Danilov and Martynov in an institution that had a tradition of obscuring the ingenuity of artillerist-fireworkers. Yet Lomonosov did not discover Danilov's secret, and it was another fireworker who figured out the recipe for green fire and replicated it. Around 1759, a young officer of the Noble Cadet Corps, Petr Ivanovich Melissino, began studying pyrotechny with Sarti and earned a position at the arsenal fireworks laboratory alongside the Italian.[146] According to Stählin, Melissino then employed the skills of another chemist, St. Petersburg's chief apothecary, Johann Georg Model, a corresponding member of the Academy of Sciences, to successfully divine the recipe for green fire.[147] Thereafter, during the final years of Elizabeth Petrovna's reign, Sarti, Martynov, and Melissino shared the responsibility of executing fireworks, which now incorporated both Italian machines and impressive color compositions.[148] Sarti left Russia in 1760, but both mechanical and chemical innovations continued thanks to the successful experiments of other artillerists.[149] Indeed, to judge from Stählin's account, brilliant colors were a characteristic feature of Russian fireworks from midcentury. Melissino showed his version of the green fire in fireworks for Empress Catherine II's coronation in Moscow in 1762 (fig. 24).[150] The artillery major Mikhail Alekseevich Nemov executed fireworks in the 1770s, perhaps with Melissino, and was "noted for his experiments in chemical labors" and for producing bright colors, including green and a "gold brilliance."[151] Stählin claimed that audiences appreciated "the beautiful colors of different fires."[152] It would be another forty years before French artificers would eventually learn to replicate this Russian color chemistry and import it into Western Europe, making all fireworks colorful thereafter.

24. Fireworks for the coronation of Catherine II in Moscow, executed by Petr Melissino, in 1762. They included "deux rangs de Palmiers d'un feû verdatre" of Melissino's recipe. Studio of E. Vinogradov, *Izobrazhenie feierverka dlia vysochaichshei koronatsii eia imperatorskago velichestva Ekateriny Alekseevny . . . v Moskve predstavlennago Sentiabria . . . dnia 1762 goda* (St. Petersburg, c. 1762). Research Library, the Getty Research Institute, Los Angeles, California.

For now, the travels of Sarti to Russia had prompted further debates and philosophical pursuits and a skillful culture of experiment among Russian artillerists working outside the state's official scientific institution. But, as in the 1730s, academicians sought to downplay the significance of such skills, maintaining the separation of artisans and scholars that characterized Russian pyrotechny.

## Conclusion

The travels of the Italian pyrotechnists thus prompted a wide range of responses from the public, men of letters, natural philosophers, and artisans in the middle decades of the eighteenth century. This chapter turned from the rising institutions of natural philosophy and pyrotechny to the circulation of fireworks skills, following the activities of Italian pyrotechnists as they moved between Paris, London, and St. Petersburg, taking with them spectacular *macchine* and ingenious effects of moving fire. In each location, similar performances prompted different reactions and responses, reflecting and furthering regional variations in pyrotechny's relation to the sciences. Thus, Servandoni and the Ruggieri's displays led to a flourishing literature on fireworks in Paris, where debates took place over the role of secrecy and theory in art and over the value of the different contributions made by architects, artificers, and men of letters to pyrotechny. In London, fireworks prompted a unique public debate over the value of courtly fireworks and led to experimental investigations intent on making fireworks useful and profitable commodities. In Russia, by contrast, the significant credit that spectacle continued to raise with the court ensured that novel Italian fireworks were met with more spectacular fireworks, leading to competition between academics and artificers over the "Columbus's egg" of green fire.

Simultaneously, the spread of Italian pyrotechny began to transform this geography of art and science. As Servandoni and the Ruggieri's techniques traveled, so they homogenized the nature of courtly fireworks displays across Europe and Russia. The same fireworks, managed by a network of artful Italians, now graced the courts of London, Paris, and St. Petersburg. In all these locations, class increasingly dictated the nature of audiences' reactions, with lowly artificers everywhere resisting the incursion of foreign fireworkers, whose exotic *macchine* and pyrotechnics were everywhere praised and patronized by the aristocracy. Men of letters and natural philosophers stood somewhere between these two positions, more ambiguous about the status of foreign artisans, but everywhere keen to assert the value of their own skills in the hierarchy of pyrotechnic labor. Subsequent decades would see further development of this new geography. However, the most immediate consequence of Italian pyrotechny was to produce an unprecedented fashion for fireworks in Europe during the 1740s and 1750s, with consequences for the performance of natural philosophy.

CHAPTER SIX

# The *Encyclopédie* and the Electric Fire: Pyrotechnic Contexts for the Arts and Sciences

The rise of Italian fireworks across several European states in the mid-eighteenth century prompted a homogenization of pyrotechnic culture and a new transformation of Europe's pyrotechnic geography. As spectacles staged by Servandoni, Sarti, and the Ruggieri became familiar sights in London, Paris, and St. Petersburg, so the local distinctions of European courtly fireworks declined. To be sure, reactions in different cities varied considerably, from the literary debates inspired in Paris to the indignation of the English and the fierce competition over spectacle engendered in St. Petersburg. But family connections and the Italian vogue spread *macchine* across the Continent, bringing a more standardized form to courtly spectacle. Other processes began at this time to create new forms of distinction and division in pyrotechnics, and these will be the focus for the remainder of this book. In London, Paris, and St. Petersburg, fireworks were witnessed by a growing class of "middling" sorts, deeming themselves neither vulgar nor courtly like the audiences of seventeenth-century fireworks, and promoting values of domesticity, education, economy, politeness, and piety that would transform pyrotechnics in the second half of the eighteenth century. In the hands of such a "polite and commercial people," the great spectacles of Servandoni and the Ruggieri would be transformed and domesticated as they passed into new venues and contexts catering to the values of this audience.[1] Critical among these values was an enthusiasm for the sciences and their power to serve utility, education, and entertainment. Fireworks would transform the sciences as they moved European audiences to a fascination with spectacle in the mid-eighteenth century, while the vogue for science would shape pyrotechnics into new forms and novel uses.

This chapter considers the significance of fireworks for two of the most vital theoretical and practical enterprises of the middle decades of the eighteenth century: the debate over art and science in Denis Diderot and Jean Le Rond d'Alembert's *Encyclopédie* and the contemporary growth of a new culture of performing experiments with electricity. The spectacular growth of interest in fireworks prompted by the Italian travels of the 1730s–50s is identified here as a vital context for these developments, producing new ideas about the relations of science and art, new scientific practices that exploited the credit of pyrotechny, and novel forms of fireworks. In the *Encyclopédie*, fireworks were recast by savants in terms of a new, productive relation of art and science championed by Diderot. Diderot's views on art and science are clarified when set in the context of the debates between Frézier and Perrinet d'Orval, on which the *Encyclopédie*'s articles on fireworks drew extensively. While Frézier maintained that the man of letters' discourse on fireworks should serve only as a mathematical recreation with no bearing on practice, Diderot offered an innovative twist on this tradition. For Diderot, the theoretical abstractions devised by the learned should no longer be merely pleasant diversions but instead should become tools for improving labor, wedding theory and practice in a new hierarchy. Now the gaze of the philosophe should manage the skill of the *artificier*.

Simultaneously, this philosophical vision relied heavily on the existing skills and knowledge of artificers, and, at midcentury, the vogue for Italian fireworks provided a critical context for the rise of a more spectacular and public science. In new contexts such as *cabinets de physique* and drawing rooms, striking phenomena, above all the remarkable spectacles of electricity, were exhibited to polite audiences and for the first time fixed experimental science in the public imagination. Fireworks provided critical resources for these new scientific spectacles, informing the way philosophers articulated the unfamiliar phenomenon of the "electric fire" in language, theory, and performance. Fireworkers offered their own spectacles of sparks, notably a remarkable "Chinese fire" shown from the mid-1750s, and, as the Parisian literary debates over fireworks interacted with new ideas about relations of art and science in the *Encyclopédie*, so pyrotechnic performance techniques helped engender the new spectacular science of electricity. Simultaneously, the work of commodifying, domesticating, and disciplining spectacle—characteristic both of Diderot's theories on the arts and of the practices of electrical performers—reshaped pyrotechnics.

## Theory and Practice in the Arts: The Frézier-Perrinet Dispute

The *Encyclopédie* is often taken as the quintessential source for discussing relations of art and science in the eighteenth century.[2] The philosophes saw in the *Encyclopédie,* with its myriad articles and plates illustrating the techniques of the arts, an opportunity to abolish the separation of the mechanical and the liberal arts and, thus, a means to social, philosophical, and economic progress. Under the Baconian claim that the essential value of art was to mobilize natural materials to human ends, Diderot and d'Alembert called on philosophers to assist the progress of the arts. Such claims appeared to celebrate the arts but equally rendered them docile. The *Encyclopédie* is easily read as a disciplinary project, making transparent otherwise obscure artisanal practices, rendering them capable of being reduced to orderly principles by philosophers and, thus, subjected to management. Far from assisting the artisans, the *Encyclopédie* replaced divisions of labor based on the mechanical-liberal distinction with a division of labor between rational management and mechanical work.[3] Hence the twist in d'Alembert's superficially respectful argument that, "while justly respecting great geniuses for their enlightenment, society ought not to degrade the hands by which it is served."[4] As in fireworks, there should be a division of labor between those who "invented" and those who "executed" the arts.

The positions of the *encyclopédistes* on arts and sciences were by no means homogeneous or static, but their novelty and significance can be assessed by setting them in the context of reactions to the Italian spectacles of the 1740s and 1750s. As we saw in the previous chapter, the response to these spectacles in France was distinctively literary. Before turning to the *Encyclopédie,* it is worth revisiting Frézier and Perrinet d'Orval's treatises to bring out their different attitudes toward the relation of practice and theory in the arts.

In his *Traité,* the man of letters Frézier not only revealed the secrets of the Ruggieri and criticized Servandoni but also attacked Perrinet's artifices, writing of a long dispute between the two men that is revealing of contemporary practices in pyrotechny. The dispute concerned the classification of rockets. Rockets were the "most noble" of fireworks, and every treatise gave them particular attention. When gunners had begun to use mathematics to articulate pyrotechnic craft in the previous century, it was applied above all to ordering the different sizes of rockets. These were classified, not according to their weight, but according to the weight of a lead ball that fitted the orifice of the mold used to make the rocket's case since the weight of any actual rocket might vary with its garniture. Measures were based on the "caliber scale," used by gunners to determine the weight of a ball or bullet needed for any particular diameter of cannon muzzle. The scale, described in detail in Siemienowicz's *Great Art of Artillery* of 1650, gave the progressively larger diameters of balls, ranging from under one to over forty pounds in weight.[5] Using cubic roots and the "rule of three," weights, volumes ("solidities"), and diameters of different-sized balls of various materials could also be ascertained. Siemienowicz described the arithmetic procedures but noted that many gunners preferred to use tables rather than learn the mathematics, which was "irksome and difficult" to master.[6]

In the 1706 edition of the *Traité*, Frézier followed Siemienowicz in using the caliber scale to designate rockets. Dividing the diameter of a one-pound rocket into a hundred parts, the diameters of all other rockets of different weights could be calculated as the roots of the diameter cubed, multiplied by an integer corresponding to the weight of a rocket.[7] A rocket mold might then be constructed on the basis of this measure.

In his *Essay* of 1745, the artificer Perrinet criticized this method, which he took to be Frézier's invention, and presented an alternative: "[Frézier] pretends that one may name rockets by the weight of a lead ball that just fits the inside of the mold . . . he goes into this subject in great detail . . . and all this to find the name of a rocket. . . . Is it not more simple, to name a rocket as of so many *lignes* than of so many ounces, which has to be figured out by calculations, which is a false denomination, and which leads one to think that it must be the weight [of the rocket]?"[8] To Perrinet it seemed self-evident to use a simpler designation of diameter in place of the laborious calculations of weight used in the caliber method. Instead, he chose the diameters of rockets measured in lines, or twelfths of an inch, as a means of classification and gave names to the resulting rockets.[9] This was another instance of Perrinet asserting his ingenuity. He viewed his new idea as conducive to good practice for artisans since it did away with the obscurities of old fireworks treatises. Significantly, Perrinet claimed that the mathematical calculations demanded by Frézier were really *social* limits on who had the right to practice pyrotechny: "This practice [the caliber scale] has been introduced for no other reason than to embarrass those who do not know the rules of caliber and is of little use to those who do know them."[10]

Frézier disagreed. Two years later, in the second edition of his *Traité*, he objected to Perrinet's criticisms of the caliber method.[11] Rejecting Perrinet's assertion that the technique originated with him, Frézier instead claimed that it was traditional among artificers "for more than a hundred years" and noted Siemienowicz's use of the technique.[12] Frézier thus deferred to past experience as justifying the rule and explicitly referred to the method as "ancient" in opposition to Perrinet's "modern" system.[13] Frézier's main criticism, however, centered on the proportionality inherent in the caliber scale. He claimed that the caliber method was more "learned" than Perrinet's and "most convenient to men of letters a little initiated in geometry because it presents at a glance the agreement of the volume and the weights of rockets to each other and to their calibers."[14] Perrinet's method produced a series of rockets, but, since they did not follow the caliber scale, different-sized rockets did not fit the proportions it set in place.

Evidently, this mattered to Frézier because he claimed to have visited Perrinet's workshop to demonstrate the usefulness of the caliber method. On July 7, 1745, he wrote, "I called in to the balances of M[onsieur] P[errinet], who . . . after having received me very politely without knowing me except by my name had the kindness to weigh in my presence in his shop at Port Louis three rockets of his choice."[15] Frézier then compared the weights and performance of the rockets with those calculated using the caliber scale. When Perrinet's rockets were fired, one burst after beginning to incline back toward the ground. Frézier claimed that the caliber method predicted this since it showed that the rocket was top-heavy, "overcharged with its garniture of stars."[16] The other was undercharged. This episode bears on the discussion of secrecy because, evidently, contra Frézier's own claims, artisans were happy to reveal their workshop practices to the man of letters if asked. But Frézier did not go to the artisan's workshop to learn about artisanal techniques—he assumed that he knew the best method already.

Perrinet then renewed his argument that the caliber scale, and mathematics in general, was superfluous to pyrotechny. In the 1750 edition of his *Essay*, he responded to Frézier's criticisms in the *Traité*: "He attacked me vigorously regarding Geometry, in which he is very well versed, and which he seemed to regret was lacking in my *Essay*. Perhaps M. Frézier ignored the fact that one may succeed very well, without Algebra, and without much erudition, in making a rocket and in demonstrating the practice; yes, I have at least experience as a guarantee . . . the science of calculations is applicable to an infinity of things, but there are many where one may have rigor without it [*ou l'on peut à la rigeur s'en passer*]."[17]

Mathematics was useful but unnecessary in the artificer's workshop, a claim that supports the assertion in chapter 1 that the geometry in pyrotechnic treatises of the seventeenth century was primarily a language for communicating to the nobility rather than a widespread practice among gunners. Significantly, Frézier made the same argument. His treatise, in fact, shared much with the tradition of books of *récréations mathématiques*, which were examined in chapter 2. In the seventeenth century, these works presented fireworks recipes and mathematical problems as an object of pleasing contemplation for the leisured virtuoso. Explaining artisanal methods would educate readers so that they would not be left in wonder by ingenious effects and gave a pleasant order to nature and art that might lead to higher spiritual and philosophical reflection, but there was no intention that readers should manufacture fireworks themselves or use mathematics to do so. Frézier claimed that his book was also largely written in this way: "I have not had as my principal aim the instruction of artisans who learn their art from a living voice and long practice, in which they are without doubt better versed than I; rather, I have written for those people of letters, curious about this art, in whom I suppose more knowledge than simple laborers [*Ouvriers*] without education, and who are happy to see certain principles made clear."[18] Frézier's exacting mathematical proportions were, thus, in principle more agreeable than Perrinet's methods. But, in practice, the artisan knew best, though he might do well to adopt the methods approved by men of letters. For Frézier, the real use of mathematics was to make artisanry appealing and intelligible, a use that gunners had themselves pioneered in the sixteenth century and that now permitted the *gens de lettres* to "read without boredom and understand without a master."[19]

Situating Frézier in the tradition of mathematical recreations also solves some puzzles regarding his other works. He has been identified as a pioneer in the mathematization of engineering in the eighteenth century.[20] His better-known *Traité de stéréotomie* offered geometrical techniques for analyzing variables associated with architecture and fortifications—in this case stonecutting and the projection of complex solids—later significant as the basis of Chastillon and then Gaspard Monge's science of descriptive geometry.[21] This work has been taken as a historical puzzle for being ahead of its time since Frézier's mathematical techniques were not applied to practice for several decades. Strangely, the author was a master of applied science yet seemed to defer to tradition and artisanal experience in judging works, despite supplying masterly mathematics for solving engineering problems. But Frézier's work was not meant to be applied science. Like books of mathematical recreations, Frézier's works presented art as a source of astonishing regularities, expressed in mathematics, for the contemplation of educated elites. Although mathematical techniques were posed as potentially useful for artisans, their main audience consisted of a higher and more liberally educated class, for whom they would offer a source of pleasant reading.

## Making Recreation into Regimentation: Fireworks in the *Encyclopédie*

The first volumes of the *Encyclopédie* followed soon after the appearance of Perrinet d'Orval's and Frézier's treatises on pyrotechnics, and, indeed, Diderot excerpted these texts for the *Encyclopédie*'s articles. Perrinet sent a copy of the latest version of his *Essay*, the *Manuel de l'artificier* (first edition, 1755;

second edition, 1757) to the editors of the *Encyclopédie*, and it is probably from them that Diderot excerpted the articles "Fusée (artifice)" and "Feux d'artifice (artificier)," which both appeared in the sixth volume of the *Encyclopédie*, published in 1756.[22] These extracts included Perrinet's claim to reveal the Ruggieri's secrets to the public. Frézier, now working in Brest as the director of fortifications for Brittany, did not contribute, but another man of letters, the librettist Louis de Cahusac, added thoughts to Perrinet's article, which appeared alongside it in the *Encyclopédie*.[23] The range of perspectives on the relations of art and science in these texts helps contextualize the better-known position of Diderot on this subject.

Cahusac's article reiterated the hierarchical division of fireworks labor presented by Frézier but chose to make even Frézier subservient to yet another tier of abstract aesthetic contributions to the art. Cahusac agreed that artificers like Perrinet dealt only with the "physical" part of pyrotechny but also claimed that Frézier only contributed ideas on the "geometry that distributes" fireworks—a remark indicative of how Frézier's *Traité* was being read. He then dismissed the interests of Perrinet and Frézier: "I do not believe I have to touch on these objects. I have sought to know them only insofar as they seemed related to the great spectacles that the king, the cities, the provinces, etc. offer to the people on solemn occasions; there seemed to me in this case a need to submit them to general laws, which were always the rule of all the arts."[24] In the case of pyrotechny, Cahusac proposed that these "general laws" made fireworks analogous to painting: "Those fireworks that only represent a kind of repetition through the play of different colors, movements, and brilliant effects, and the decoration on which they are placed, no matter how cleverly designed, will never amount to anything more than the frivolous charms of paper cutouts. In all the Arts it is necessary to paint. In the one that we call Spectacle, it is necessary to paint with actions."[25]

Cahusac did not mention the Italians in the remainder of his article and went on to attack French artificers for failing to achieve the successes of China and Russia in pyrotechny: "Could we not, by adopting all that these foreign nations have already discovered, invent the means . . . to broaden the boundaries of an art whose effects are already most pleasant and that could also become most honorable for inventors, as for the nation?"[26] Despite acknowledging the inventiveness of Russian and Chinese fireworks, to be considered further below, Cahusac chose Servandoni's 1730 fireworks for the dauphin when he turned to describing model displays, though he did not mention Servandoni. Then came Perrinet's article, revealing the secrets of the Ruggieri.

Cahusac, then, reiterated the distance between practice and theory promoted by Frézier, who took pyrotechnic knowledge as an opportunity to delight in geometrical order or "general laws." Perrinet, in contrast, sought a rapprochement of theory and practice to restore the ingenuity of artisans lost by institutionalized pyrotechny. Both these arguments were utilized in Denis Diderot's comments on art in the *Encyclopédie*, and both help contextualize Diderot's position.[27] Addressing a readership similar to Frézier's, Diderot sought to encourage in his audience an interest in studying the arts. He thus began by highlighting the aesthetically pleasing regularities in artisanry. Studying the arts was

> well worth the trouble whether we think of the advantages they bring us or of the fact that they do honor to the human mind. . . . What mathematical demonstration is more complicated than the mechanism of certain clocks or the different operations to which we submit the fiber of hemp or the chrysalis of the silkworm before obtaining a thread with which we can weave? What projection is more beautiful, more subtle, and more unusual than the projection of a design onto the threads of a simple and from there onto the threads of a warp? . . . I could never enumerate all the marvels that amaze anyone who looks at factories, unless his eyes are closed by prejudice or stupidity.[28]

As for Cahusac and Frézier, the orderly principles inherent in the arts were the locus of pleasurable contemplation. But Diderot went on to emphasize what was of less significance to these men of letters—the use of such principles to reform artisanal labors. Now Diderot's views echoed those of Perrinet, as he argued for the value of practice in the improvement of art: "Let them [artisans] carry out experiments, and let everyone make his contribution to these experiments: the artist should contribute his work, the academician his knowledge and advice . . . soon our arts and our manufactures will be as superior as we could wish to those of other countries."[29] Unlike Frézier, Diderot insisted that the abstractions observed in art should be used to bring discipline to practice.

To emphasize this point, Diderot turned to Francis Bacon, whose philosophy of inventions he greatly admired.[30] Like Bacon, Diderot thought it critical that the intellectual should preside over the process of invention in the arts because, if left to themselves, artisans would be prone to make discoveries only by chance or trial and error, not systematic inquiry.[31] Furthermore, the learned should never refuse to adopt inventions out of skepticism over their long-term success, a point Diderot illustrated with the example of pyrotechnics. Montaigne, he said, had dismissed the invention of firearms as a fad. "Imagine Bacon," he wrote, "in the place of Montaigne: you would see him study the nature of the agent and prophesy, if I may say so—grenades, mines, cannons, bombs, and the entire apparatus of military pyrotechnics."[32] Diderot also drew on Lemery's pyrotechnic experiments to promote his arguments, noting how few in previous centuries would have believed that the present one would use "a dust that breaks rocks and overthrows the thickest walls from an unbelievable distance, that a few pounds of this dust, enclosed in the depths of the earth, shake the earth."[33]

Diderot thus exploited both the traditions of mathematical recreations and ingenious invention in his article on art, but now he synthesized them to propose that the abstract principles derived from studying art could be applied, not only to the entertainment of *gens de lettres*, but also to the improvement of art itself. It was not just that the man of letters stood above artificers in the hierarchical division of pyrotechnic labor, but the knowledge at the man of letters' command should shape the artificers' practices. This was a novel argument, reflecting neither Perrinet's nor Frézier's position exactly, but containing elements of both. The application of abstract principles to art was then neatly illustrated in the plate accompanying the article on fireworks in the *Encyclopédie*. Diderot took it from Perrinet's *Manuel de l'artificier*, but, while he kept the artificer's images laying out tools for making rockets, Perrinet's picture of a busy artificer's workshop was replaced by an empty, mechanical, and schematic space more in keeping with other images of work disciplined by rational analysis in the *Encyclopédie* (figs. 25–26).

Such images have often been identified as means by which the philosophes sought to prize open artisanal secrets in order to render them transparent, orderly, and manageable.[34] Certainly, Diderot wrote that "the artists should know that to lock up a useful secret is to render oneself guilty of theft from society" by seeking advantage for an individual instead of the "common welfare."[35] Yet this was exactly the position that gunners had long adopted in accounts of pyrotechnic art, and it echoed Perrinet's reasons for revealing the secrets of the Ruggieri, which, of course, Diderot excerpted in the *Encyclopédie*. However, Diderot's portrayals of the arts as secretive, like those of Perrinet and Frézier, should be read more as strategies for competing with rivals than truths about the arts. Perrinet, after all, had openly communicated with Frézier despite the latter accusing artisans of secrecy in his *Traité*, and, while Perrinet claimed that the Ruggieri kept their technique of lighting moving and fixed fires a secret, a claim reiterated in the *Encyclopédie*, this too was likely a construction. Thus, in the second edition of his *Manuel de l'artificier* of 1757, Perrinet changed his text to state: "The secret of this

25. The artificer's view of the artificers' atelier. Plate 1 from J. C. Perrinet d'Orval, *Manuel de l'artificier. Seconde edition, revue, corrigée & augmenté* (Paris, 1757). Courtesy of the Hagley Museum and Library, Wilmington, Delaware.

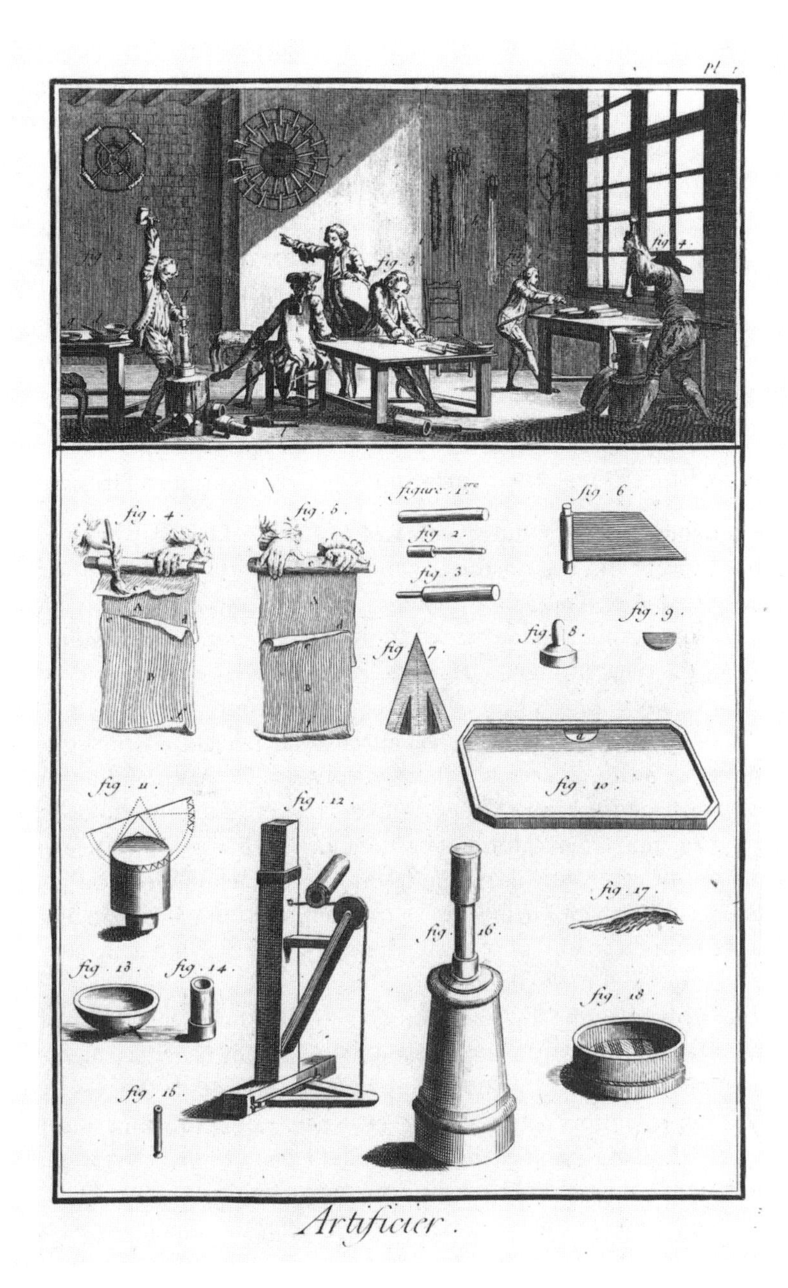


26. The *Encyclopédie*'s view of the artificer's atelier, adapted from Perrinet d'Orval's *Manuel de l'artificier*. The first of seven plates under the title *Artificier* in *Receuil de planches, sur les sciences, les arts liberaux, et les arts méchaniques avec leur explication*, 11 vols. (Paris, 1762–1772), vol. 1. These plates accompanied the article by J. C. Perrinet d'Orval, "Artifice," in *Encyclopédie; ou, dictionnaire raisonné des science, des arts et des métiers* (17 vols.), ed. Denis Diderot and Jean d'Alembert (Paris, 1751–65), 1:740–44. From the author's collection.

communication of fire was brought from Bologna to France in 1743 by the brothers Ruggieri. . . . This secret, which they have very obligingly communicated to amateurs of the art, . . . consists of a thing so simple and of an easy execution that the machine is at once made."[36] The meaning of *secret* here shifted from "intentional concealment" to "technique."[37] Evidently, when they were asked, the Ruggieri, like Perrinet himself, explained how it was done. Later, the Ruggieri informed the public that they would give explanations of their techniques at their atelier.[38] Perhaps Perrinet was just hasty, or the Ruggieri altered their position in the face of Perrinet's revelations. Whatever the case, accusations of secrecy were not facts about artisans but efforts to gain advantage in the competitive world of pyrotechny.

## The Chinese Fire: Appropriating New Inventions in Fireworks

It is notable that, consecutive with Frézier and Perrinet's debates and the publication of the *Encyclopédie*, natural philosophers joined artificers in seeking to appropriate, imitate, and gain control over pyrotechnic spectacles. This was explicit in the case of another innovation shown by the Ruggieri on the stage of the Comédie-italienne, the Chinese fire, and clearly evident in a new innovation by philosophers themselves, in spectacles of the electric fire. In the former case, Parisian academicians adopted the role of revealing artificers' secrets, echoing Diderot's call that the philosophes should manage and improve the arts. In the case of electricity, philosophers devised an entirely novel form of brilliant fire that brought spectacle under their control while exploiting the credit of fireworks in electrical performances. Both activities helped make natural philosophy increasingly fashionable in the second half of the eighteenth century, giving it a credit that pyrotechnists in turn would later draw on.

In March 1755, the Ruggieri brothers showed a new Chinese fire for an intermezzo following the play *Camille esprit folet*.[39] The secret of Chinese fire was much sought after by European pyrotechnists, and the Ruggieri's display prompted further revelations by Perrinet. Its exhibition fitted European tastes for chinoiserie and changing attitudes toward Chinese pyrotechnics, expressed in Cahusac's enthusiasm in his article for the *Encyclopédie*.

Before the end of the seventeenth century, Europeans enjoyed access to travelers' accounts of Chinese fireworks but had not set out to imitate them. Dutch merchants and Jesuit missionaries were ambiguous about the status of Chinese fireworks. The Jesuit Matteo Ricci, resident in Peking from 1601 to 1610, wrote that the skill of the Chinese in manufacturing fireworks was extraordinary, producing ingenious displays, but took the alleged preference of the Chinese for festive rather than military fireworks as a sign of their lack of belligerence and, hence, military weakness.[40] Others claimed that Chinese fireworks were inferior to European pyrotechnics. Johann Nieuhof's account of the Dutch embassy to China in 1665 noted the view of Athanasius Kircher: "Though we acknowledge they had the Invention of Gun-powder before us . . . yet they never arriv'd to our Perfection, being unskill'd in Fire-works."[41] Attitudes changed in the second half of the seventeenth century as the taste for chinoiserie developed in the courts. Now, the imitation of nature by local artificers was supplemented with the imitation of the arts of distant artificers. Chinese lacquers, porcelain, filigree, the *façon de la chine* textiles, and silk were all imported and emulated, while courtly artists and architects designed pavilions, palaces, and gardens *à la chinoise*.[42] More positive assessments of Chinese fireworks followed. Peter the Great's ambassador, Lev Izmailov, reported from China: "They make such fireworks that no one in Europe has ever seen."[43] In 1696, the Jesuit Louis Le Comte claimed that the Chinese made "the finest Fireworks in the World."[44] Amid this growing enthusiasm, Europeans began seeking out recipes for Chinese fireworks. Of particular interest were pyrotechnic "trees" or "vine arbors" that used brilliantly colored fire to represent flowers, wood, leaves, grapes, and fruits. Joseph de la Porte, the abbé de Fontenai, recalled witnessing fireworks in Canton in 1744: "One sees entire trees, covered with leaves and fruits; grapes, apples, oranges, with their particular color. . . . This is something our French artificers have not yet been able to do."[45] The desire to emulate Chinese colors may have been one motivation behind the search for green fire discussed in the previous chapter, though evidence of this is scarce.[46] In any case, in 1735 another Jesuit, Du Halde, also described a brilliant "Shower of Fire . . . the Light of which was like Silver, and which in a moment turned Night into Day."[47] This was the Chinese fire, whose brilliancy prompted several efforts across Europe to discover its secret. Already in 1712, Peter the Great requested Lorens Lang, the Russian commissar in Peking, to "obtain and dispatch the secret of the composition of Chinese firework fountains."[48] In 1721, the exiled Lord Bolingbroke sought the "secret of Chinese fireworks" in Paris, though no details of his efforts are known.[49]

Then, in 1755, the Chinese fire was shown by the Ruggieri, in what must have been a remarkable spectacle of brilliant white sparks. The earliest record of these performances explained the source of the brothers' recipe: "The sieurs Ruggieri, authors of fireworks that have been seen up to now in the *Théatre Italien*, . . . bought from an artificer who for a long time exercised his art in China the secret of this composition."[50] Evidently, Perrinet was

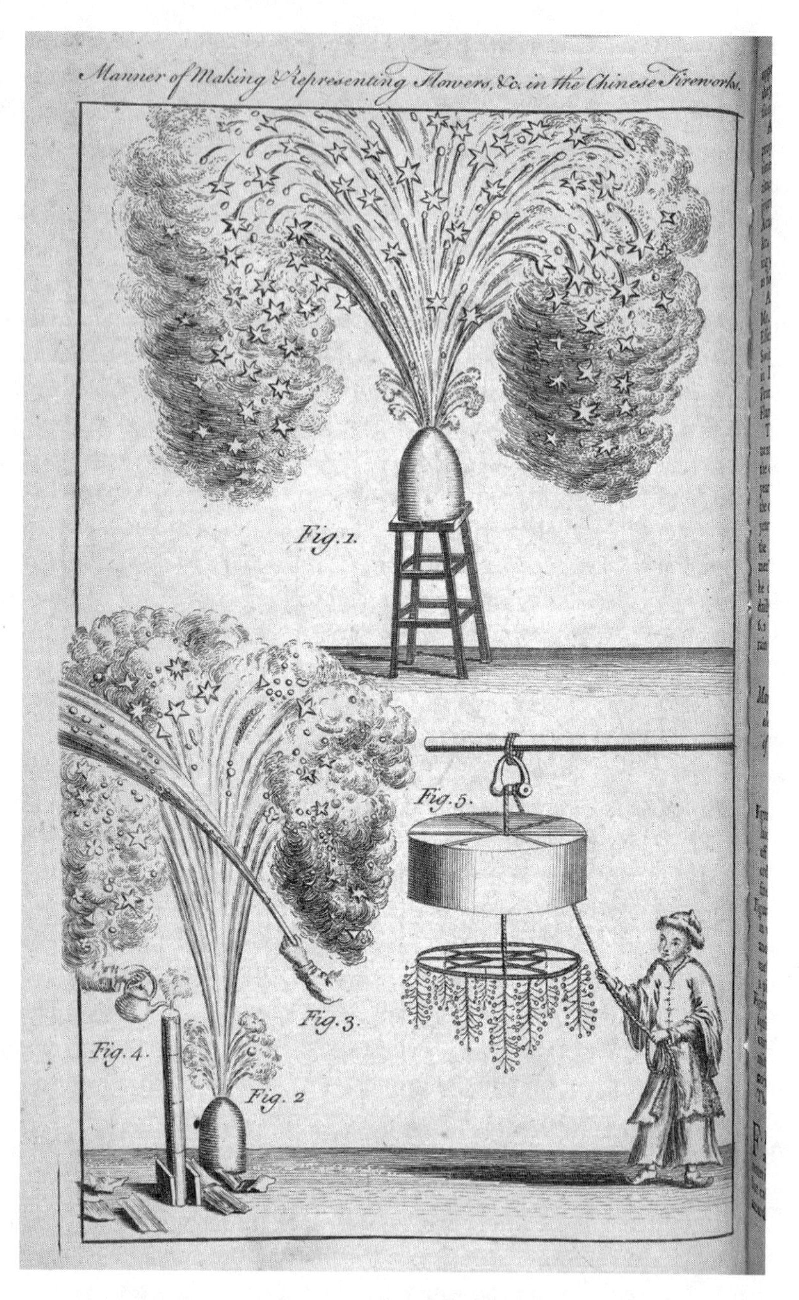


27. Chinese fireworks, from "Manner of Making Flowers in the Chinese Fire-Works, Illustrated with an Elegantly Engraved Copper-Plate.—From the Fourth Volume [just published] of the Memoirs Presented to the Academy of Sciences," *Universal Magazine* 34 (1764): 20–23. University of Chicago Library.

also determined to discover the secret, but he learned it from Pierre Nicholas le Chéron d'Incarville, another Jesuit missionary, resident in Peking, who often sent intelligence on Chinese arts back to Paris to assist local producers of chinoiserie.[51] Although the details of this transaction are unclear, once Perrinet had the secret, he published it, again revealing the method behind the Italians' spectacles. Indeed, he described how to make Chinese fire in the same *Encyclopédie* article as his revelations of the Ruggieri's other secrets.[52] However, Perrinet kept his own secret in this article because he did not reveal the recipe for Chinese fire, just the method for making it. It was only a year later, in the second edition of his *Manuel de l'artificier*, that the recipe was first given. He no doubt kept the secret to ensure the sales of his book, another strategy to compete with the Italians.

Eventually, the secrets of many Chinese fireworks were revealed by the Paris Academy of Sciences. Incarville, whose missions took him to Quebec and Peking, was an important figure in French imperial networks, regularly supplying the academy and the Jardin des plantes with seeds and news of artisanal inventions from China. In 1758, he submitted a long report on Chinese fireworks to the academy, which published it five years later.[53] In his report, Incarville explained that the Chinese fire was made by crushing old iron pots and scraps into sand and adding the sand to gunpowder. This produced what the Chinese called *thieh ê* (iron moths), *thieh sha* (iron sand), or *thieh hsieh* (iron granules).[54] Incarville also provided instructions for making Chinese "flowers" from the iron sand composition, colored "grapes" from a sulfur paste, and the "Chinese Drum," a remarkable effect in which a series of illuminated figures dropped in succession from a drum (fig. 27).[55] Jesuit accounts and the Ruggieri's performances were probably enough to raise Cahusac's admiration of the Chinese in the *Encyclopédie*, and, after Incarville's text was translated in 1765, Chinese fireworks would become popular across Europe, prompting further efforts to discover their secrets.[56] As for Incarville, he claimed: "I have the secret of the emperor's artificer, who comes to see me from time to time."[57]

## Electricity and the New Fashion for Natural Philosophy

Academicians appropriated the techniques of artificers in a culture where fireworks had reached the height of fashion. In the wake of Ruggieri's and Servandoni's displays and Diderot's proposals that the philosophes should be the legitimate managers of art, it made sense for Parisian savants to seek out intelligence on spectacular effects. But, if philosophers appropriated artificers' fires in order to manage spectacle, they could also achieve this by developing their own. The remainder of this chapter suggests that the fashion for fireworks was similarly important for another novel form of philosophical spectacle in the mid-eighteenth century—the electric fire, which became the hallmark of a new public science and the basis for an abundance of new theories and practices in natural philosophy.

Two cultures of sparks flourished in the mid-eighteenth century—fireworks and the new spectacle of electricity. It is well-known that electricity, a phenomenon recorded since ancient times but examined systematically only from the end of the seventeenth century, provided a critical resource for the success of the new experimental science in the eighteenth century.[58] Prior to the 1740s, experimental science was the province of a small community of virtuosi and academicians, promoted to limited audiences (such as London traders or Parisian aristocrats), but little known to a broad and growing public. However, after natural philosophers learned to harness static electricity using static-generating electrical machines in the first decades of the eighteenth century, new experiments emerged that quickly captured the public imagination, launching science into the public sphere. From the 1730s, experimenters began dispensing shocks and sparks to surprised audiences, electrifying couples who received a shock when they kissed, or suspending electrified boys from silk threads who then attracted bits of paper and straw. After the invention in 1745 of the Leyden jar, which could store a large electric charge, these experiments grew more and more spectacular, the most famous being the abbé Jean-Antoine Nollet's use of the Leyden jar to make some two hundred Cistercian monks holding hands jump into the air on being shocked. These experiments were spectacular and theatrical and, most important, made science fashionable and appealing for the eighteenth-century public. It was during the 1740s that science thus shifted from being a relatively esoteric pursuit to a widely celebrated public commodity. As the German naturalist Albrecht von Haller wrote of natural philosophers: "From the year 1743 they discover'd phenomena, so surprising as to awaken the indolent curiosity of the public, the ladies and people of quality, who never regard natural philosophy but when it works miracles. Electricity became all the subject in vogue, princes were willing to see this new fire which a man produced from himself, and which did not descend from heaven. [People] resorted from all parts to the publick lectures of natural philosophy, which by that means became brilliant assemblies."[59]

If electrical science was spectacular, the question arises as to why it became spectacular and theatrical at this time. Historians have usually assumed that the reason for this lay in the inherent properties of electricity itself, its power to shock and attract and repel matter lending itself to spec-

*jets de feu* that move like serpents with lightning rapidity."[71] Nollet imagined electric fire as streaming jets of electric fluid, represented in illustrations resembling the sparks streaming from pyrotechnic devices in contemporary fireworks engravings (fig. 28; cf., e.g., fig. 24 above). Indeed, Nollet gave a pyrotechnic explanation of electricity. Thus, he thought that particles of electric matter were enveloped in "some fatty, saline or sulphurous matter" that was ruptured to release the elemental fire. This set off a reaction creating the electric fire, "a little like a grain of powder ignited and lighting others placed nearby."[72] On this basis, in his electrical demonstrations Nollet dipped metal wires in sulfur, believing that this enhanced the sparks emitted when they were electrified: "It is a very well-known fact that electric matter, which divides into diverging rays as it emerges into the air, lights up much better when it is required to traverse a fatty or sulphurous matter."[73] For Nollet, electric fire *was* a form of pyrotechnic fire.

Others made the comparison. The Parisian *ingénieur-opticien* and *cabinet de physique* proprietor Charles Rabiqueau argued in *Le spectacle de feu élémentaire*, published by the Librarie pour l'artillerie et le génie in 1753, that electrical effects were caused by an extremely fine "powder of electric fire" analagous to gunpowder.[74] Alternatively, the English electrician and painter to the Board of Ordnance Benjamin Wilson noted similar links when discussing the question of whether electric fire might have an impulse in resistance to the air. He took a brass or iron wire, bent the ends at right angles to the wire, and suspended it at the middle on a metal wire placed on an electrified body. The wire turned with a "great velocity, moving always in a direction contrary to that in which the electric fluid issues from its points." It did so because "the electric particles, by their elastic force, issue directly forwards from the points, and endeavour to expand themselves; but meeting with some resistance from the air, force the wire to move backward in a contrary direction, much in the same manner that a Catherine-wheel is made to turn round in a direction, contrary to that in which the small rockets affixed to its periphery discharge themselves."[75]

Both electric fire and gunpowder fire could exert an impulse on the air. Thus, in its early years, electricity could be understood as a special form of powder fire, which is significant because electricity is better known as a phenomenon that *superseded* gunpowder explanations of meteors in the eighteenth century, after Benjamin Franklin showed the equivalence of lightning and the electric fire. Already in 1770, John Mills exclaimed how, "since the invention of gunpowder," the causes of thunder and lightning had been "ascribed to a mixture of nitrous and sulphureous vapours by some means set on fire in the air and exploding like that powder. . . . Franklin's soaring

28. Electricity streams like *jets de feu* above an illustration of the Leyden experiment. Plate 4 from Jean-Antoine Nollet, *Essai sur l'électricité des corps* (Paris, 1746). University of Chicago Library, Special Collections Research Center.

genius has realized the fable of Prometheus in bringing fire from heaven and furnished us with a better theory."[76] Yet, before this happened, it was the gunpowder theory that explained electricity, as philosophers applied the familiar explanation of natural fires to the new electric fire, and the same gunpowder theory continued in explanations well after Franklin.

The place where pyrotechny was, perhaps, most important for electrical science was in its performances. In a period when fireworks were in vogue, the electric fire was presented above all as a public spectacle, shown before an audience as a series of arresting and entertaining effects that appealed to the senses and could be varied according to the ingenuity of the performer. The spectacular nature of electricity was made to seem self-evident: "Electricity seems to furnish . . . Phenomena so various and so wonderful . . . as must have been designed by the Almighty Author of Nature for the production of very great Effects."[77] But it was a choice on the part of experimenters to construct electrical science as one of dramatic demonstrations, and they chose to do so at a time when fiery spectacles were in the public imagination. Competition to exploit the demand for fiery spectacles was fierce, as the proprietors of Mulberry Gardens learned: "The . . . Fireworks at this place have gain'd so general applause, that a splenetick, envious temper has lately prompted one or two neighbouring publicans to attempt the like amusements."[78] It is unsurprising, then, that electrical performances of this time shared a number of elements with pyrotechnic displays. As the physician Pierre Massuet wrote in 1752: "An electrified man touches a pile of gemstones or a silver vessel with his hand; one sees fire come out from all sides. It is the most brilliant firework I have ever seen."[79]

Electricity was used to ignite fires, imitate artillery, and set off explosive powders. At the ceremonial reopening of the Berlin Academy of Sciences in January 1744, the physician Christian Friedrich Ludolff ignited spirits in a spoon using an electric spark generated from the finger of a person connected to an electrical machine. Johann Heinrich Winkler repeated the experiment in Leipzig and then used electricity to ignite lycopodium, flax, *aurum fulminans*, and gunpowder.[80] Winkler also proposed artillery uses for electricity, while the North American lecturer Ebenezer Kinnersley showed gunpowder and a "Battery of eleven Guns" ignited by electric fire, in addition to an "Electric Mine sprung."[81] In philosophical lectures at York, Adam Walker used the electric fire to ignite fireworks.[82] Nollet's electrified jumping monks recalled the animated figures of pyrotechnic monks in fireworks displays (plate 1).

In fact, electricians were quite explicit about the analogy between pyrotechnic and electrical performances because, from the early 1740s, they

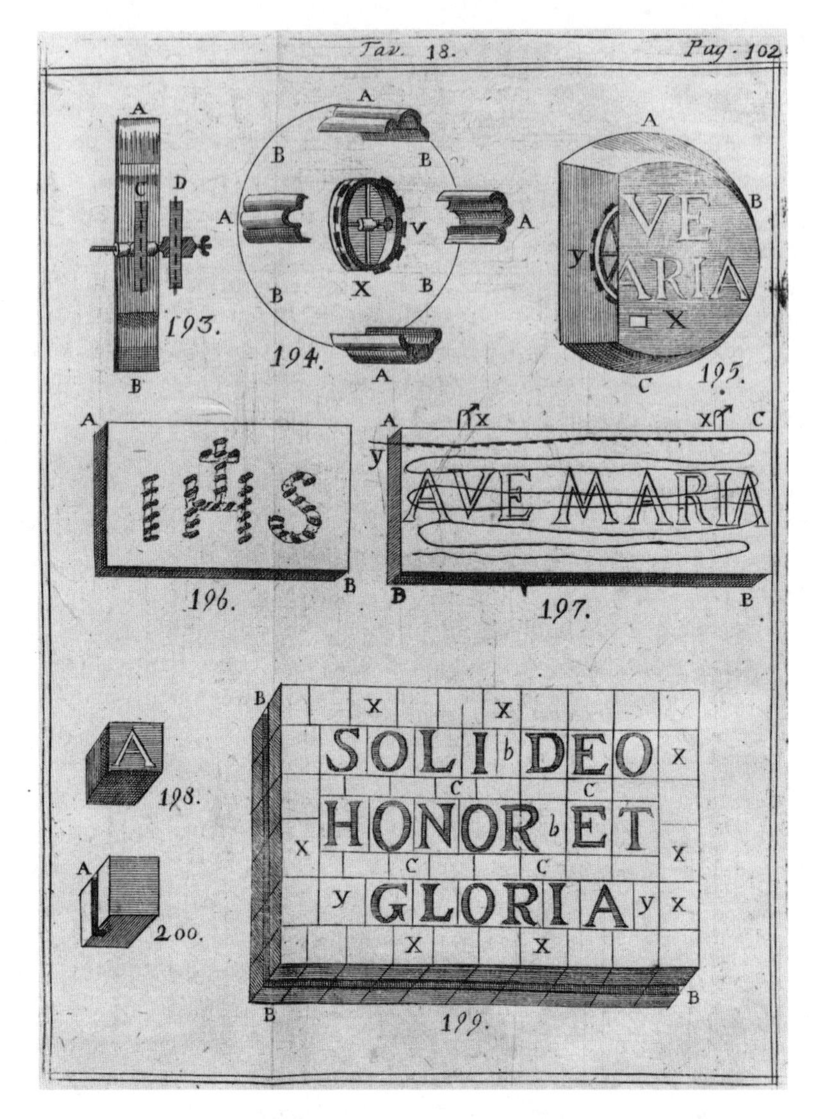


29. Various means for illuminating names with fireworks. Plate 5 from Giuseppe Francesco Antonio Alberti, *La pirotechnia o sia trattato dei fuochi d'artificio* (Venice, 1749). Courtesy of the Hagley Museum and Library, Wilmington, Delaware.

sought to imitate fireworks and illuminations using the electric fire. One of the most common elements of pyrotechnic performances was the presentation of names, figures, mottoes, and royal monograms in fire, using either *lances à feu,* small spark-emitting cartridges, or illuminating lamps, arranged in patterns on a wooden frame or board so that, when lit, they spelled out

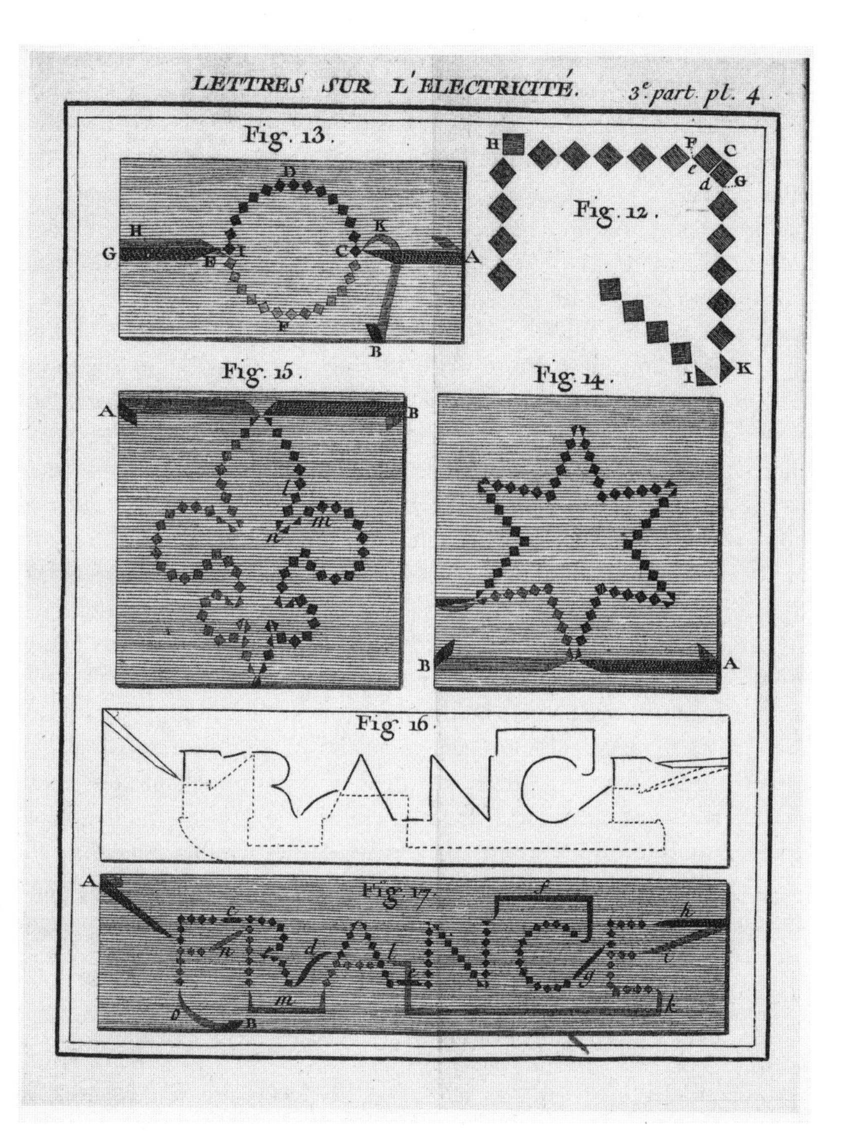


30. Electric illuminations. Plate 4 from Jean-Antoine Nollet, *Lettres sur l'électricité: Dans lesquelles on examine les dernières découvertes qui ont été faites sur cette matière, & les conséquences qu' l'on peut en tirer*, 3 vols. (Paris, 1753–67), vol. 3.

words or showed decorative figures in fiery outlines (fig. 29).[83] Already in the 1740s, Winkler used electricity to celebrate royalty and described how to spell out the words AUGUSTUS REX with electricity.[84] In 1767, the abbé Nollet wrote to Laura Bassi in Bologna explaining how to elaborate this effect by passing the discharge of a Leyden jar through figures of letters or pictures made from lines of tinfoil squares fixed between sheets of glass (fig. 30).[85] Nollet was direct about the value of pyrotechnics for giving electrical science credit with the public: imitating them would "make this science tasteful and captivate their attention."[86] One could "varier à l'infini" these "inventions ingénieuses" or "petites illuminations" according to the taste of the performer, much as fireworks were varied to demonstrate the skill of their performers and to ensure the approbation of audiences.[87] Nollet showed these illuminations in an entirely dark room, where he presented them as wonders.[88] He claimed to have shown them many times and was not anxious about hiding their secret in order to raise wonder, "making my spectators believe for a time that the electric fire will pronounce everything I choose on the same glass."[89] Since audiences could not see the plates containing different words being exchanged by the performer, fiery figures seemed to appear by themselves. Fireworkers also relied on the darkness to hide their actions. Indeed, Nollet also proposed making "pyramids" for these performances, another figure often shown in fireworks, using spinning electrified needles of decreasing length made to spark at the ends so that their revolutions created circles of electric fire. For this, Nollet mixed pyrotechnic and philosophical ingredients, suggesting that the end of the stem on which the needles pivoted be dipped in melted sulfur to produce a greater spark, following his argument that sulfurous matter enhanced electric fire.[90]

## Electrical Meteorology

Eighteenth-century electricians also used the electric fire to make artificial meteors. The state of the weather was important for both fireworks and electrical demonstrations, and performers of each attended carefully to the suitability of the weather for showing effects, assuming that warm and dry conditions were ideal for good performances.[91] Performers then manifested meteors with ingenious effects. Nollet thought that the techniques of "petites illuminations" could imitate lightning by discharging an electrified rod through zigzags of tinfoil squares attached to glass in an entirely dark room.[92] This was just one of a multitude of electrical imitations of meteors that proliferated in demonstration lectures in the eighteenth century.

Chapter 2 argued that the flourishing of pyrotechnic imitations of meteors and bodies in sixteenth-century court culture helps explain why nitro-sulfurous accounts of natural meteors and physiological states proliferated in that period. Now, electrical imitations of the same effects coincided with another shift, it being at this time that nitro-sulfurous accounts of meteors and medicine were replaced by electrical ones. Phosphorus provided a precedent, and, indeed, study of the electric fire emerged from the same series of late-seventeenth-century pneumatic experiments with which Boyle had explored the properties of phosphorus. In Frederick Slare's hands, phosphorus was offered as an alternative to gunpowder theories in natural philosophy, though his project faltered. The same was not true in the case of electricity. As electrical imitations and the popularity of electrical displays grew, so the electric fire superseded nitro-sulfurous fire as a candidate for explaining a variety of meteorological phenomena. As gunpowder theories additionally applied to human physiology and the *flamma vitalis*, so they were superseded by an electrical medicine and notions of an electric "spark of life," echoing the parameters of the gunpowder theory.[93] Nevertheless, it would not be until the nineteenth century that the gunpowder theories finally disappeared.

Electricity was used to imitate a whole range of celestial and earthly meteors in the eighteenth century. When Bose first tried the experiment of beatification using electric fire, he did so in the armory of Leipzig, where metal-wire trimmings attached to an electrified suit of armor were made to "Sparkle and glitter with Tails like Comets," recalling the pyrotechnic illumination of armor earlier described by Nicolas Lemery.[94] Joseph Priestley used electricity to imitate earthquakes, while Stephen Gray and Granville Wheler built an electric planetarium.[95] The Philadelphia lecturer Ebenezer Kinnersley made "Air issuing out of a Bladder set on Fire by a Spark from a Person's Finger, burning like a Volcano" as well as "Various Representations of Lightning."[96] As Benjamin Franklin explained: "We represent lightning, by passing the [electrified] wire in the dark, over a china plate that has gilt flowers."[97] Thunder houses with lightning rods and discharges between glass plates subsequently multiplied lightning effects in demonstrations.[98] John Brice Becket and others imitated the aurora borealis. An exhausted and sealed glass tube was rubbed with a woolen cloth or applied to an electrified body to produce vivid luminosity: "If the hand be apply'd to the end of it, a bright spark will sometimes dart along the tube like a meteor in the air."[99] All these artificial meteors were displayed as sublime marvels, as Joseph Priestley commented: "The discharge of a large electrical battery is rather an awful than a pleasing experiment, and the effects of it, in firing gunpowder . . . and in imitating all the effects of lightning, never fail to be viewed with astonishment."[100]

The new electrical meteorology emerged alongside performative contexts such as these. Philosophers continued to make imitation epistemologically significant, with the result that electrical artifices, like those made with phosphorus and chemicals earlier, constituted *models* and, thus, explanations of natural effects. This had been philosophy's twist on pyrotechnic artifice's courtly meanings, and it was now applied to electricity. "Lightning," Nollet wrote, "is, in the hands of nature, what electricity is in our hands, so that the marvels that we deal with at present on our scale are little imitations of the great effects that terrify us and depend on the same mechanism."[101] Tiberius Cavallo reckoned of the thunder house: "By this experiment the true state of the earth, when covered by electrified clouds, may be represented exceedingly well; and several meteors . . . may be imitated; such as water-spouts, and whirlwinds, besides . . . thunder and lightning."[102] Winkler observed: "All the effects, which human art has hitherto discovered by machines in electrifying, manifest themselves in thunder-storms."[103] Franklin's kite experiments designed to secure the equivalency of lightning and electricity emerged from this culture of meteoric imitations. The experiment hinged on analogizing lightning and the explosive discharge of the Leyden jar, caused by the completion of a circuit between the positively charged interior and negative exterior of the jar. The same was supposed, and, with the kite, shown to take place on a larger scale, when lightning redistributed charge from the positively electrified earth to negatively charged clouds (or vice versa).[104] Such analogies drew on pyrotechnic traditions. Indeed, fireworks linked kites, fire, and meteors long before and after Franklin's celebrated performance. The "fiery dragons" described in chapter 2 were often figured with sails and banners flown in the air and attached to crackers or lanterns. The term *kite* was first used in English to describe one version of the dragon in John Babington's *Pyrotechnia* of 1635.[105] Philosophers recognized the kite as an ancestor of the flying dragon. In 1777, John Anderson's *Institutes of Physics* explained how electricity could be drawn from clouds using "a boy's kite or dragon."[106]

Kites were not exclusively used in fireworks, but they were routinely associated with pyrotechny and meteors. In December 1741, Captain William Gordon wrote to the *Philosophical Transactions* from Vauxhall, site of ordnance experiments and pleasure gardens, describing a fireball in the sky, "nearly in the Form of a large Boy's Kite, projecting a long Tail towards the North-west, not unlike those Slips of Paper set on Fire." Gordon was unsure whether this was "the Work of Art or Nature" but noted that the fireball made "a Noise like a Clap of Thunder."[107] Philosophers also flew pyrotechnic kites into clouds. In 1749, at Camlachie in Scotland, the Glasgow professor of astronomy Alexander Wilson used a series of kites linked to one another to measure the temperature of the air at high altitudes. To release the thermometer at a height of about three thousand feet involved attaching it to a kite with a slow-burning match, which scorched through cords to send the thermometer plummeting to the ground, where Wilson quickly read off the temperature.[108] Kites would continue to be linked to fireworks in experiments, as we shall see below.

Franklin's assertion that lightning and electricity were one prompted a widespread use of electrical explanations in meteorology after 1752.[109] William Stukeley already assumed that "fireballs, thunder, lightning, and coruscations" all proceeded from the electric state of the atmosphere.[110] By 1787, the abbé Pierre Bertholon's *De l'électricité des météores* could propose that every manner of meteor, from fiery exhalations to volcanoes, earthquakes, thunder, and lightning, was the result of electrical causes, many capable of being modeled with electrical apparatus.[111] His assertions relied directly on making and performing imitations of pyrotechny and meteors using electricity. In 1776, Bertholon published an article in Rozier's *Journal de physique* explaining how these remarkable performances were shown "in my *cabinet,* to the majority of electrical physicists, [and] to different persons curious about the electrical machine."[112] Bertholon developed the "petites illuminations" that Nollet described in his letter to Bassi, consisting of electrified figures of metal foil and wires placed between glass plates and divided into segments. Sparks ran between the segments to create illuminated outlines in imitation of pyrotechnics. Bertholon's "electric illuminations" included the standard words and names figured in fire described by Winkler and Nollet in addition to portraits and illuminated physiognomies. Bertholon also showed a whole series of illuminated figures of human and animal bodies, celestial and meteorological imitations, including an electric rainbow and orrery or solar system and figures of plants and minerals, equations from mathematics and physics, and complex machines. This extensive application of electric illuminations to so many natural and philosophical phenomena reflected Bertholon's willingness, in *De l'électricité des météores,* to apply electrical explanations to all meteors. Indeed, the techniques of imitating pyrotechnics and those of imitating meteors were indistinguishable since Bertholon claimed that figures could also be illuminated by electrifying differently shaped evacuated glass tubes creating a "barometric glow" identical to that created with the apparatus used to model the aurora borealis a few years earlier.[113] Bertholon drew his readers' attention to electric illuminations in *De l'électricité des météores*.[114] Here, he rejected the gunpowder theory of meteors, or "causes pyrotechniques," by estimating the quantity of inflammable material necessary to move a great mass of earth and concluding: "All the gunpowder made since its invention would not be capable of moving this solid mass."[115] He then asserted that only electricity was capable of such action, owing to the rapidity of its conduction and the distance it could travel in a short time. Earthquakes were, thus, "underground lightning" (*tonnerres souterrains*), no different from lightning in the atmosphere.[116] Electrical explanations of meteors followed, in which electric illuminations served as philosophical models. For example, falling stars and fireballs were exactly equivalent to electric sparks and traces made between wires on a glass plate in the electric illuminations.[117] Bertholon thus presented both new forms of pyrotechnic theater using electric fire and new theories of meteors founded on these performances.

## Hybrids of Electrical and Gunpowder Theories

Electrical pyrotechnics helped transform meteorology. Nevertheless, electricity's victory over the gunpowder theory was far from conclusive in the eighteenth century, and gunpowder and nitro-sulfurous accounts endured in electrical and meteorological science into the nineteenth century. Just as performances were often hybrids of pyrotechnic and electrical effects, so theories often mixed both kinds of causes to account for phenomena or else took the gunpowder theory in novel directions related to other new kinds of fire emerging in the eighteenth century.

Winkler accepted Franklin's equation of lightning and electricity but did not take it that lightning was exclusively electric. In 1754, Winkler proposed that the ignition of various materials by electricity in his university rooms was replicated in clouds, where accumulated electric matter ignited sulfurous particles, that is, gunpowder, to produce "hot claps" of lightning. If sulfur was not present in the air, "cold claps" resulted, which struck, but did not burn, "trees, men, and houses," an effect that Winkler thought contradicted, but did not eliminate, the standard gunpowder explanation of lightning.[118]

English philosophers took a similar view. In 1757, Erasmus Darwin's first scientific publication attacked the Reverend Henry Eeles for promoting an electrical theory of meteors.[119] Eeles rejected "an analogy between thunder and fired gunpowder."[120] For example, while gunpowder exploded in all directions, lightning acted in rectilinear lines. It seemed impossible that, even if sulfur and niter existed in the atmosphere, they could somehow combine with a charcoal substitute and, in the proportions needed, make gunpowder-like explosions. Even if they did, they would fall to the ground before they ever ignited, while water vapors in the clouds would stifle the ignition. Electricity, in contrast, could be seen in experiments to fly off bodies it pervaded with a spark and a crack and, thus, seemed a better candidate for "analogy." Eeles proposed that electricity permeated vapors that ascended to form clouds, which, colliding, released a "body of fire two or three hundred yards in length."[121]

Recalling the strategies of the seventeenth-century Royal Society, Darwin lampooned Eeles's "enthusiasm" for "his idol goddess Electricity."[122] The miraculous status of electric fire was, here, turned against its credit: "There is ever such a charm attendant upon novelty, that be it in philosophy, medicine, or religion, the gazing world are too often led to adore, what they ought only to admire."[123] Darwin's alternative explanation of meteors passed from the gunpowder theory to the steam engine, an idol better suited to his industrializing Midlands milieu. He began with the theory of gunpowder by noting the "immense rarefaction of explosive bodies by heat," as when "Nitre . . . in detonation . . . emits great quantities of air."[124] Steam shared this power of expansion but, unlike air, condensed again on cooling to a liquid. Thus, water could "by heat alone become specifically lighter than the common atmosphere," rising and falling according to the heat of the sun. If clouds were electrified, as Eeles contended, this was only "an accidental circumstance, and not a common one," and the ascent of vapors needed no electrical explanation. Darwin also offered thoughts on expansion due to heat, caused, he maintained, by the rotary motion of particles equivalent to the "verticillary motion given to charcoal-dust thrown on nitre in fusion, or the wonderful agitation of the parts of burning phosphorus."[125] Thus, a modified form of gunpowder meteorology might challenge the electricians.

## The Lightning Rod Debate

Pyrotechnic challenges to electrical meteorology continued into the early nineteenth century. A key site where this relation endured was in debates about lightning rods. Just as in the 1740s theories explained electricity in terms of gunpowder, so solutions to the destruction caused by lightning also hinged on pyrotechnics. Already in 1745, William Watson contrasted the electric fires issued from blunt and pointed metal objects in terms of different artificial fireworks. While the nitro-sulfurous account of lightning and fiery exhalations endured, it was not obvious that lightning rods, of either shape, were the best means to manage aerial explosions. Thus, in 1758, the aging Newtonian Stephen Hales, whose fireproofing techniques had been used by the Ordnance in the Green Park fireworks of 1749, proposed further experiments linking meteors, fireworks, and kites. In his *Treatise on Ventilators*, he offered "a proposal to try if the Force of Hurricanes can in any degree be abated: As also to abate the Noxiousness of some Kinds of sultry sulphureous Airs in hot Climates."[126] Hales thought that hurricanes were caused by a "great ferment" of rising sulfurous vapors and descending cold and pure air. When this destroyed the elasticity of the descending air, its expansibility was restricted and created a vacuum, with the result that air "rushing into the Vacuity with great Velocity and Violence . . . thereby makes those tempestuous, rebuffing, whirling Hurricanes."[127] The ferment intensified the mixture until it exploded as lightning. Hales concluded that it might be possible to stop hurricanes, and, hence, lightning, by igniting the sulfurous vapors before this process began. To do so, he proposed using "Bombs, or Grenadoes, shot to great Heights out of Cannon, or . . . Rockets."[128] As it did Erasmus Darwin, local industrial experience offered Hales an analogy. Just as a candle flame ignited sulfurous vapors in mines, so a rocket burst of "Wild-fire" would do the same in the air. Hales also adapted the dragon to a novel purpose: "Paper-kites may be raised to a very great Height, by fixing the line of the first Kite to the body of the second Kite, and so on to the third, fourth, or fifth, &c. Kite; the lower Kites to be larger and larger than the preceding upper ones: And a Rocket might be fired by means of a Match at the upper Kite when at its greatest Height."[129] This technique borrowed directly from pyrotechnics and from Alexander Wilson, who had been the first to propose linking kites together as part of his experiments in high-altitude thermometry. Hales also proposed suspending pots of iron filings, spirits, or phosphorus attached to kites or rockets to ignite the vapors and diffuse the winds and lightning.[130]

Such pyrotechnic solutions to meteorological puzzles did not decline for some time. In 1776, Giambattista Beccaria fired rockets carrying threads into clouds and fog to determine whether they discharged electricity.[131] John Lyon presented another elaborate defense of the gunpowder theory against electrical meteors in 1780.[132] And, in August 1814, the Berlin engineer Blesson deployed rockets to prove that the *ignis fatuus* was chemical, not electric, as most believed by then.[133] However, for philosophers convinced that lightning was electric and not pyrotechnic, the lightning rod offered an alternative means to control the heavenly fire. In 1750, Franklin proposed, using an artificial meteor, that a rotating, electrified brass scale dish descending toward a metal instrument would spark violently into it, unless a pin were placed on the instrument, in which case the electricity silently discharged without sparking.[134] Artificial meteors were, thus, given a new use, taking on the role of small-scale models with which to test large-scale effects. Long metal rods attached to tall buildings might serve as "pins" to dissipate aerial electricity and prevent damage from lightning—but what was the best shape for the rod to achieve this? Philosophers debated the merits of blunt and sharp points until the controversy came to a head. Even now, gunpowder was at the center of the debate. From 1763 to 1768, the Board of Ordnance transferred its powder magazine from Greenwich, a traditional site of fireworks manufacture, to Purfleet and, to protect the powder, sought advice on lightning rods from Benjamin Franklin and Benjamin Wilson. Wilson, an advocate of blunt rods, argued that tall, thin pointed rods would "invite" lightning too readily and take a massive discharge from a cloud creating an explosion. Franklin advocated just such tall pointed rods, the higher the better.[135] Under the supervision of Sir Charles Frederick, the director of the Green Park fireworks in 1749 and now the surveyor-general to the Ordnance, Franklin and his Honest Whig allies in the Royal Society won the argument against the Tory outsider Wilson. But pointed rods at Purfleet failed to stop a lightning strike in May 1777, prompting another round of debates. Wilson, the society painter, mobilized spectacular artificial meteors to make his argument, presenting a huge electrical conductor and model of the Purfleet Arsenal primed with gunpowder before King George III at the Pantheon on Oxford Street, the site of theatrical performances and princely spectacle.[136] Wilson was well versed in theater, having introduced his friend David Garrick to the Duke of York, for whose personal theater Wilson painted stage sets and acted.[137] As the model and conductor converged, sparks ignited the gunpowder to create a dramatic explosion, which Wilson's lightning rods supposedly avoided. Such spectacle was intended to resolve the lightning-rod debate, though royal approval did not stop Wilson's marginalization by the Royal Society.

## Conclusion

While Italian fireworks delighted and offended different European audiences in the mid-eighteenth century, their effect on literary debates and philosophical performances proved consequential. As they had many times previously, pyrotechnic discourse and techniques offered resources for new learned visions of nature and ideas of work. For the Parisian artisan Perrinet, fireworks needed improvement, through the artisan's observation of his own labor and the application of theory to practice. To Frézier, observing pyrotechnic art yielded satisfying regularities and mathematical order, which might assist practice, but, more important, served pleasure. Diderot, drawing on artisanal and lettered accounts of fireworks for the *Encyclopédie*, offered a third alternative. The pleasant order discernible in art by the man of letters should be turned to the improvement of practice. Theory should drive progress in art, perhaps through the hands of the artisan, but ideally under the gaze of the philosopher. Similarly, the secrecy of the arts laid bare by competing artisans might also be revealed by the savant. The learned should manage the marvels and secrets of art, for the profit of all.

Diderot's remarks reflected a context in which the learned were appropriating spectacle to their own ends, revealing secrets, and producing new marvels whose effects were the subject of both theoretical and practical elaboration. After a variety of actors sought the secret of Chinese fire, the Paris academy exploited its networks of traveling savants to learn and publicize the recipe. The hunt for Chinese fire reflected a widespread desire among midcentury Europeans to discover arresting shows of brilliant sparks and their high regard for the remarkable spectacles of the Italian pyrotechnists. It was within this context that the Leyden experiment and shows of electric fire flourished. Natural philosophers exploited the credit of fashionable pyrotechnics to shape new public spectacles of sparks, utilizing the language and performance techniques of fireworks to articulate the behavior of this unprecedented effect. Diderot linked the philosophers' interventions in art with progress, and, to many natural philosophers, this cultivation of electrical effects led to rapid advances. The seventeenth-century science of nitro-sulfurous particles, explaining meteors and physiology, now set the parameters for a dramatic electrical science of meteorology and the body, though gunpowder theories, fireworks, and electricity remained integrated in the sciences for several decades after Franklin's kite experiments.

Despite its debts to pyrotechnics, however, the electric fire was unique. Its incomparable properties were critical to its success as a philosophical phenomenon and spectacular effect. With it, philosophers succeeded in generating an image that showed that *they* were the masters of fire: "Franklin's soaring genius has realized the fable of Prometheus." In the remainder of the eighteenth century, both natural philosophy and pyrotechny would be celebrated as sources of astonishing fiery effects, and, in this new environment, where scientific and pyrotechnic spectacles became equally fashionable, a remarkable new genre of philosophical fireworks would arise.

CHAPTER SEVEN

# Philosophical Fireworks: Domesticating Pyrotechnics for a Polite Society

Fireworks provided an important context for electrical performances in the eighteenth century. At the same time, electricity was itself a marvelous spectacle. If the rise of electrical demonstrations owed much to pyrotechnics, they were also creditable because the electric fire was unique—it was *not* the same as pyrotechnic fire, and no existing pyrotechnic culture could accommodate it. Electric fire was fascinating and hard to comprehend because it could repel and attract bodies, pass through materials invisibly, and produce shocks in the body. Although it might be used to set objects on fire, it did not consume them like pyrotechnic fire. As Albrecht von Haller and many audiences in Europe recognized, electricity was something fundamentally novel and a powerful source of credit for natural philosophical performances.

Consequently, as the fashion for philosophical effects grew after the 1740s, so the performances of gunpowder fireworks and other effects converged. Sometime in the late eighteenth century, a Dutch clockmaker turned his hand to philosophical fireworking, producing a cabinet whose interior mechanism sent colorfully painted transparent patterned discs revolving on a spindle. Illuminated from behind with a candle, the whirling spirals and patterns on the discs mimicked the swirling flames of fireworks. Other discs in this cabinet of "optical fireworks" illuminated a pyrotechnic temple, a coat of arms, and spiraling stars, fiery columns, and the image of a Montgolfier balloon (plate 8). This was one example of a new genre of "philosophical fireworks" that flourished in the second half of the eighteenth century, a collection of variously optical, electrical, hydraulic, mechanical, and chemical imitations of pyrotechnics and fireworks imitating nature's spectacles. This chapter surveys the culture of philosophical fireworks and explores a new pyrotechnic geography to which they gave rise.

Critical to the growing fashion for both fireworks and natural philosophy, and especially for their hybridizations, was the rise in the eighteenth century of a polite and commercial public that traversed all three sites we have been considering: London, Paris, and St. Petersburg.[1] Audiences for fireworks in the seventeenth century tended to be divided, in the literate imagination at least, between the nobility and the vulgar, one knowing and enjoying spectacle, the other fearing its powers. But, with the growth of a "middling sort" in the eighteenth century, such an image disappeared. Now polite society increasingly made its presence felt, using modest wealth to rise above the vulgar and imitate on a smaller scale the conspicuous consumption of the aristocracy. These people promoted good manners as a means to smooth social intercourse and valued self-control and self-discipline, manifested in an orderly appearance and dress, to be contrasted with the smells and dirt associated with vulgarity.[2] They preferred intimate or domestic gatherings and commercial venues for entertainment to the bustle and crowds of grand fireworks displays, with the result that the period witnessed a flourishing of coffeehouses, salons, theaters, and pleasure gardens, all paying venues, where, to be sure, different classes mingled, but where the lowest orders tended to be absent. Literacy and a taste for reading further defined the public, and a preference for education and improvement mixed with the desire for entertainment. It was to these tastes that philosophical fireworks appealed, and, while grand pyrotechnic performances continued in the second half of the eighteenth century, they were accompanied by a flourishing of this new genre of effects. In pleasure gardens, theaters, and cabinets, performers domesticated fireworks, transforming them into miniaturized commodities, devoid of soot and smoke, and suitable for polite consumption.

This new genre also had consequences for artificers, whose careers increasingly depended on and converged with those of natural philosophers. The history of art and science is as much about competitions between artisans as it about debates between artisans and scholars. Indeed, identities were fluid and defined as part of this competition. This chapter also considers practitioners who mobilized claims to scientific authority to compete with rivals in the art of fireworks, doing so in new spaces for pyrotechnics, the pleasure gardens. Giovanni Battista Torré, an Italian instrumentmaker in Paris, made artificial volcanoes the centerpiece of his philosophical fireworks, while the Ruggieri family, continuing to show fireworks in Paris, mobilized the new chemistry of Antoine Lavoisier to attract larger audiences. The scenes of performance were important for making fireworks philosophical. Both protagonists established pleasure gardens in the 1760s, and these sites became important locations for entertaining and improving spectacles, tied into scientific culture as much as any *cabinet de physique* or laboratory. Commercial pressures drove innovation here, competition leading to new collaborations and inventions whose urgency was amplified as the market for royal and state fireworks declined in Paris in the 1770s and 1780s. Throughout the period, however, a tradition endured, one that saw men of letters and architects considered above artificers in the hierarchy of pyrotechnic labor. Torré and Ruggieri's spectacles challenged this hierarchy, and, after the French Revolution, a second generation of Ruggieri set about overturning it for good. In the age of Napoleon, Claude-Fortuné Ruggieri mobilized the new chemistry of Lavoisier and Chaptal to claim authority in pyrotechnics and to banish the architect and the allegorist from fireworks. The result was an enduring image of pyrotechny as an art inescapably linked to the science of chemistry, a connection that Ruggieri consolidated by turning to the Russians and their colored fires.

## Combustion and the World of Goods: Making Fireworks for a Polite Society

An important difference between electrical and fireworks displays was their venue. Although electrical displays could take place outside, many occurred in the intimate and domestic settings favored by polite society. Nollet depicted his experiments taking place in *cabinets de physique*, drawing rooms, and parlors before audiences of a few well-heeled individuals or family members. The latter was also the subject of Joseph Wright of Derby's well-known painting of 1768 depicting an experiment with the air pump in a domestic setting (plate 9). Such venues appealed to a class that increasingly disdained the crowd for its boisterous and unpredictable behavior. Crowds for pyrotechnics had always been rowdy, a fact that court society tended to find amusing, watching at a distance from balconies and boxes. But, in the eighteenth century, polite society, which did not have the luxury of separation from the masses, started to complain. In 1706, a London freeman asked the lord mayor to discontinue illuminations, which threatened his domestic tranquility: "It gives ye Rude Rabble Liberty to Doe what they list . . . they Breake windows with stones, fire Gunns with Pease in our Houses to ye Hazard of Peoples life or Limbs."[3] This kind of criticism grew steadily, particularly in England, where the middling sort flourished. In 1761, following fireworks for the king's birthday on Tower Hill in London, there were more complaints: "As soon as the Tower Guns are fired off . . . the Populace here consider it as a Signal for every kind of Outrage. . . . We reproach the Romans with the Barbarity of their public shews. . . . But is a Rabble let loose to commit the most shocking Violences in the Heart of a City like this an instance of more Refinement?"[4]

Artificers and philosophers understood the implications and increasingly created smaller, more exclusive venues for performances. These reinterpreted outdoor spectacles such as fireworks for indoor sensibilities and mixed them with new small-scale wonders such as electricity. Already with the advent of the Ruggieri's performances at the Comédie-italienne, fireworks were shown in more intimate and exclusive settings. It is hard to imagine that the Ruggieri's introduction of Chinese fire, with its brilliant white sparks, occurred without reference to electrical demonstrations. Servandoni also brought fireworks to his stage shows and included electrical effects in these enclosed and elite spaces. After his great firework of 1739, Servandoni took premises in the Salle de machines in the Tuileries, where together with the king's *artificiers* François-Noel Lemarié and Guérin he showed stage machines depicting Roman vistas and allegorical fables. These centered on themes such as the *Spectacle de Pandora*, telling the story of Pandora's creation by Vulcan as Jupiter's punishment for Prometheus. An architectural ensemble, the machinery glittered with "rains of fire, overturned boulders, thunder, lightning."[5] In 1747, Servandoni and the royal sculptor Blaise Lagrelet developed these shows by creating a pretty electrical display in the Saint-Laurent Fair in Paris, combining optical devices, automata, and elder-pith figures made to jump on metal plates by "terrifying sparks."[6] Such philosophical fireworks proliferated in theaters and *cabinets de physique* thereafter.

Polite society here included women in addition to men, and philosophical fireworks were aimed at audiences of both sexes. Female rulers such as the Russian empresses participated in approving fireworks designs, and the practice extended to the aristocracy. For the recovery of the dauphin in 1752, the Marchioness of Pompadour composed an allegorical firework at Bellevue in which monsters attacked a "luminous dolphin" (i.e., the dauphin) before Apollo, hurling fireworks, destroyed them.[7] Women also set off fireworks. Ladies figured as "gallant fireworkers" in a midcentury German print that showed fireworks as a more intimate and small-scale event, played off by well-heeled individuals (plate 10). Furthermore, in the culture of philosophical fireworks, women could also be artificers and played an important role in establishing the vogue for new effects. Women assisted their artisan husbands in ateliers crafting fireworks, managing enterprises if their husbands died. Sometimes, only their own deaths by accident led to public notice.[8] However, in Paris in the 1760s, one Mlle Saint-André advertised her "laboratory and cabinet" at no. 25, rue Fauxbourg S. Denis, above the Saint-Laurent Fair.[9] Saint-André, an *artificière du roi,* claimed credit on the basis of the sciences, owing "her progress to twenty years of work": "Her principles are taken from the sciences from which pyrotechny naturally derives, such as physics, mechanics, and geometry. This in effect has been the sole means leading to the success that connoisseurs admire in the work of this lady."[10] Her wares were predominantly geared toward polite consumers and included the first "indoor fireworks." Thus, one could buy "artifice for apartments": "At times when visits and compliments are reuniting friends and family, one likes to relax and vary the amusements of society by small fêtes in which one includes some fireworks."[11] Pyrotechnic details were not given, but Saint-André offered instruction in their use, to avoid "the embarrassment that these executions ordinarily cause."[12] A few years later, Michael Höckely, a military and festive fireworker of Auxonne, Burgundy, described the production of such miniature fireworks.[13] Tiny cartridges filled with powder, rammed with iron wire, and rolled up into spirals or "pastilles" could be set on needles, adorned with colored paper flowers, and set spinning. Others were fired in bottles or buckets, set off inside tobacco boxes or needle boxes. There were even miniature rockets and a flying dragon. Like Nollet's electric illuminations, such "table fireworks" should be shown in a darkened room, with the windows and doors open "in order not to incommodate the society in the room by the smell of powder and sulfur."[14] In this way, "a gathering of people in their rooms may entertain themselves without great cost and in an easy manner."[15]

Another important venue for philosophical fireworks was print. Books provided further consumable forums for fireworks, and literacy offered more opportunities for distinction from the vulgar. Accompanying the growth in public venues for pyrotechny was a growing taste for printed literature that, like earlier pyrotechnic treatises, revealed and disseminated the techniques behind novel fireworks and offered recipes for domestic pyrotechnic amusements and entertainment. Some of this literature dealt in traditional fireworks. Several short works appeared in Paris detailing pyrotechnic recipes for amateurs in the 1770s and 1780s.[16] In Russia in 1777, Mikhail Danilov, now a major, published Russia's first treatise on fireworks "as . . . in the provinces and certain towns . . . many people [were] enthusiastic to present fireworks and illuminations" but did not have "the opportunity to see the works at the laboratory."[17] Included were recipes for traditional fireworks, stars, rockets, fountains, and fireballs, though the recipe for green fire was notably absent. In London, Godfrey Smith published a substantial collection of recipes for making fireworks in his *Laboratory; or, School of Arts,* which went through seven editions between 1740 and 1810. Verses on the frontispiece made clear the distinguishing features of such knowledge:

> No vulgar Eye enjoys a fond Delight
> In Naturs Beauty and Productions bright;
> This nursing Mother, is ye Second Cause
> Of Plenty, Life, and uncontroling Laws;
> When Art doth court her, She unveils her Face
> And shews her Charms to her adopted Race.[18]

Danilov and Smith listed traditional fireworks recipes, but other books described new philosophical fireworks. These were books of "rational recreations," extending the seventeenth-century tradition of mathematical recreations. Just as Hanzelet and others exploited the flourishing of new artificial effects in the 1620s, mediating them to noble audiences in print, so the same occurred with the advent of philosophical fireworks. New and greatly expanded editions of the *Récréations mathématiques* thus appeared from 1769, edited by Jean-Etienne Montucla, a friend of Diderot's and Blondel's, and then by the geographer, physician, and postal service director Edme-Gilles Guyot.[19] Pretty experiments in physics, electricity, and philosophical fireworks were added to the original contents, with Montucla explaining Nollet's trick of showing illuminated electric figures in the dark and offering instructions for making electric *gerbes* and fixed suns.[20] Links to gunners remained since Charles Hutton, the professor of mathematics at the Royal Artillery Academy in Woolwich, produced the English translation.[21] There were many such works, whose recipes circulated among different authors and editions throughout the era. William Hooper's *Rational Recreations* passed through numerous editions, drawing on Guyot and others to explain how to make artificial rainbows, thunder, lightning, volcanoes, and earthquakes in the home. All these effects were easy to make, small in size, and suitable to domestic entertainments. Hooper appealed to natural philosophers to make the sciences suitable for polite entertainments: "Should we not endeavour to render useful learning, not dull, tedious, and disgustful, not rugged and perplexing, not austere and imperious, but facile, bland, delightful, alluring, captivating? that Philosophy . . . decked in all the glowing ever-varying colours of the skies may gain admittance to the parties of the gay and careless?"[22]

An important dimension of books of rational recreations and displays of philosophical fireworks was that they should be both educational and entertaining. In the seventeenth century, English natural philosophers had used the rhetoric of utility and learning to avoid their experimental performances being equated with enthusiasm and Catholic incendiarism. As the eighteenth century progressed, similar rhetoric served to distinguish the polite from the vulgar, assigning politeness to those who presented or consumed shows as educational and improving and rudeness to those who merely showed wonders or gaped at marvelous things.[23] In this context, it was easy to interpret grand public fireworks as no more than spectacles for their own sake, as some did of the Green Park fireworks of 1749. Verses in the satirical *Green-Park Folly* contrasted the politeness of science and the rudeness of fireworks. Because of the fireworks, the river Thames,

> Whose Banks afford a pleasing walk
> For Meditation or for Talk;
> Where the young Sons of Learning rove,
> And chat of Science, or of Love,
> Is forced here to fill the Show,
> And a dire Penance undergo,
> Is scorch'd and stench'd with Sulph'rous Fire
> To make a Crowd of Fools admire.[24]

Philosophical fireworks, however, could be presented as more illuminating. "The spectacle is instructive as well as curious," wrote the witness of a *spectacle pyrique et hydraulique* in Paris in 1768. "What image is not presented to the imagination?"[25] In *Julie; or, The New Heloise*, Jean-Jacques Rousseau's treatise on the education of women, the author made education entertaining by having Julie entertained with "rockets . . . from China, which made quite an impression."[26] Hooper similarly used an enduring trope to explain how rational recreations "enlarged and fortified" the "mind of man" by training the senses to understand the artifice behind ingenious effects, which the vulgar failed to comprehend. The learned man's "heart is incessantly rapt with joys of which the grovelling herd have no conception, compared with whose ignorance, the insensibility of the blind and deaf . . . are but trifling imperfections."[27] Similar morals were no doubt raised at shows of electric illuminations and table fireworks.

## Inoffensive Fireworks and the Pleasure Gardens

Another effort by the public to distinguish themselves from the vulgar rested on avoiding the smoke and smells associated with fireworks. In the *Green-Park Folly*, the stench of sulfur was allied with the crowd and contrasted to pleasing riverbanks and learning. Philosophical fireworks avoided such impurities and associated them with vulgarity. Thus, a distinctive feature of the new pyrotechnics was a concern to render fireworks cleaner, safer, and devoid of the inflammatory risks and sooty by-products of traditional fireworks. Ideally, pyrotechnics should be "brilliant and inoffensive," as one reviewer put it.[28] In addition, such attitudes placed a new value on the "natural" in experience and behavior.[29] Court society had once made the cultivation of artifice a sign of its power, at a time when magnificent art was its exclusive prerogative. Now, in an age of growing consumption and easier access to artificial products, the natural was to be prized as much as art. The shift was reflected in audiences for fireworks. If, according to the courtesy protocols of theatergoing, spectators of the 1730s ignored fireworks, they now remarked on the naturalness of displays. The witness of the Parisian "spectacle pyrique et hydraulique" in 1768 exulted imitations of thunder, lightning, and "the phenomena of nature . . . they are pieces of the latest taste: everything is rendered with such art, that the images are equal to Nature."[30]

The electric fire was considered valuable on such terms, as a natural light devoid of the smoke and smells that even miniature fireworks still produced. Performances continued through the end of the eighteenth century. In 1784, Nollet's student Joseph Sigaud de la Fond explained how to make electric *jets de feu* in his textbook on the use of a *cabinet de physique*.[31] Across the channel, George Adams reproduced Nollet's techniques, illuminating the word LIGHT with electrified tinfoil.[32] Such performances were directly allied with social graces. Electric sparks could be presented as natural, honest, pure, and enlightened, while pyrotechnic sparks were artificial, noisy, and sensual. In 1798, John Bennett thus contrasted the true and false graces and virtues appropriate to young ladies by reference to electrical and pyrotechnic effects. False grace was "loud and noisy" like "an humble fire work, which cracks and sparkles; the other is that lightning, which, in an *instant*, electrifies and shocks; this is the offspring of heaven; that, the artificial creature of the world."[33]

Novel pyrotechnics were designed to eliminate offensive elements. The Ruggieri's Chinese fire, with its brilliant white sparks, was considered more polite than other fireworks: "It gives more brilliance to artifice, is less subject to smoke, and, since it goes out very promptly, it is communicated without danger."[34] "Hydraulic" fireworks made with water also claimed to eliminate smoke and smells.[35] At the Panthéon in Paris, and then at the London Lyceum, the Dutch physicist Charles Diller presented "Diller's Philosophical Fireworks," made with colored jets of inflammable air that sprang from spinning copper pipes supplied with gas from bladders.[36] Diller's shows, which proved immensely popular, were advertised as safe and hygenic, "destitute of smell or smoke, yet inflammable in the closest apartment, and incapable of detonation by coming into contact with atmospheric air."[37]

In the age of *lumières*, some of the most inoffensive philosophical fireworks relied on optical effects. In his *récréations mathématiques*, Guyot described the "manner of imitating real fireworks naturally, solely by means of light and shadow," the instructions appearing in a number of other works.[38] Pieces of thin paper should be painted white, red, blue, or yellow and rubbed in oil to make them transparent. Black or blue paper might then be fashioned into cones, wheels, globes, and pyramids and spiral patterns cut out from the shapes. When another spiral drawn on transparent colored paper was turned around behind these figures and illuminated from behind, the crisscrossing patterns would "resemble sparks of fire that incessantly succeed each other" (plate 11).[39]

Like table fireworks and electric illuminations, these fireworks were small and suitable for domestic consumption. Craftsmen produced a variety of ingenious cabinets and boxes inside which miniaturized visions of fireworks and illuminations could be seen. Besides the Dutch cabinet mentioned above, peep shows and *vues d'optique* showed brightly painted engravings of court displays to be observed through a lens enhancing their perspective and depth (plate 12).[40] In shadow theaters, pyrotechnic decorations were painted on glass plates and illuminated inside a wooden box frame like stage scenery (plate 13). By the 1790s, Diller's theater shows had been commodified as numerous instrumentmakers sold tabletop apparatus showing inflammable air fireworks with jets emitted from spinning copper wheels and pipes (plate 14). Pyrotechnics thus became domesticated into inoffensive commodities, transformed from grand incendiaries into polite performances the size of a carriage clock.

Not all performances were miniaturized, however. William Hooper reckoned of optical fireworks: "These pieces have a pleasing effect when represented of a small size, but the deception is more striking when they are of large dimensions."[41] He claimed to have made wire and paper figures of the sun up to six feet wide, a suitable figure for enlightened pyrotechny.[42] In the pleasure gardens, artificers also developed philosophical fireworks on a larger scale and transformed pyrotechnics into a profitable business. In England, gardens expanded their pyrotechnic repertoires after the Green Park fireworks of 1749, offering spectacles to anyone able to pay the entrance fee. Expansion encouraged more Italians to travel to England, including several who also made a living selling scientific instruments: "At BUNN's Pantheon, On Monday, June 4, 1781, (being his Majesty's Birth-day) will be performed a Concert of Vocal and Instrumental Music . . . To conclude with an elegant Display of Fire-works, by Sig. Baptista PEDRALIO, in a Variety of Designs, in Brilliant, Chinese, Rayonant, Gold, Blue, Red, and Yellow Fires, particularly a curious Sun-Piece, forming a brilliant Glory to the Letters G.R. Note: Admittance One Shilling."[43]

Pedralio also made barometers and was probably an apprentice of Giovanni Battista Torré, another Italian instrumentmaker and fireworker, whose career will be discussed below. Pedralio was one of many artists in fireworks, both men and women, regularly performing at the gardens. At Cuper's and Marylebone, Benjamin Clitherow was another resident pyrotechnist, assisted by his wife. These artificers made a commercial career out of pyrotechny, making and selling fireworks to the public, their wares including the new table fireworks and the Chinese fire, now widely known thanks to translations of Incarville's reports. Clitherow's elaborate trade card, which advertised him as a "Fire Worker" who had performed fireworks before the "Prince of Wales and the Rest of ye Royal Family" (fig. 31), thus explained: "Furnishes Fireworks after Ye Italian & China method . . . Has the Real True & Genuine China fire yt Represents a Beautifull Fruit tree in full bloom will Throw its flowers from 10, to 30 feet High The small ones may be fired in Rooms, without Danger . . . Sold by no one else in England."[44]

Clitherow marked his card with a royal coat of arms to prevent "counterfeits," and his fireworks could be ordered and sent to any part of England. One person who bought fireworks in this way was the natural historian Joseph Banks, who spent £5 2s. 6d. on pinwheels and Roman candles in 1772. Others preferred to order fireworks from France.[45] Alternatively, the less well-off could go to the pub, and public houses provided further new venues for small-scale shows of natural philosophy and pyrotechnics. Clitherow thus showed fireworks at the King's Arms in Moorfields, while a Mr. Flockton displayed an illuminated "Chinese Bridge" and "Chinese Temple" at the White Lyen pub in Highgate.[46]

Sarah Hengler was another well-known artificer, the "Starry Enchantress of the Surrey garden!"[47] Born in Surrey around 1765, she performed fireworks with her German immigrant husband Johann Michael Hengler, the couple setting up workshops in their three-story house south of Westminster Bridge in 1795. After John's death in 1802, Hengler worked with her daughter Magdalen Elizabeth Jones, and later her grandson, performing fireworks. Constantly touring, she visited the many venues for fireworks available by this time—the Royal Circus and Philip Astley's Amphitheatre in Surrey, the Prospect Hotel in Margate, the Olympic Circus in Hull, the Market Place in Leicester, and elsewhere. By 1821, Hengler was identified as

31. Trade card of Benjamin Clitherow. © The Trustees of the British Museum.

"the celebrated pyrotechnic to his Majesty."[48] She died in 1845, killed by an explosion in her atelier.[49]

Commercial pyrotechnics thus proliferated in the second half of the eighteenth century and soon spread across Europe. In Paris, the Boulevard du Temple and Palais royal provided the location for numerous *cabinets de physique* and theaters where scientific lectures and polite pyrotechnics were put on show.[50] A Monsieur Aubin simulated a rocket ascent with multicolored smoke and flames in his *cabinet.*[51] Another showed "Gifts of Fire [*Etrennes en Feu*], all sorts of possible figures, geometric, likewise astronomical, such as satellites, tourbillons, &c."[52] In 1778, the Parisian *artificier du roi* Delavarinière showed pyrotechnic solar eclipses at the Cirque royal on the Boulevard du Luxembourg and "a grand firework . . . preceded by the Transit of Venus across the Sun, and the combat of Salamanders in colored fire, carried by *Courantins,*" that is, rockets on a line, a term predating electric currents.[53]

By the 1770s, commercial entertainments were also growing in St. Petersburg and Moscow. In Russia, as elsewhere, the public (*publika, obshchestvo*) was identified as all those distinct from the common people (*chern'*), usually by their consumption of plays, books, and art. Much the same as in Paris and London, there were salons, theaters, opera, literary circles, and learned societies in the Russian cities. However regional distinctions remained as the Academy of Sciences continued to dominate natural philosophy while Jacob Stählin continued to compose court fireworks into the 1770s.[54] The court and the nobility were the principal patrons of commercial spectacle, and their estates and palaces became significant sites for pyrotechny. Count Lev Aleksandrovich Naryshkin gave fireworks for a visit of Catherine II to his country estate in July 1772 and printed an *opisanie* to commemorate it.[55] The noble Boris Petrovich Sheremetev built an "illuminations theater" on his estate of Kuskovo near Moscow, while his descendants turned the estate into a pleasure garden in the 1760s, complete with a theater of serf actors, architectural follies, music, dances, and fireworks.[56] Moscow and St. Petersburg's first pleasure gardens were also located in noble gardens belonging to Prince P. N. Trubetskoi in Moscow and Count Naryshkin in St. Petersburg.[57] Entertainments were similar to those elsewhere and were often performed by artists from abroad.

## Sublime Fireworks: Vauxhall, Volcanoes, and the Rise of Mr. Torré

Two of the most successful pyrotechnists of late-eighteenth-century Europe were Giovanni Battista Torré and Claude-Fortuné Ruggieri. Both men opened pleasure gardens in Paris, where they took philosophical fireworks in novel directions. While artificers stressed the lack of danger posed by philosophical fireworks, there was also room to appeal to a growing taste for the sublime in pyrotechnics, emphasizing the *pleasure* of danger in fireworks. Edmund Burke's *Philosophical Enquiry into the Origin of our Ideas of the Sublime and Beautiful* identified the sublime as "whatever is fitted in any sort to excite the ideas of pain, and danger . . . productive of the strongest emotion which the mind is capable of feeling."[58] Real dangers created terror, "but at certain distances, and with certain modifications, they may be, and they are delightful."[59] The volcano provided the preeminent vision of the sublime for many in the late eighteenth century, thanks in part to the shows of Torré, who made artificial volcanoes the centerpiece of his pleasure garden performances. Ruggieri, alternatively, chose the new French chemistry of Antoine Lavoisier and the colored fires of the Russians to make fireworks more appealing to his audiences. His innovations proved enduring, making Russian colored fire and pyrotechny's link with chemistry last well beyond the eighteenth century.

Torré and Ruggieri's activities also challenged existing hierarchies of pyrotechnic labor, forging a new identity for the artificer that equally endured for many years—the idea of the pyrotechnist as an artist in "applied science." If the middle decades of the eighteenth century saw much celebration of Italian fireworks, they also witnessed, as chapter 5 showed, a contempt for the Italian artificers who set them off. French competitors revealed the Ruggieri's techniques for communicating fire and the recipe for Chinese fire in the 1750s. English audiences scoffed at their fireworks, and French men of letters placed them at the bottom of the pyrotechnic tree. Resistance and competition from local artificers continued to cause problems for the Ruggieri. By the early 1760s, they still enjoyed regular commissions from the French court, serving as *artificiers du roi*, and preparing displays with substantial budgets.[60] But, in the same decade, the number of artisans claiming the title *artificier du roi* swelled to twenty-eight, while artificers continued to compete with the Italians in displays (plate 12).[61] In addition, another enterprising Italian, Giovanni Battista Torré, opened a new venue in Paris for fireworks, the "Waux-Hall" pleasure garden, whose spectacular pyrotechnic volcanoes rapidly caught the attention of the town. Thanks in part to Torré's shows, the study of volcanoes, like that of electricity, emerged in the 1760s and 1770s as another science mixing natural philosophy and pyrotechnics.

Torré's career is indicative of the multiplying intersections of pyrotechny and natural philosophy that marked the second half of the eighteenth century. Giovanni-Battista Torré hailed from the Lake Como region, where he was trained by his Swiss-Italian father as a maker of barometers and thermometers.[62] Leaving his village, Torré traveled widely, to "Switzerland and many provinces of France." One of the earliest notices of the lightning rod, read in translation by Benjamin Franklin, identified a "Gio. Batista Torrè, Dimonstratore di sperienze curiose di Fisica" in Brussels in June 1752, raising an iron rod on his house, but no further details are known.[63] In any case, Torré settled in Paris, where he made instruments for the academician René-Antoine Réaumur, whose laboratory the abbé Nollet had managed previously.[64] He also studied with Servandoni, and, in 1753, after Servandoni's disgrace in London, Torré was invited there to stage fireworks in the new pleasure gardens of Marylebone.[65] By 1760, Torré was back in Paris embarking on a different venture, with the opening of a *cabinet de physique experimentale,* a shop selling books and instruments, including Dolland telescopes, solar microscopes, and electrical machines.[66] Styling himself a "physicist," and residing at the court of the Quinze-vingts, Torré offered public scientific lectures on physics and optics in rooms above the Temple of Peace coffeehouse near the Porte du Temple. Soon after, however, his attention again turned to fireworks, and, by the summer of 1764, advertisements appeared in the Paris papers promoting a new venture:

> Mr. Torré has managed to obtain from His Majesty permission to exercise the profession of artificer in Paris. . . . His spectacle will be open in the course of August, and the public will be informed by notices three days in advance. He will not give a detailed description of this spectacle. The public will judge it. But one may be assured that the multitude, and the variety of pieces of artifice and colors of these fires, will give to this spectacle, which lasts at least an hour, the merit of its novelty. The author owes his talents to a long study of experimental physics and above all to chemistry, the soul of artifice.[67]

Torré was about to open Paris's first pleasure garden on the Boulevard Saint-Martin, which he named the "Waux-Hall d'Eté" after Jonathan Tyer's Vauxhall in London.[68] With a nominal entrance fee to keep out the *bas peuple,* but affordable to middling shopkeepers, clerks, civil servants, and their like, the gardens offered walks, refreshments, illuminations, and polite fireworks.

Torré's performances followed the same line as his advertisements. At the Waux-Hall, he presented "pyri-pantomimes"—fireworks integrated with live action amid perspective scenery on a stage. Like the Ruggieri at the Comédie-italienne, Torré thus recast fireworks as a smaller-scale and more intimate performance, similar to electrical and philosophical lectures that Torré himself had presented. The most celebrated of these performances was *Les forges de Vulcain sous le Mont Etna*, which Torré based on the episode in Virgil's *Aeneid*. Following a display of transparencies or "tableaux d'artifice," a curtain opened to reveal an artificial volcano. Present at the show, Louis-Petit de Bachaumont explained how

> one saw in the interior of the mountain Vulcan and his Cyclops all dressed in the customary manner. One saw Venus descend, who came to ask her husband for arms for Aeneus. The palace of Vulcan occupied the bottom of the den [*antre*] and formed a most rich and deep perspective. The work of the Cyclops produced effects in artifice most advantageous and that could still be multiplied further. But above all the public seemed struck by the effects of the volcano, effects taken from the nature of the thing itself. One feels the subject could not be better chosen, and the artifice appeared to have been invented expressly in order to imitate these sorts of natural phenomena.[69]

Torré's fireworks thus appeared as a natural, meteorological imitation, much like the electrical imitations of lightning or the aurora borealis discussed above. Indeed, Bachaumont claimed that Torré had obtained "some weight [*consistance*] with people of letters" for these shows.[70]

The niceties of Torré's volcanoes were not self-evident to everyone. When, in 1772, Torré traveled to London to present the *Forge of Vulcan* at Marylebone gardens, the politeness of philosophical fireworks was challenged. Invited to England by David Garrick, and residing in the Haymarket, Torré showed the *Forge of Vulcan* at Marylebone for two years. But performances began only after a protracted dispute with locals, who challenged the garden proprietors in court over the shows.[71] When a cargo of Torré's machinery arrived on carts at Marylebone, the "extraordinary appearance and declared use of these machines occasioned universal alarm." Locals claimed that Torré's fireworks would be *offensive*, likely to wake up and frighten children, upset the sick and infirm, and threaten locals with explosions. But the campaign, headed by a Mrs. Fountain, the mistress of a school for young ladies, collapsed at court, and Torré's shows went ahead, though rockets and maroons were removed to allay the locals' fears.[72] Reviewers pointed to the now exemplary *politesse* of the performance: "Altho' there were present many blooming maidens, many beautiful wives, and many handsome widows [at Torré's first Marylebone display], not one of them had a sensation of fear during the firework; the safety of which was so self-evident, that the whole might without injury be exhibited and played off in a ladies dressing-room, were it capacious enough for the purpose."[73] Soon, the fireworks were widely celebrated in London society, and Torré finally received royal honors, performing the *Forge of Vulcan* for King George III's birthday in 1772.[74]

If the politeness of Torré's shows was in doubt, their philosophical credentials were not, hinging on the volcano, a central motif in fashionable tastes for sublime and natural spectacles. Volcanoes had long been imitated in fireworks spectacles, and, by the 1760s, they figured as exemplary instances of the sublime. Both pyrotechnic and natural philosophical treatises alike included Lemery's recipe for imitating a "Vesuvius," while fireworks and volcanoes were regularly compared.[75] The painters Joseph Wright of Derby, Francesco Piranesi, and Peter Fabris all produced pairs of paintings depicting the fireworks at the Castel Sant'Angelo and the eruption of Mount Vesuvius. Wright, who witnessed both on the grand tour, explained: "The one is the greatest effect of Nature the other of Art that I suppose can be."[76] A reviewer called his pictures "wonderful examples of sublimity."[77]

In its celebration of pleasurable terror, the category of the sublime, like the category of the natural, contributed to the distinction of the polite classes from the vulgar by inverting the traditional power hierarchy of spectacle. In the sixteenth and seventeenth centuries, elites were marked by knowledge of artifice, which permitted them to avoid being afraid of artificial effects deemed terrible by the vulgar. Now, at a time when the artifices involved in fireworks were more widely understood (or at least experienced), elites distinguished themselves by a cultivated fear. *Not* to be struck with some degree of terror by fireworks was now vulgar. Nevertheless, as William Hooper's comments on the "grovelling herd" suggest, the vulgar could still be equated with fearfulness when appropriate.

Volcanoes were equally valued as "natural artifices," and Torré's effects were praised as "natural" imitations of meteors. When, in 1798, there were no longer any volcanoes on display in London, Isaac Disraeli lamented: "Is the national genius too austere to be pleased with these striking representations of some phenomena in nature?"[78] As in the case of electricity, however, artifice still provided the vocabulary for articulating even the most natural of experiences, and, as Wright's comments suggest, art retained great value even as nature gained it. This is especially evident in the labors of William Hamilton, the British envoy extraordinary to the Spanish court of Naples from 1764 to 1800 and a preeminent observer of volcanoes. At his residence in the British embassy at the Villa Sessa in Naples, Hamilton, of noble birth, combined the study of Vesuvius with the pursuit of polite collecting and connoisseurship of art and antiquities.[79] He celebrated the natural and thought that a real volcano always surpassed the spectacle of its imitations: "It is impossible to describe the beautiful appearance of these girandoles of red hot stones, far surpassing the most astonishing artificial fire-works."[80] Nevertheless, he regularly articulated the natural by reference to art. Hence the format of his remarkable *Campi Phlegraei, Observations on the Volcanos of the Two Sicilies* of 1776. Printed at a personal cost of £1,300, this sumptuous book set the science of volcanoes in a presentation volume reminiscent of splendid festival books, incorporating fifty-nine hand-colored copper engravings based on paintings of Vesuvius by Peter Fabris, also a painter of fireworks, together with descriptions of Vesuvius's geology and eruptions.[81] The book was literally a spectacle since Hamilton included with an edition sent to the Royal Society an elaborate theatrical apparatus representing a series of illuminated scenes.[82] In his paintings, Fabris showed streaks of lightning in the sky around plumes of black smoke and lava erupting from the volcano, reflecting Hamilton's observation that the smoke issuing from volcanoes was "pregnant" with the electric fire.[83] But Hamilton did not invoke electricity among the causes of volcanic activity, and, even when he did discuss electric fires, it was in pyrotechnic terms: "He had, with many others, seen balls of fire issue [from volcanoes] which bursting in the air, produced nearly the same effect as that from the air-balloons [a form of bomb] in fire-works, the electric fire that came out having the appearance of the serpents with which those fire-work balloons are often filled."[84]

Hamilton's pyrotechnic analogies were typical of philosophical accounts of the volcano, which, like accounts of other meteors, continued but modified the gunpowder theory of the seventeenth century. In the 1630s, Castelli had argued that volcanoes erupted like fireworks, but, by the eighteenth century, philosophers proposed more specific fuels and ignition mechanisms than simply the lighting of nitro-sulfurous vapors.[85] Martin Lister's suggestion that pyrites provided volcanic fuel, supported by Lemery's imitations, revised the gunpowder theory.[86] Nevertheless, in the eighteenth century eruptions of volcanoes were regularly accounted for in pyrotechnic terms. When Vesuvius erupted in 1737, the *Philosophical Transactions* included reports such as the following: "There were many great Flashes of Lightning darted through this Pillar of Smoak [ascending from Vesuvius], and frequent Discharges as of Cannon or Bombs, which were followed by falling Stars, such as we see from well-made Rockets . . . great Quantities of Sulphur and Nitre are, to be sure, the Operators of these great Explosions, Lightnings, Bombs . . . and Nature can certainly make much stronger and more elastic Gunpowder, than Mankind."[87]

Philosophical accounts also rehearsed the poetics of Torré's sublime spectacles in order to appeal to the public. In the verses of his natural historical poem *The Botanic Garden*, Erasmus Darwin presented the volcano in more or less the terms of Torré's pantomimes, with Cyclops and Vulcan working the forge inside Etna:

> Thus when of old, as mystic bards presume,
> Huge CYCLOPS dwelt in Etna's rocky womb,
> On thundering anvils rung their loud alarms,
> And leagued with VULCAN forged immortal arms;
> Descending VENUS sought the dark abode.
>
> .................................................
>
> With radiant eye She view'd the boiling ore,
> Heard undismay'd the breathing bellows roar . . .
> With smiles celestial bless'd their dazzled sight,
> And Beauty blazed amid infernal night.[88]

As hybrids of natural philosophy and sublime pyrotechnics, Torré's volcanoes thus helped ensure an enduring use of fireworks in the comprehension of nature. Torré's shows also prompted a variety of imitations and elaborations. Catherine the Great was entertained with a grand volcano and *Forge of Vulcan* in 1772, while volcanic performances multiplied in the London and Paris pleasure gardens for several decades (fig. 32).[89] Influential among these displays was the *Eidophusikon* of Philip James de Loutherbourg, Garrick's set designer at Drury Lane, an admirer of Servandoni's and a traveling companion of Torré's.[90] From 1781, Loutherbourg created fiery scenes from Milton's "Pandaemonium" with the *Eidophusikon*, a stage set of moving parts illuminated with lamps tinted by stained glass of different colors.[91] Like Torré, Loutherbourg was praised for the naturalism of his scenery, as a man "who could create a copy of Nature to be taken for Nature's self."[92] Subsequent shows depicted Etna and Vesuvius in eruption; these, in turn, inspired painters such as Thomas Gainsborough and Caspar David Friedrich, becoming instrumental in a new fashion for landscape painting.[93] For many, pyrotechnic volcanoes were the epitome of enlightened spectacle. Hence the construction, in 1788–94, of the Stein, an artificial volcano that spewed pyrotechnic lava in the gardens of Leopold III Friedrich Franz in Wörlitz-Dessau. An inspiration to Goethe, the Stein was taken to represent enlightenment itself, as a constructive force shaping the environment for the benefit of mankind.[94]

32. Fireworks in an artillery camp for Empress Catherine II of Russia, August 1773. *Vulkanova blagodarnost' za vysochaishee eia imperatorskago velichestva poseshchenie, na ego pole, izobrazhennaia v feierverke pered artilleriskimi lagerem, avgusta, "-" dnia 1773 goda* (Moscow [?], 1773 [?]). Research Library, the Getty Research Institute, Los Angeles, California.

## Appealing to the Chemists: Competition in the Pleasure Gardens

Torré's Paris Vauxhall and naturalistic volcanoes posed significant competition for the Ruggieri. To rival Torré, the family opened its own pleasure garden and also turned to the sciences for innovations. A year after Torré's Waux-Hall opened, Pietro Ruggieri established the Jardin Ruggieri, located near Torré's gardens at 20, rue Neuve-Saint-Lazare, in the Porcherons district to the northwest of Paris. With the new venture, the Ruggieri extended their appeal to the polite public. Ruggieri was described as "the dignified rival of Torré," his gardens offering a "beautiful and brilliant promenade" with "poèmes pyriques" or pantomimes closely imitating Torré's volcano shows.[95] However, the Ruggieri also sought out innovations from a different source. These came from a young Antoine Lavoisier, chemist, geologist, and a recent entrant in the competition to devise new forms of lighting for Paris streets, for which he would gain membership in the Academy of Sciences and an interest in the workings of combustion.[96] In 1766, Lavoisier composed a report describing his efforts to make colored fireworks for the Ruggieri brothers and particularly green fire, the "Columbus's egg" of eighteenth-century pyrotechnics, which Danilov had succeeded in making in Russia a decade earlier, and which the Russian chemist Lomonosov had been unable to replicate.[97] Lavoisier now tried his hand—and also failed.

In the report, Lavoisier explained his efforts to make yellow, blue, and green fire. Of these, green was presented as by far the most important:

> It is unnecessary that I expound here on the great advantages that the introduction of colored fires could spread to artifice. It is above all in the decorations in *feux de lance* [small cartridges used to make fiery outlines of pictures and buildings] that the effect would be most marvelous. If it is possible for us one day to produce fires of all the different colors, one could execute then decorations of artifices that would form profitable tableaux. One will be able to use them to represent gardens, small woods, arbors; and one will be able to place [in them] parterres enameled with flowers, temples, and palaces. The columns, palasters, and the different orders of architecture of which they will be composed will seem to be made of serpentine, jasper, amethyst, emerald, ruby. . . . Green is one of the principal colors that we must achieve now to execute these marvels.[98]

Like Lomonosov, Lavoisier knew that "different salts with a cuprous base" dissolved in alcohol would yield a green flame, but he noted: "None of these salts succeed in artifice. They give absolutely no color and are for the most part so delicate that at the end of a few hours the composition moistens and is no more able to be lit."[99] Thinking sulfur might be the agent that diminished the color, Lavoisier tried making compositions without it.[100] This, however, did not work. Like Lomonosov, Lavoisier assumed some quality-bearing substance to be responsible for color, and he next proposed that green might be produced by mixing yellow and blue compositions. Using techniques developed by Johann Heinrich Pott and Louis-Claude Bourdelin, he experimented to make yellow and blue fireworks, succeeding in both, using a purified form of zinc dissolved in vinegar to make blue.[101] But, again, when the two were mixed: "The flame was yellow on jumping from the cartridge and blue in its extremity, but the two colors were never combined."[102] Like Lomonosov, Lavoisier also failed to make green fire. In the age of *lumières*, academic chemistry might still flounder in comparison with artisanal ingenuity.

Whether the Ruggieri displayed colored fires of Lavoisier's recipes is unknown, but this was the beginning of a new taste for the sciences in the Ruggieri brothers' works.[103] The Jardin flourished, attracting the public with suitably refined commodities—dances, illuminations, cafés, and boutiques.[104] In the meantime, however, the Ruggieri's traditional source of income, the sooty spectacles of grand court fireworks, was declining as revenues from royal and city fireworks increasingly diminished, particularly after a tragic accident in 1770. The city of Paris, which paid the Ruggieri for fireworks, was running out of money. While fireworks for the taking of Mahon in 1756 cost 11,500 livres, there were only 4,500 livres to pay for celebrations of a more important event two years later, victory over the English. After 1760, the city declined to pay for any fireworks except for royal weddings, and even these proved disastrous.[105] In spring 1770, celebrations began surrounding the wedding of Marie-Antoinette and the dauphin, the future King Louis XVI. At first, things appeared to go well. The future queen traveled about the country and witnessed many fireworks displays, even assisting in one performed by the *artificier* Dominique Bray at Châlons-sur-Marne.[106] Torré showed a splendid firework at Versailles: "It is said they will give him the medal of St. Michel [for it], an honor that is quite suitable to his talents."[107] There was even a philosophical firework, as the *receveur général des finances*, M. Varennes de Beost, presented an illumination arranged so that its lamps would all be lit at once, a project that Lavoisier reported on, advising on its safety.[108] But the grandest fireworks were reserved for Petronio Ruggieri. On May 30, Ruggieri planned to show a vast Temple of Hymen on the Place Louis XV with a 600,000 livre budget for fireworks.[109] The event, however, proved disastrous. The fireworks illustrated the dangers of crowds of spectators thronging the streets, typical of great courtly

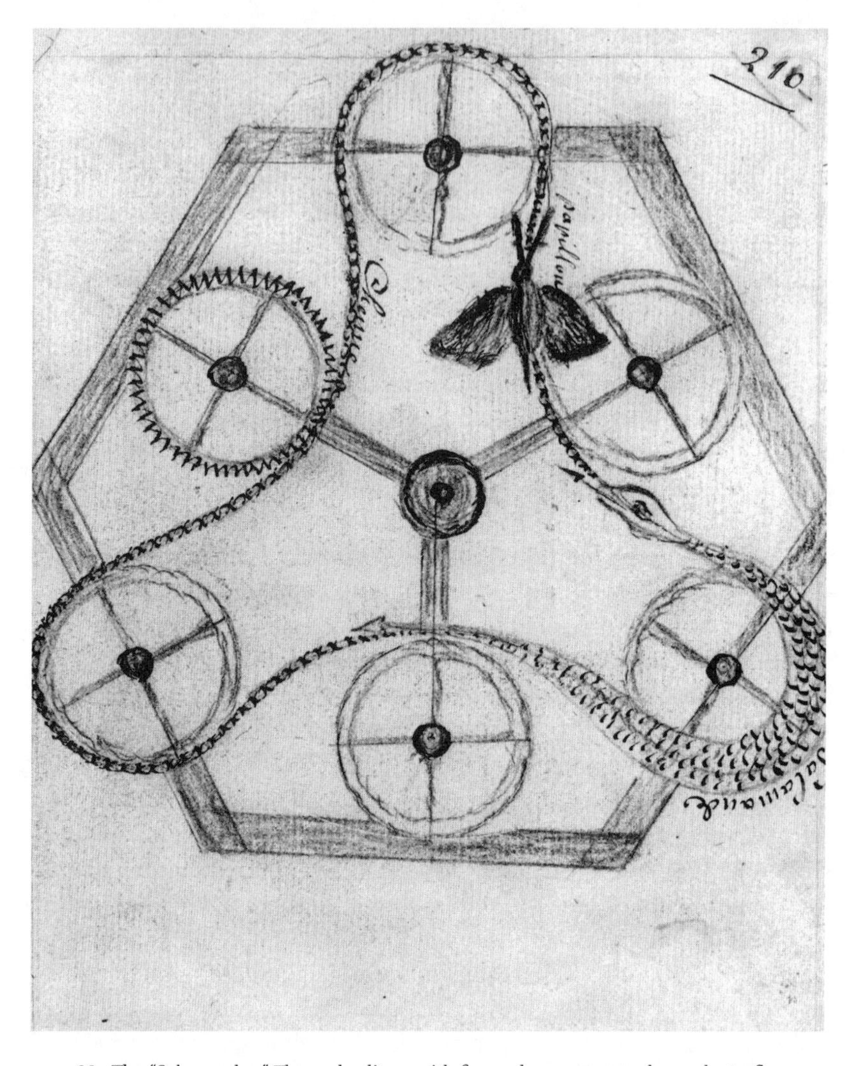


33. The "Salamander." The snake, lit up with fireworks, appears to chase a butterfly, using hidden machinery. From *A Colbert: Traité des manoeuvres et de l'artif[i]ce de guerre et de joie pour l'artillerie, dédié à messieur[s] du Corps royal de l'artillerie, regiment de La Fère,* Guttmann Collection on Explosives, Firearms, and Military Science, 1450–1905, Hagley Museum and Library, Wilmington, Delaware.

displays. During the performance, a stray rocket ignited Ruggieri's cache of fireworks, setting off a huge explosion, burning the decorations.[110] Many spectators panicked and were trampled or pushed into the Seine. Several hundred people were probably killed, and scorn was soon poured on the dauphin and his *artificiers*.[111] Indeed, such were the feelings of the city, polite and vulgar alike, that, when the dauphin, subsequently King Louis XVI, was executed, his body was interred in the churchyard of St. Magdalen so that "his ashes repose between the people who were stifled in the throng on the 19th of April, 1770."[112]

The city of Paris did not pay for another grand firework until 1782, and, in the meantime, royal performances at Versailles were reduced in scale and executed by Torré, whose reputation was not hurt by the Ruggieri's misfortune.[113] Indeed, when the comte d'Artois and Marie-Thérèse of Savoy were married in 1773, they were entertained, like George III in England, with a performance of Torré's *Forge of Vulcan*.[114] Torré's philosophical fireworks were now thought fit to serve as courtly displays.

In the wake of the 1770 disaster and Torré's success, the Ruggieri sought to restore their reputation. Some of their efforts hinged on shows of ingenious machinery. Exploiting their mechanical skills at the Jardin Ruggieri, the brothers presented the remarkable "Salamander," which used a system of wheels and chains hidden from view to represent "the Serpent with the head of a dog, rushing the nest of birds to flush them out, preceded by many *Pièces Pyrriques*" (fig. 33).[115] Then came Torré's death in 1780, which was fortunate for the Ruggieri since "the crowd returned to their spectacles."[116] Over the next decade, the Jardin was remodeled, incorporating a growing number of philosophical fireworks. Once again, the Ruggieri turned to the increasingly fashionable science of chemistry to improve pyrotechnics. The family had already collaborated with Lavoisier, and, after a wave of discoveries of new "airs" or gases led to Montgolfier's and Charles's popular balloon ascents, chemistry became the focus of the Ruggieri's attentions. The aeronaut Blanchard was invited to perform balloon ascents at the Jardin, while the brothers devised fireworks made with inflammable air. Appealing to the financial sensibilities of the polite classes, the Ruggieri also collaborated with a physicist named Le Roux to produce "flame-proof clothes," "by means of which one may, in a considerable fire, be saved with one's family and fortune, without being incommodated by flame or by smoke."[117] Nevertheless, the Ruggieri struggled with family deaths and continuing competition. Pietro died in September 1778, passing ownership of the Jardin to Petronio. In 1783, Bachaumont thought that the success of Torré and another new venue, the Colisée, "has left [the Ruggieri] lost from view and absolutely fallen."[118]

## Science, Spectacle, and the French Revolution

The next generation of Ruggieri set about restoring the family fortunes and succeeded admirably. In 1794, Petronio also died, the last of the original brothers, leaving the Jardin Ruggieri to the guardian of his two sons, Michel-Marie and Claude-Fortuné.[119] Claude-Fortuné Ruggieri spent his career forging links between pyrotechny and the new chemistry. He did so to distinguish his skills from those of his rivals and as a challenge to hierarchies of pyrotechnic labor that still insisted that artificers were beneath the talents of architects and men of letters. Chemistry offered credit, as the science itself underwent new transformations in the revolutionary era.

An important consequence of the rapid growth in the public's taste for philosophical fireworks and scientific demonstrations in the 1770s and 1780s was a change in the nature of the sciences, particularly in Paris. Just as polite society sought to distance itself from the vulgar by enjoying domestic and commercial spectacles of natural philosophy and pyrotechnics, so too scientific elites began to distance themselves from the public spectacles in which they had participated since the late seventeenth century. The 1780s witnessed a transformation of the rowdiness of the crowd into revolutionary action. Spectacle was politicized, with attacks on hated ministers and the return of the exiled Parlement in late September 1787 accompanied by pyrotechnic riotousness: "They fire a great many rockets, firecrackers and all sorts of fireworks, which lasts until almost four in the morning. All the windows in the place Dauphine are illuminated."[120] The following year, crowds on the place Dauphine lit firecrackers and bonfires and "burned the effigy of Monseigneur de Lamoignon, the current French minister of justice, after having him do public penance for his wrongdoing." Locals "had to close all the shops and illuminate all the facades of all the houses."[121] Like religious enthusiasm, radicalism was regularly associated with fiery passions, inflammatory acts, and uncontrollable incendiarism. As Thomas Carlyle said of the meeting of Etoile before the Revolution, it was the "first spark of a mighty Firework."[122]

In this context, scientific elites in Paris increasingly steered away from overt spectacle. They faced the same paradox as the Royal Society in the previous century. While spectacle appealed to the crowd and promoted natural philosophy, appeals to the crowd could look dangerous in times of political unrest. The Royal Society solved the problem by altering its methods and carefully managing spectacle, promoting it or avoiding it when appropriate. The Royal Society also claimed to police the very forces that threatened the state, with a science explaining the physiology of enthusiasm and overheated passions. The Paris academy deployed similar strategies, reforming methods to make them less theatrical, and withdrawing from the arena of public science when suitable. Academicians also cultivated a police role, identifying popular practices that it deemed suitable or unsuitable to the values of science and the state. In methodological terms, this change produced novel approaches to experiment. Experimenters such as Lavoisier, Laplace, and Coulomb shunned the theatrical and sensuous, instead emphasizing experiments that by their nature could not be performed in public and that demanded a mathematical expertise. Quantitative approaches and precision measurement thus grew increasingly important, in Lavoisier's measures of heat or Coulomb's use of the torsion balance, for example. Sensuous measures in chemistry were replaced with more instrumental methods.[123] Similarly, the cosmopolitanism of theatrical science gave way to a more exclusively male pursuit. Metaphysics was dismissed as "feminine" in favor of a "masculine" calculating reason.[124] Interest in the particulars and details of nature's mechanism, the observational strategy entailed in appreciating the "Theater of Nature," was also rejected in favor of abstract, generalizing "systems," a panoptic perspective that only an expert eye could achieve.[125] Thus, as polite spectacles grew popular and crowds more political, the techniques of elite science insisted on special expertise and retreated deeper into exclusive venues.

The academy did not shun all theater, however, and, when occasion suited, academicians endorsed or participated in spectacle.[126] The academy also took on a growing police role, authorized by the crown to decide *which* Boulevard theatricals were dubious and which desirable. Academicians had always enjoyed a police function. Since the seventeenth century, the academy took on the responsibility of assessing new inventions for the crown, and academicians regularly joined local experts to approve or reject requests for *brevets* or *privileges* to safeguard inventions.[127] But, during the eighteenth century, this role extended to commissions assessing novel, and often spectacular, practices. Lavoisier was instrumental here, as in transformations of scientific method. Commissions to assess the animal magnetism of Franz Anton Mesmer and the dowser Barthelemy Bléton in the late 1770s and the 1780s have been well documented for their assaults on popular medicine and practice. However, assessments of pyrotechnic innovations were more sympathetic.[128] In the 1760s, Mlle Saint-André kept certificates from academicians, including the abbé Nollet, Alexis-Claude Clairaut, and Jacques Vaucanson, on the door of her atelier.[129] After Lavoisier worked with the Ruggieri in the 1760s, he reported favorably on a number of other pyrotechnic projects, especially those publicizing the new chemistry of airs such as Diller's philosophical fireworks made with gas.[130]

Some spectacle thus suited the new chemical and physical sciences otherwise distanced from the theatrical, and, like the Royal Society earlier, the Paris academy and its successor institutions could exploit or retract from spectacle as appropriate. Similarly, the new sciences served pyrotechnists. Although the Revolution would sweep the academy and Lavoisier out of existence, the credit of science and its new methods remained. Claude-Fortuné Ruggieri then made his career fashioning spectacles out of the new chemistry. At the turn of the nineteenth century, he used the creditable science of Lavoisier's disciples to transform the image of pyrotechnics and secure an intimate association of fireworks and chemistry that has endured ever since.

## A Revolution in Fireworks: Claude-Fortuné Ruggieri and the New Chemistry

Claude-Fortuné Ruggieri remains one of the most celebrated pyrotechnists in the history of fireworks, principally for his remarkable *Elémens de pyrotechnie,* first published in 1802, and revised in a third edition of 1821.[131] The son of Petronio and Jeanne-Elizabeth Ruggieri, he was probably born about 1777 and inherited the Jardin Ruggieri on Petronio's death in 1794, which reopened under the direction of his guardian sieur Ducy in the same year. Claude worked there with his brother, Michel-Marie, himself an accomplished artificer, until 1805, when he established his own atelier at no. 3, rue de Clichy.[132] In the interval, the Revolution swept away the grand fêtes of the ancien régime and, with them, the post of *artificier du roi.*[133] The Revolution did not, however, abolish fireworks, whose popular and commercial associations were by now as strong as their associations with the elite. In the era of Jacobin power, the Ruggieri brothers thus gave fireworks for the first anniversary of the fall of the Bastille and, in 1796, opened a new pleasure garden, the Jardin Boutin or Tivoli.[134] Revolutionary fireworks were variously philosophical. Claude collaborated with the aeronaut and inventor of the parachute André-Jacques Garnerin and, later, with Etienne-Gaspard Robertson, for whose shows of phantasmagoria he provided the fireworks.[135] Michel-Marie produced spectacular fireworks in the first years of Napoleon's regime, including a volcano, perhaps following the words of Napoleon to the Council of Ancients on the 18 Brumaire: "You do not now meet under common circumstances; you are upon a volcano" (plate 15).[136] Both Claude and Michel-Marie showed balloon ascents, displays of Chinese fire, the salamander, and other spectacular pyrotechnics in these years. Garnerin, Ruggieri claimed, then incited him to prepare a book on fireworks, and the *Elémens* was the result.[137]

Both Garnerin and Claude Ruggieri shared an interest in the new French chemistry of the revolutionary era and the works of Lavoisier, Pierre-Simon

Laplace, Antoine François Fourcroy, and Jean-Antoine Chaptal, no doubt following the earlier collaboration between his father and Lavoisier. The *Elémens* was thus dedicated to Chaptal, Napoleon's minister of the interior, the former director of the Grenelle saltpeter works, and the author of the Lavoisian textbook *Elémens de chimie* (1791), whose title evidently inspired Ruggieri.[138] Chaptal had much to say about the state of artisanry in France. In the last years of the eighteenth century, he promoted a vision of French prosperity rooted in the demands of philosophes such as Diderot for the application of the sciences, especially chemistry and mechanics, to manufactures. Against what Chaptal perceived as a debilitating divide between the arts and the sciences in the old regime, he offered the image of a revolutionary "new man" who would combine theoretical, ideally chemical, knowledge and artisanal practice to accumulate private wealth and national prestige.[139]

Chaptal's egalitarianism and determination to apply chemistry to art clearly appealed to Ruggieri, whose family had experienced precisely the social distancing identified by the chemist, as architects and men of letters regularly placed them at the bottom of the pyrotechnic hierarchy of labor. Despite the rise of philosophical fireworks and a lull in grand fireworks during the Revolution, such men continued to claim jurisdiction over pyrotechnic designs. After the Revolution, Napoleon reasserted magnificent state spectacles. He took a copy of Frézier's *Traité des feux d'artifice pour le spectacle* on his Eastern campaign in 1798, and subsequent fireworks employed painters and architects' *macchine* as grandiose as any of those of the ancien régime.[140] For his coronation in December 1804, a vast representation of the emperor crossing the Alps at the Saint Bernard Pass was constructed, after Jacques-Louis David's portrait utilizing the same theme (plate 16).[141] In 1810, for Napoleon's marriage to Marie-Louise of Austria, the architect Pierre Bénard designed a grand *macchina*, with a Temple of Hymen on a rock, eighty feet high and illuminated with colored lances.[142] Ruggieri subjected this continuing reliance on architects to criticism in the *Elémens*, writing, he claimed, because "the architects who have been in charge of the beautiful fêtes have always sought to emphasize the architecture, even at the expense of the effects of the fire. This vice, of which they are not the only guilty ones, still occurs."[143]

In the *Eleméns de pyrotechnie*, Ruggieri then utilized the same tactics his adversaries had employed earlier in the century, claiming his own skills as authoritative. In a world of domesticated fireworks often figured in books appealing to polite audiences, Ruggieri chose print as the medium in which to forward his claims. Up to now, the Ruggieri had kept their techniques outside the realm of print, but commercial interests and the popular appeal of books on fireworks likely led Claude to publish. Print also created a forum for shaping knowledge. In the *Elémens*, Ruggieri presented pyrotechnic knowledge in a carefully ordered hierarchy. Like the men of letters, Ruggieri identified "artifice," the manufacture and execution of fireworks, as insufficient for true pyrotechnic expertise. Artificers must also have a knowledge of architecture and art to warrant any higher status. But, against the men of letters, Claude claimed that artificers must additionally be skilled in physics and chemistry:

> Artifice, by itself, is nothing more than mechanical work, which demands no more than the measure and name of the materials that it employs; but to merit the name of an artist in this genre, it is also necessary to be a physicist [*physicien*], in order to foresee without having to resort to trials the effects of any operation; a mechanic, for the perfection of a piece that one has invented; an artist and architect, because the pyrotechnist must know how to make the effects of fire agree with all the rules of architecture. . . . Knowledge of chemistry is also of absolute necessity, to combine with certainty the materials that one employs and to make these compositions with the most economy possible.[144]

Contra Frézier and Cahusac, who had argued for allegory, architecture, and painting as the principal elements of pyrotechny, Ruggieri now posed his own skills as essential and frequently referred to the new chemistry throughout the book. Ruggieri's grasp of the new chemistry was not always precise. While Lavoisier replaced the traditional notion of phlogiston with a new substance called *caloric* to explain heat and combustion, Ruggieri confused the two:

> Saltpeter is . . . nothing other than the combination of oxygen gas and others, condensed and reduced to the smallest volume that it may occupy. It is not only the heat named *caloric* that suffices to dissipate the saltpeter and evaporate it into the state of a gas but the fire itself, named *phlogiston*, in action, that is, in a state of causticity. Thus are all the parts of charcoal divided to infinity, which catch fire as fast as they are consumed. It is in this instant . . . that the active fire ignites the oxygen of niter, evaporates it, and reduces it to the state of a gas.[145]

Nevertheless, a vocabulary at the end of the treatise presented much new chemical terminology drawing on the work of Lavoisier and Chaptal.

Ruggieri thus redefined pyrotechnic language just as Lavoisier had recently sought to redefine chemistry through linguistic and methodological reforms.[146] Through strategies such as these, Ruggieri asserted a new proximity of chemistry and fireworks. Allegorical inventions, architecture, and art were necessary in pyrotechny but not sufficient: "Pyrotechny . . . is a dark chaos which one cannot penetrate without the torch of chemistry."[147]

## Green Is the Color: The Chemistry of Colored Fires

The era of philosophical fireworks had given rise to a multitude of new spectacles blending natural philosophical and pyrotechnic effects: optical, mechanical, hydraulic, electrical, volcanic, and aerial. The practical effect of Claude's new assertion of chemical skill in pyrotechnics was to produce another new philosophical firework: colored fire. In both the *Elémens* and his performances, Ruggieri cultivated a new form of pyrotechnics whose novelty centered on the production of brilliantly colored fires using chemistry. In so doing, he transformed the innovations made by Russian artificers in the eighteenth century into a central feature of European fireworks, one that has endured ever since.

In the revised edition of the *Elémens*, Ruggieri thus presented a series of pyrotechnic "discoveries" resulting from his "research" into new effects. Principal among these was what Ruggieri called "Green Fixed Fire for Palm Trees . . . [whose] discovery is due to studious investigations which have led me to imitate the method which the Russians used to make this beautiful fire, whose color rivals that of nature."[148] Green fire was, indeed, a Russian invention, one that Ruggieri began to imitate following reports from his friend Garnerin, who had visited St. Petersburg and Moscow with his wife, Jeanne-Genevieve, in 1803 and 1804, performing there the first balloon ascents in Russia.[149] Garnerin and Ruggieri evidently assumed that Russia was a nation superior to France in its fireworks since Garnerin subsequently explained, "that the Russian artificers were not as skilled as some wanted to believe, and that he had seen nothing in St. Petersburg that he had not already seen in Paris. Only one object had appeared remarkable to him; namely, the representation of a palm tree with large leaves, and in a fire whose color rivaled that of nature."[150]

Thus, the French now discovered the effect that had been a hallmark of Russian pyrotechnics since the middle of the eighteenth century—color. As we saw in chapter 5, after Danilov's creation of a green fire in the 1750s, the recipe had been divined by the artillerist Melissino and the chemist Model. The technique remained prominent in displays. In 1824, the Russian artillery lieutenant Fedor Cheleev published a fireworks manual at St. Petersburg in which he celebrated the Russian discovery of green fire in the eighteenth century: "What nature has up to this time hidden from us Artillery General Melissino . . . discovered—a spiritous green fire—and composed from it a representation of palm trees."[151] Danilov's name was not mentioned, but Cheleev revealed Melissino's recipe, which used "6 pounds of Venetian verdigris [*iari Venitseiskoi*], 2 pounds of sal ammoniac [*nashatyriu*], 1 pound of mercuric chloride [*sulemy*], 2 pounds of saltpeter, and 20 *shtof* [24.6 liters] of alcohol [*spirta*]."[152] Cheleev indicated that this composition produced a "lively and bright but very pale" green that faded to green-blue. Claiming that Melissino's recipe was still secret in 1792, he tried to replicate and improve it himself, using equal parts of verdigris and copper sulfate, with sal ammoniac dissolved in alcohol. Cotton fuses coated in the resulting solution were then laid on hooks inserted in the wooden figure of a palm tree, which could be hoisted up by ropes during a display, when it would ignite in a brilliant green "palm fire" (*pal'movyi ogon*).[153]

Garnerin may have witnessed such palm fires at fireworks for the wedding of Grand Duchess Maria Pavlovna in August 1804 since Cheleev recalled that one General Major Makoveev had displayed them using an improved version of Melissino's recipe during the celebrations.[154] Certainly, Ruggieri's recipe was similar to the Russians', using four parts crystallized verdigris, two parts copper sulfate (or "blue vitriol"), and one part sal ammoniac. Ground into a paste with alcohol, Ruggieri explained, the mixture was diluted with alcohol half an hour before it was due to be shown. Cotton threads soaked in the mixture were hung on nails outlining the shape of a palm tree cut from a wooden board, and, when ignited, this produced "a superb green flame."[155]

Ruggieri claimed to have first shown his green fire in June 1810, at the Neuilly château of the princess Borghèse, for the wedding of Napoleon and Marie-Louise. This was the same firework at which the architect Bénard had displayed a great Temple of Hymen, and Ruggieri's green fire was likely intended to show the force of chemical innovations against architectural marvels. Thus, in addition to the green palm fire, Ruggieri showed transparent portraits of the emperor and the empress, "illuminated by the light of oxygen gas in which phosphorus was burned."[156] He was assisted by one citizen Beyer, an *artiste-physicien* and Paris's superintendent of lightning rods who owned an atelier near Ruggieri's on the rue de Clichy. Beyer and Ruggieri also imitated "the noise of artillery, by using a detonating salt discovered by Berthollet."[157] This was likely oxymuriate of potash, or potassium chlorate, which became the principal ingredient in colored fireworks

Plate 1. Pyrotechnic monk and nun. From the manuscript of Johann Moritz I von Nassau-Siegen (1606–79), *Etliche schöne Tractaten von aller-/handt Feüerwercken und deren Künstlichen Zubereitung. . . Anno 1610*. Bildarchiv Preussischer Kulturbesitz/ Art Resource, New York.

Plate 2. Joseph Furttenbach the Younger (painter), *Fireworks in the Garden of the Merchant Johann Khoun, in Ulm, 1644*. Oil on canvas, ca. 1645. Germanisches Nationalmuseum, Nuremberg.

Plate 3. Mikhail Vasil'evich Danilov (painter), fireworks in St. Petersburg to celebrate the birthday of the empress Elizabeth on December 18, 1741. Watercolor, gouache, and gilding on paper. National Library, St. Petersburg, Russia/the Bridgeman Art Library.

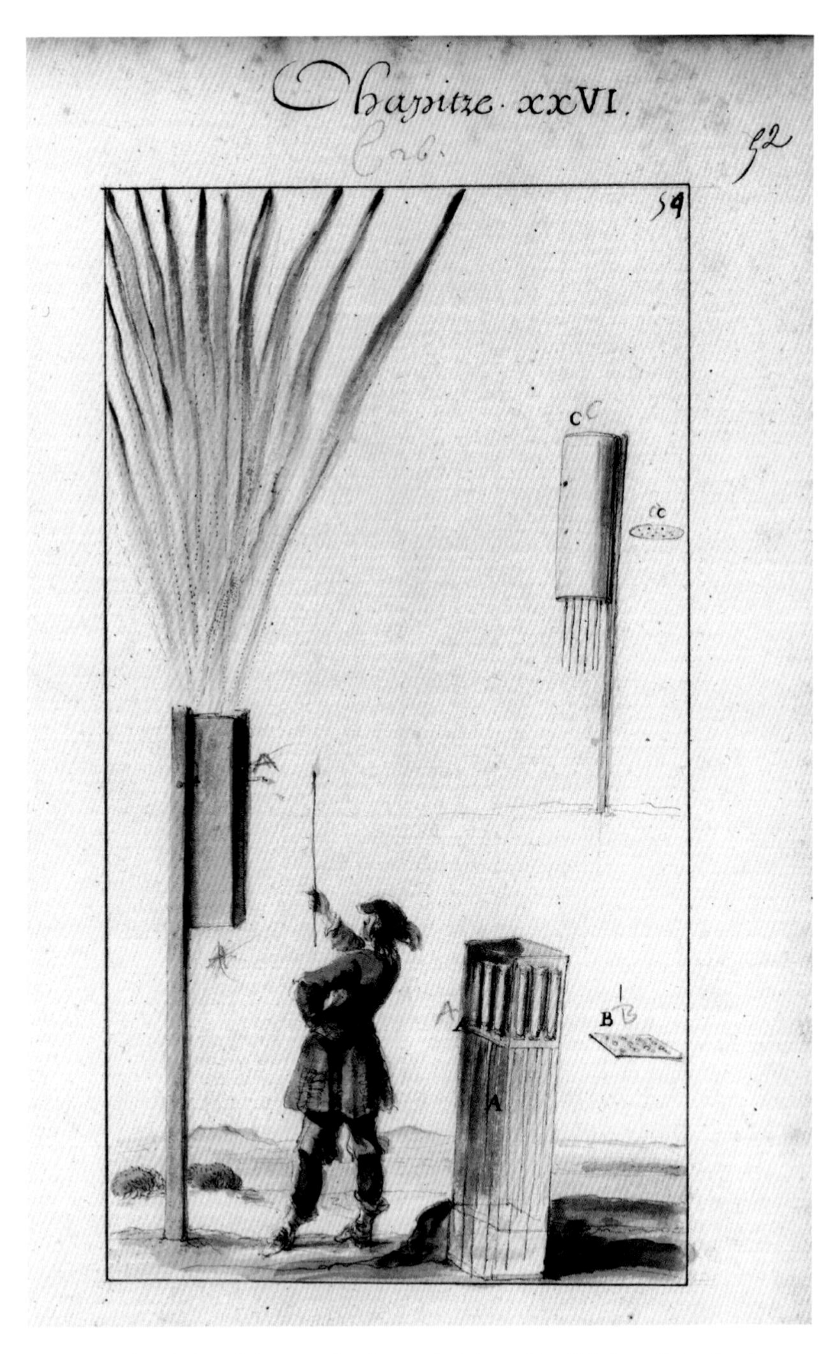


Plate 4. Stefano della Bella's watercolor shows a Parisian artificer lighting a flight of rockets, or *gerbe,* from a chest illustrated beside the *artificier.* From *Traicté des feux artificielz de joye & de recreation* (Paris, ca. 1649), fol. 54 (52). Bibliothèque nationale de France.

Plate 5. *Feu d'artifice donné à Versailles par Louis XIV*. Oil on canvas, n.d. Musée Lambinet—Versailles.

Plate 6. J. Blondel (engraver), *Veue générale des décorations, illuminations et des feux d'artifices, de la feste donnée par la ville de Paris sur la riviere de Seine en presence de leurs majestés le vingt neuf auost mil sept cent trente neu à l'occasion du marriage de Madame Louise Elizabeth de France, et de Dom Philippe Infant d'Espagne* (Paris, 1739). Erich Lessing/Art Resource, New York.

Plate 7. A hand-colored engraving of Servandoni's fireworks in the Green Park, published by H. Overton. *A View of the Public Fire-Works to Be Exhibited on Occasion of the General Peace Concluded at Aix la Chappelle, October 1748* (London, ca. 1749). Bibliothèque nationale de France.

Plate 8. Dutch optical cabinet, late eighteenth century. Photograph by the author, courtesy of Marlborough Rare Books, Ltd.

Plate 9. The domestic setting of philosophical performance. Joseph Wright of Derby, *An Experiment on a Bird in the Air Pump*. Oil on canvas, 1768. © The National Gallery, London.

Plate 10. *Gallant Gentlemen and Lady Fireworkers* (Augsburg [?], ca. 1746–55). Germanisches Nationalmuseum, Nuremberg.

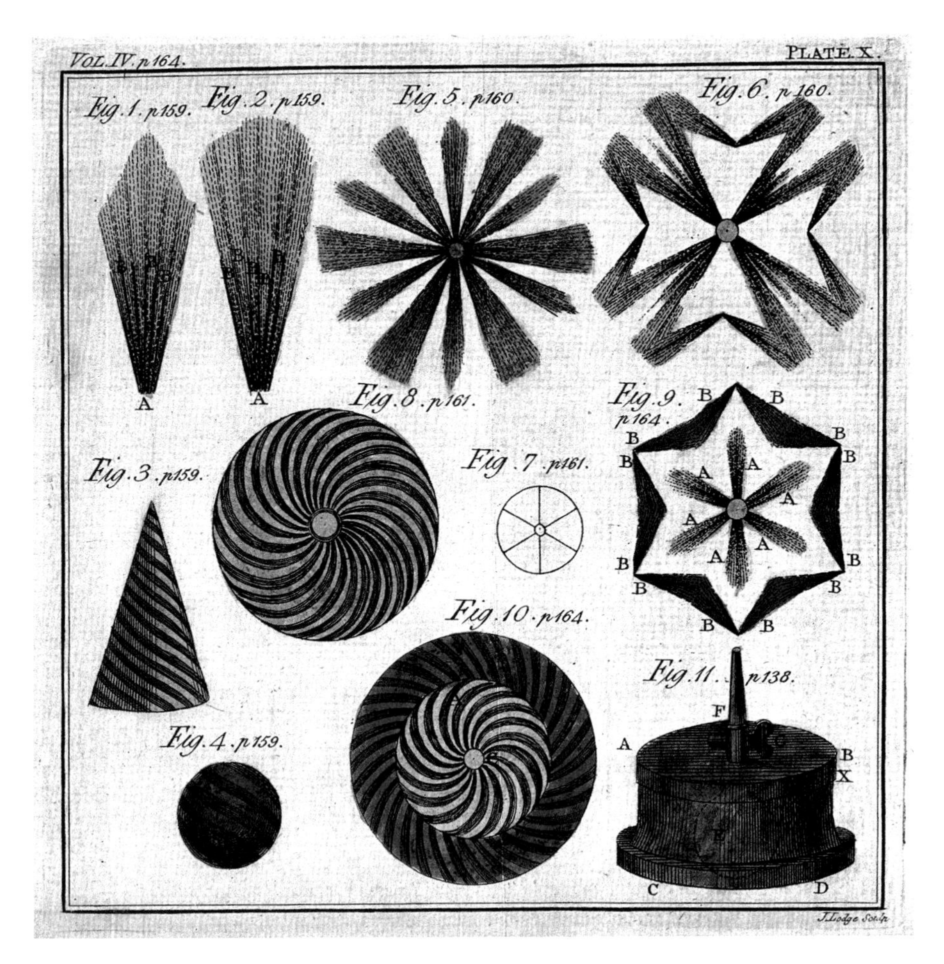


Plate 11. Hand-colored engraving of optical fireworks by J. Lodge, showing *jets de feu*, crosses, spirals, cones, and stars. Plate 10 from William Hooper, *Rational Recreations, in Which the Principles of Numbers and Natural Philosophy Are Clearly and Copiously Elucidated*, 4th ed., 4 vols. (London, 1794), 4:opp. 164. Wellcome Library, London.

Plate 12. *Vue d'optique* by the Basset family of rue Saint Jacques, Paris, showing fireworks celebrating the Treaty of Paris. French and Italian artificers competed at this display, captured in the engraving in typically bright colors for viewing through a lens. *Vue perspective du feu d'artifice qui doit etre tiré sur l'eau pres la Place de Louis XV en rejouissance de la paix le vingt deux juin 1763* (Paris, ca. 1763). Bibliothèque nationale de France, Paris.

Plate 13. Shadow theater with scenes of fireworks. Painted glass in oak frame, late eighteenth century. © Sotheby's, Amsterdam.

Plate 14. Apparatus for imitating fireworks with inflammable air, ca. 1790s, consisting of brass burners in a mahogany box, manufactured by W. & S. Jones of Holborn, London.

Plate 15. Francesco Piranesi produced a series of very large gouache paintings of fireworks, including this one of a display by Michel-Marie Ruggieri in the Jardin de Luxembourg, 1800. Francesco Piranesi, *Représentation du feu d'artifice donné par le Sénat, le XXV brumaire an XIII*. © Roger Viollet/The Image Works.

Plate 16. A hand-colored lithograph of fireworks for the coronation of Napoleon, December 16, 1804. *Superbe feu d'artifice représentant le Mont Saint-Bernard, erigé et tiré sur l'eau en façade de l'Hôtel de ville de Paris. A l'occasion du glorieux avenement de Napoleon Ir . . . le dimanche 26 frimaire de l'an 13*. Bibliothèque nationale de France.

in the nineteenth century. Discovered by Claude-Louis Berthollet in 1786, the compound had been used for artillery and gunpowder since the late 1780s, having earlier been considered too volatile to be used in fireworks.[158] Although it would be some years before potassium chlorate was used to produce colors, a proposal first made in print by the Philadelphia chemist James Cutbush in 1822, Ruggieri was evidently experimenting with it a decade earlier.[159]

Following the Neuilly display, Ruggieri regularly showed the green palm fire at the Tivoli gardens and elsewhere.[160] There were other colors too with similar histories to green. Early pyrotechnics made pale colors, but the addition of potassium chlorate and other volatile metal salts made them brilliant and intense from now on. Zinc filings produced a fine blue flame, as did crude antimony; amber, pitch resin, and purified salt ("Muriate of Soda") produced yellow; lampblack and lycopodium produced red. In 1822, a year after the appearance of Ruggieri's recipes for these colors in the *Elémens*, Cutbush described the use of nitrate of baryta (barium nitrate), metallic arsenic, and oxymurate of potassa (potassium chlorate) as a means to produce green fire, which should be burned "in a reflector" to make a "beautiful green light."[161] Cutbush's 1825 treatise *A System of Pyrotechny* explained his intention of imitating green fire with potassium chlorate, which he continued to associate with the Russians: "This salt, mixed with other substances, will produce the green fire of the palm-tree, in imitation of the Russian fire."[162] Thereafter, potassium chlorate became a common ingredient in fireworks, while other color compositions using metallic salts multiplied, until a systematic presentation of variously colored fires was offered by the artificers F. M. Chertier and Paul Tessier in the 1830s.[163] These provided the basis of color compositions for the remainder of the nineteenth century. Thus the trajectory of green fire from the hands of the Russian artillerist Danilov to the widespread use of color in the art of pyrotechny. It is, however, Claude-Fortuné Ruggieri's name that today is associated with the introduction of color into fireworks.[164] His efforts to secure credit using chemistry evidently worked.

## Conclusion

Claude-Fortuné Ruggieri's colored fires emerged from the fashionable culture of philosophical fireworks that arose in the second half of the eighteenth century. After natural philosophers captured the attention of the public with spectacles of electric fire, philosophy and fireworks both enjoyed a fashionable status. Appealing to a growing middle-class audience, performers, whose careers often took in both pyrotechnic and philosophical pursuits, turned to an array of new venues such as theaters and pleasure gardens, to present spectacles imitating or mixing traditional fireworks with natural effects. Displays were varied. Some appealed to polite sensibilities with miniaturized and sanitized commodities suitable for consumption in the home. Others, such as Torré's artificial volcanoes, appealed to sublime sensibilities and a new taste for thrilling spectacles. But all philosophical fireworks sought to offer distinction to polite consumers from both the grand courtly fireworks of tradition and from the vulgar and increasingly volatile crowds who frequented them. The court was by no means absent from the scene of philosophical fireworks, and, indeed, royal patronage remained a powerful resource for artificers, but philosophical and commercial fireworks eclipsed the court's great displays toward the end of the eighteenth century. Philosophical fireworkers also sought to provide both education and entertainment and did so in a competitive market culture where artifice provided a living for authors, savants, artificers, and instrumentmakers. In the *Encyclopédie*, Diderot had urged the learned to reform the arts, prompting historians to identify this era as one when savants brought artisans under a disciplinary gaze. But, in the marketplace, artificers were principally concerned with the activities of other artificers, and, while savants certainly policed their works, they equally offered attractive new techniques and effects that could be exploited to compete with rivals. In the theaters and pleasure gardens, natural philosophy was as much a resource for artisans as it was a force for controlling them.

Claude Ruggieri's fortunes lay in just such an exploitation of philosophical credit. In the nineteenth century, the great *macchine* and allegorical decorations of architects and painters typical of courtly firworks would be replaced by the abstract patterns of colored fires that continue to feature in fireworks displays today. This change was due in part to Claude's efforts to reshape the art of pyrotechny into a form more sympathetic to the skills of the artificer. The Ruggieri brothers' links to the chemists provided the initial expertise for making colored fire, a move prompted by intense competition in the new marketplace for fireworks, but a young Antoine Lavoisier failed to meet their demands. Only later, in the age of Napoleon, did color emerge as a viable component of pyrotechny, intended to exemplify Claude's assertion that a knowledge of chemistry and physics was absolutely necessary for legitimate pyrotechny. As for the gunners of the seventeenth century, appeals to the sciences could raise the status of the fireworker. Ruggieri thus succeeded in overturning the hierarchy of pyrotechnic labor as it had stood since the age of Louis XIV. After Ruggieri, the pyrotechnist would appear regularly as an artist skilled in "applied chemistry," while fireworks recipes appeared in books of "chemical technology."[165] Ruggieri also relied on the Russians in making his colored fire. His efforts highlight the high regard with which Western Europeans held Russian fireworks in the eighteenth century and the importance of the Russian tradition in shaping, not only interactions of art and science in the age of enlightenment, but also a form of pyrotechnics that continues to delight audiences today.

CONCLUSION

# The Geography of Art and Science

Fireworks played an important role in the life of Europeans from the Renaissance to the age of Napoleon. Besides the pleasure and entertainment that they provided for countless spectators, they offered tools of power, distinction, and diplomacy to nobles and courtiers and means of distinction and expression to the lower and middle classes. Fireworks provided an arena for articulating notions of politics, religion, economy, and history in brilliant spectacles and ingenious displays. They also constituted a vibrant practice around which Europeans debated the identities of science and art, their relations, definitions, and antagonisms. This book has followed the career of fireworks to reveal these varying and complex histories of science and art.

The first two chapters followed the rise of "artificial fireworks" in the courts of Europe and the gunners' efforts to raise the status of their art. Pyrotechny was one of various mechanical arts that gained a higher status during the Renaissance as practitioners utilized *scientia*, the liberal knowledge of rhetoric, mathematics, and natural philosophy, to mediate their skills to courtly audiences. Gunners transformed both the knowledge and the practice of pyrotechny in this manner, reshaping a military, alchemical craft into a discourse of ingenious invention and a practice producing abundant new pyrotechnics. Fireworks then became a forum where reading, patronage, and attendance at the scenes of performance brought together communities of courtiers, gunners, and natural philosophers, prompting a substantial importation of pyrotechnic knowledge and practice into the realm of learning. Pyrotechny, like other arts, thus came to play a significant role in the reconstruction of natural philosophy traditionally identified as the Scientific Revolution. As gunner-artificers were obtaining a new level of fame and credit for their endeavors, so their fireworks became resources for new sciences. These ranged from speculative philosophies that likened meteors, celestial bodies, and physiology to the actions of pyrotechnics to new magical and experimental philosophies engaging with nature through practical interrogations of artifice and the variation of inventions and artificial effects to novel ends and the comprehension of nature. Fireworks played a significant role in these enterprises, informing the work of figures such as Paracelsus, Descartes, Bacon, Boyle, Newton, and Leibniz.

The history of fireworks thus affirms the Zilsel thesis that crafts made a significant contribution to the creation of a "new science" in the sixteenth and seventeenth centuries. But pyrotechny's career also shows the limits of circumscribing this exchange within the Scientific Revolution. The traffic of skills between fireworks and natural philosophy continued after the seventeenth century and always entailed a complex, shifting, and reciprocal exchange of skills and ideas. Moreover, as the practices of science and pyrotechny became increasingly institutionalized, so a distinctive regionality developed in their relations, producing a geography that traditional accounts of arts in the Scientific Revolution ignored. Chapters 3 and 4 thus examined two of the institutions of the new science, the Royal Society and the St. Petersburg Academy of Sciences, to reveal geographic distinctions in their relations of art and science.

In London's Royal Society, the use of spectacular effects in science was closely linked to shifting political and religious circumstances in the wake of the English Civil War. Tracing the history of pyrotechnic experiments and state festivals during the Restoration and in the aftermath of the Glorious Revolution showed how intimately these activities were connected. Experimental programs reflected the shifting values attributed to fireworks, as Londoners variously celebrated and condemned them. When suspicions were high against Jesuit incendiarism, philosophers claimed that experiments could control fiery spirits but steered away from pyrotechnic associations. When tempers cooled, philosophers felt freer to invoke spectacle in their experiments, and, by the close of the reign of Queen Anne, men like Desaguliers could be both experimenters and fireworkers. Suitably reformulated as evidence of divine, rather than princely, powers, fireworks became prominent in the English discourse of natural theology, while a rhetoric of utility and profit led natural philosophers to engage with artificers in order to turn fireworks into useful inventions.

In contrast, relations of artisans and scholars in Russia in the first decades of the eighteenth century were more distant, despite the Russian Academy of Sciences having a more formal role in the production of fireworks. Science was a relative novelty in Russia, introduced as part of Peter the Great's plans to civilize his Muscovite subjects and make Europe recognize Russian power. Peter's own enthusiasm for fireworks as a tool for teaching his people the rigors of war fed into the courtly culture constructed by his successors in the 1720s and 1730s, with grand fireworks displays becoming a regular feature of court life under Empresses Anna and Elizabeth. Simultaneously, the new Academy of Sciences in St. Petersburg found it difficult to interest the Russian nobility in the sciences, abandoning lecture programs just as scientific lecturing was flourishing elsewhere. Instead, academicians recognized that obligations to the court to design fireworks might be turned to academic advantage. In the 1730s, academicians then exploited their role as designers of fireworks allegories and scenery to bring the sciences to the attention of the sovereign and the nobles. Presenting the Russian state as one that professed "love for the sciences," fireworks obligated audiences to live up to their image, prompting the empress Anna to astronomy and increased funding for the academy. Fireworks thus eclipsed lectures as the means to promote the sciences in Russia, reflecting distinctive local conditions. Moreover, although academicians shared the task of producing fireworks with the Russian artillery, they nevertheless maintained a less interactive relationship with them than did their English counterparts in the Royal Society. On the contrary, eager to ensure that courtly attention was focused on the Academy of Sciences, Russia's academicians worked hard to make artisanal labor in fireworks displays invisible, omitting it from printed records of fireworks that they managed.

France, despite similarities to Russia in its absolutist court culture, offered a third distinctive pattern of relations between art and science, evident in the pyrotechnic culture that developed under Louis XIV. While the uses and representations of fireworks in seventeenth- and early-eighteenth-century Paris echoed those in Russia, with labor erased from view in performances and pictures, this was intended to serve, not the Paris Academy of Sciences, but the king himself, whose ruling image cast him as a unique miracle worker, responsible for all actions in the state. Literature on fireworks presented the elements themselves, not artisans or academicians, working to please the king with fireworks. While Russia joined the pursuit of science with artistic composition in its academy, in Paris the two enterprises were separate. An academy of art and court administration oversaw courtly spectacle, and, if the Academy of Sciences played a role in the king's fireworks, it was in determining and displaying the regular order and laws of nature, against which the king's miracles might stand in magnificent relief.

Three different contexts thus sustained quite distinctive relations of fireworks and the sciences by the early decades of the eighteenth century, each closely related to local circumstances. However, this geography of fireworks, art, and science was never static but changed over time as practitioners circulated about Europe, carrying skills from one place to another. Chapters 5–7 then followed this process as Italian architects and artificers brought new forms of pyrotechny to the European courts, prompting an unprecedented fashion for fireworks in London, Paris, and St. Petersburg. Responses to the Italians' fireworks varied from region to region and followed existing traditions. Parisians reacted through a literary debate, while the English offered practical alternatives to pomp and spectacle. The Russians sought to produce spectacles superior to those of the Italians, to compete for the attentions of the court.

However, these were not absolute or permanent distinctions, and, indeed, in all three cities local artificers resisted the intrusion of foreigners, while the Italians' aristocratic patrons everywhere supported them. Moreover, as the Italians traveled across Europe, they homogenized the style of courtly fireworks displays, which from now on routinely included grand, classical *macchine* and performances of several acts. In England and France, the great cost of impressive, Italian-styled courtly spectacles would lead to their decline in the second half of the eighteenth century, while commercial fireworks, appealing to more middle-class audiences, expanded during the same period. In a new context of commercial and domestic venues for fireworks such as pleasure gardens, theaters, salons, and *cabinets de physique*, a different geography prevailed. In these venues, in London, Paris, and St. Petersburg, philosophers and artisans routinely interacted, and, indeed, the careers of many performers incorporated both scientific and pyrotechnic skills. Simultaneously, performers and consumers, now consisting of both sexes, made much effort to avoid the "vulgar" audiences common to grand public displays of fireworks, with the result that access to venues was restricted, while the effects on display catered to polite tastes and exclusive values.

Fireworks thus shifted from a courtly to a commercial context in the eighteenth century, though there was never any strong division between these realms. Pleasure gardens regularly hosted entertainments for the courts, and, in Russia, commercial gardens belonged to the nobility. Princely patronage remained important in pyrotechny throughout the sixteenth to the nineteenth centuries, and even the most commercially minded artificers sought royal credentials on their trade cards. Nevertheless, audiences for fireworks also changed in this interval, with the rise of a middle-class public having a considerable effect on pyrotechnic culture. Spectators' reactions to fireworks reflected these changes and, indeed, contributed to shaping the divisions of social order. Around 1600, when fireworks were relatively rare events, the court and the nobility claimed distinction from the vulgar through experience and knowledge of fireworks, which were otherwise terrifying spectacles, indistinguishable from natural portents. By the early eighteenth century, fireworks remained fearful, but now a wider audience of nobles approached them nonchalantly, manifesting control of the passions and the body through an artful apathy to dramatic spectacle. The middle classes later in the century continued to seek distinction from the vulgar but now preferred fireworks shown in more exclusive venues and domesticated into cleaner and less dangerous forms. Simultaneously, as knowledge and experience of pyrotechnists' skills became widespread, it no longer served as a means of distinction. With the aesthetics of the sublime, distinction from the vulgar was made to turn on enjoying fear rather than surpressing it, the expression of a pleasurable fear being considered more natural than the artifice of nonchalance.

Definitions of the natural and the artificial were intrinsic to these varying social identities and were, in turn, shaped by ongoing debates over the nature of art, science, and pyrotechny. From the sixteenth century to the early nineteenth, different communities debated the proper combination and order of skills necessary to stage fireworks, rarely agreeing on which of the arts and sciences were essential. Different positions in these debates served the interests of different practitioners. Renassiance gunners took pride in alchemical and mathematical skill and ingenious invention, while men of letters in the age of the Sun King urged that knowledge of allegory and emblems was fundamental in fireworks. For much of the late seventeenth century and the eighteenth, architecture was the most prized skill in pyrotechny, with architects employed across Europe to design elaborate edifices for fireworks. Frézier and Cahusac claimed that a knowledge of letters, classics, and painting rose above architecture in this hierarchy of skills, before Ruggieri argued that physics and especially chemistry were paramount. These debates indicate the danger of claiming that pyrotechny has always been an art of "applied chemistry" and testify to the remarkable variety of scientific and artistic skills that a single practice could entail over two centuries. There is no reason to suppose that other arts and sciences were not similarly variegated in this period.

Different definitions of pyrotechny arose and declined depending on local circumstances and changing fashions, and these in turn affected debates on the nature of relations between the arts and the sciences. Gunners were among the first to claim that a knowledge of art would lead to higher contemplation of nature, in books of mathematical recreations. In these works, authors had little expectation that readers should study recipes and procedures of art in order to make fireworks; rather, they expected readers to delight in the orderly principles and regularities to be found in a body of artisanal knowledge. Such attitudes endured into the eighteenth century in works such as Frézier's *Traité des feux d'artifice pour le spectacle*, which offered mathematical rules in pyrotechny less as practical knowledge than as delightful reading. Against this tradition was another, aimed at setting down the rules of art for practitioners themselves to read. Gunners from the author of the *Feuerwerkbuch* to Perrinet d'Orval thus wrote treatises intended to improve practice by recording and communicating recipes and techniques. In this tradition, artisans themselves identified the areas of pyrotechnic art that they deemed worthy of improvement. With Diderot and d'Alembert's *Encyclopédie*, a different position was proposed, synthesizing these two streams. Now the savant was to use the orderly regularities discernible in art as a guide to art's improvement. Now the learned should manage art, proposing reforms that artisans were expected only to carry out. This was the middling sort's solution to the problem of the relation of art to science, and, as polite and commercial pyrotechny superseded that of the courts, so this philosophy of rational management grew increasingly popular.

It would be wrong, however, to see the second half of the eighteenth century as an era when the artisans were subsumed entirely under the disciplinary gaze of the philosophes. As chapter 7 showed, the principal concern of artisans in the increasingly competitive world of fireworks was other artisans. Local artificers resisted foreign intruders, revealed and imitated their secrets, or competed with alternative spectacles. In the pleasure gardens, Torré and the Ruggieri's spectacles regularly mirrored one another as each tried to outdo his rival. Natural philosophers certainly sought to appropriate the techniques of artificers and to exert control over creditable spectacles, as the cases of Chinese and electric fire indicated. But the sciences served as much as resources for competing artisans as they were constraints. Thanks to the rise of public science, whose credit depended partly on spectacles drawing on pyrotechnic tradition, artificers operated in a culture where natural philosophy was becoming increasingly fashionable. In these circumstances, artificers regularly exploited new philosophical effects to their own ends of distinction and profit, producing a remarkable variety of philosophical fireworks in the second half of the eighteenth century to capture audiences and defeat competitors. Cumulatively, by the beginning of the next century, pyrotechny would be subsumed into a branch of applied science or "chemical technology," but this was the result of many profitable exchanges between philosophy and artifice, whose principal exponent was, not a savant, but the artificer Ruggieri. At the same time, it would be a mistake to claim too much distinction between communities of artificers and natural philosophers during this period since their identities increasingly incorporated the skills of both. Nollet, Torré, Pedralio, Bertholon, and Ruggieri, to name a few, pursued careers engaging variously fireworks, physics, chemistry, and instrumentmaking, and it was only at the beginning of the nineteenth century that the identity of the pyrotechnist began to settle into the image of an artificer skilled in applied chemistry. Moreover, if any community suffered from these developments, it was the architects, whose contributions to fireworks diminished steadily in the nineteenth century and disappeared entirely in the twentieth.

Interactions of the arts and sciences in pyrotechny thus extended well beyond the Scientific Revolution and were by no means limited to a traffic of skills from art into science or even to science and art exclusively. This book has, rather, presented a vibrant culture of multiple interactions and exchanges that extended over several centuries and engaged a variety of different communities. Neither were the locations of these interactions limited to the sites of scientific and artisanal production, such as the workshop, the academy, or the laboratory. Sites of use and consumption, such as churches, city squares, theaters, pleasure gardens, and domestic spaces, were just as important in shaping exchanges between fireworks and natural philosophy. Artisanal travel likewise helped spread techniques and knowledge between communities, most notably bringing the Russian art of colored fire to Western Europe, from whence it became ubiqitous in the nineteenth century.

## Fireworks in the Nineteenth Century: War Rockets and Economic Principles

A brief look at the subsequent career of fireworks in the nineteenth century demonstrates how such interactions continued to develop and vary after the age of Claude Ruggieri. The history of European fireworks began with military pyrotechnics, and, in the early nineteenth century, it returned to them. The growth of philosophical fireworks affected not only fireworks for pleasure but also fireworks for war. In 1805, the English Tory editor and inventor William Congreve set out to apply the methods of science to produce a new form of military rocket, in response to attacks on British troops in India using iron-bodied rockets with bamboo sticks. Imitating the Indian war rockets of Hydar Ali and Tipu Sultan of Mysore, the resulting "Congreve Rockets," great solid-fueled devices weighing up to three hundred pounds, established a tradition of improving military rocketry through the application of scientific principles that has endured ever since.[1] In 1813, William Moore, Congreve's colleague at London's Royal Arsenal, published *A Treatise on the Motion of Rockets [and] an Essay on Naval Gunnery*, the first mathematical analysis of rocket motion.[2] After successful trials on expeditions against Napoleon's French fleet stationed off Boulogne in 1806, Congreve's rockets were mobilized in campaigns around the world, becoming a vital component of European artillery through the nineteenth century, and providing the starting point for developing liquid-fueled rockets in the twentieth.

Financial sensibilities continued to mediate interactions of art and science in fireworks. Like the pleasure garden fireworkers, Congreve made economy a key principle in his work, claiming that the rocket's lack of recoil made it a cheaper and more efficient weapon than traditional heavy ordnance. The same economic concerns did much to reshape fireworks for pleasure in the nineteenth century. In 1830, the Scots geologist John MacCulloch, one of Congreve's collaborators at Woolwich, first proposed the creation of pyrotechnic spectacles using different patterns of similar, mass-produced pieces. Like Congreve, MacCulloch viewed this as an application of scientific principles to fireworks, in which geometrical patterns of colored fires produced a maximum of spectacle for a minimum of cost: "Picturesque arrangement, which costs nothing, is . . . as highly conducive to beauty as it is to economy."[3] Claude Ruggieri's desire to supplant architecture with artifice was now realized with increasing frequency, as economic concerns intensified. Subsequent decades saw a decline in the use of expensive ephemeral architecture in displays, and, while *macchine* continued to appear in the nineteenth century, by 1900 they had mostly disappeared, supplanted by cheaper patterns of differently colored fires.

The nineteenth century's interest in economy might have eliminated the fireworks display altogether, as an act of pure conspicuous consumption, yet fireworks endured as useful commodities, as entertainment for the masses, and as a subject of criticism in polite society. Nineteenth-century political economists, heirs to the Green Park laments of Georgian London, regularly identified fireworks as the epitome of superfluity and waste, while polite pressure groups railed at the dangers and vulgarity of pyrotechnics to emphasize their own civility.[4] Technologies of light, new media of the nineteenth century, also continued the trend toward a commodification of inoffensive spectacles that philosophical fireworkers had cultivated in the eighteenth century. Magic lantern shows regularly imitated fireworks in the nineteenth century.[5] MacCulloch's colleague and publisher David Brewster captured the principles defining economized fireworks in another new invention, the kaleidoscope. The kaleidoscope, Brewster claimed, was designed to produce a maximum variety of spectacles for a minimum of cost: "There are few machines, indeed, which rise higher above the operations of human skill. It will create in an hour, what a thousand artists could not invent in . . . a year; and while it works with such unexampled rapidity, it works also with a corresponding beauty and precision."[6] Early kaleidoscopes were advertised as offering imitations of fireworks, while fireworks came to be seen as imitating kaleidoscopes, showing patterns of light transforming "like the varying designs of a colossal kaleidoscope."[7] Many other inoffensive light shows followed, culminating in the immersive spectacles of photographic projection and cinema whose special effects are today the heirs of the fireworkers' pyrotechnic displays.[8] In sum, the interactions of pyrotechnics and natural philosophy endured in the nineteenth century, testimony to a continuing, dynamic, and reciprocal process of exchange.

## Gone in a Flash: The Disappearing History of Fireworks

The careers of fireworks, art, and natural philosophy that this book has recounted raise a final and critical question. If fireworks were so closely related to the arts and sciences from the sixteenth century through the nineteenth, why have these interactions become so obscure in European history? Why do we hear nothing today of the poetic inventiveness of gunners, the pyrotechnic roots of electrical science, or the bustling theater of enlightened philosophical fireworks? The history of fireworks and the sciences seems to have gone up in smoke, as ephemeral as the spectacles it recounts.

It could be argued that, far from disappearing from history, philosophical fireworks, or at least the middle-class culture of domesticated and rational spectacle to which they belonged, made a deep impression on formative histories of science, creating a discipline in which these resources have become so ubiquitous as to be invisible. Polite spectacle thus helped shape some of the earliest accounts of the history of science, in works such as Joseph Priestley's *History and Present State of Electricity* (1767) and Adam Smith's *History of Astronomy* (1795). Priestley and Smith presented history as itself a rational recreation for the leisured classes that would reveal an order in the progress of ideas that was sublime, pleasant, and civilizing. Priestley wrote: "The History of Electricity is a field full of pleasing objects, according to all the genuine and universal principles of taste. . . . Scenes like these, in which we see a gradual rise and progress in things, always exhibit a pleasing spectacle to the human mind. Nature, in all her delightful walks, abounds in such views."[9]

Priestley entangled history and fashionable aesthetics since the "unlimited increase" latent in the idea of progress equated it with "a prospect really boundless, and sublime."[10] Fields, walks, and prospects all recalled the pleasure gardens, and Priestley went on to decribe the history of electricity as one of "scenes," "drama," and "actors." But, while the history of natural philosophy was a "delightful pleasure," it was also one to which only the learned should have access because supposedly only they were able to comprehend the interrelations of nature's effects and systematically consider their connections. Like philosophical fireworks, the image of science proposed by the historians should be both entertaining and improving. Singular experiments might produce arresting and pleasing effects, but these should always be joined together to form a sense of more general regularities and the progress of ideas. As Smith wrote in his *History of Astronomy*: "Philosophy is the science of the connecting principles of Nature."[11] Smith then contrasted epistemic strategies in terms familiar from the scenes of spectacle, discriminating the fearful response of the vulgar to spectacular meteors, stars, and comets from a civilized and fearless understanding of celestial phenomena achieved through the progress of astronomy. The geography of the pleasure garden was here reconstituted in the history of science but now in *temporal* terms and in a form that subsequently became ubiquitous in histories of the physical sciences. Smith wrote of progress from the savage wonder of unmediated experience "in the first ages of society" to the current refined and civilized consideration of scientific principles that underlay, and so annulled, the wonder of nature's marvels.[12]

History was, thus, part of the marketplace for rational recreations and inherited its values. But new histories, and the rising values of economy and progress, simultaneously erased the particular debts of the sciences to fireworks of the past. Market values thus contributed to a division of labor between the history of science and that of spectacle. As I have shown elsewhere, natural philosophers and entrepreneurs who exploited philosophical fireworks to useful and profitable ends in the early nineteenth century were keen to distance themselves from the spectacular ventures on which their schemes initially drew.[13] Thus, the entrepreneurs William Murdoch and Frederick Winsor both asserted originality in the invention of gaslighting while accusing their rival of merely elaborating an existing invention—the philosophical fireworks made with gas shown at the Lyceum Theatre by Charles Diller in the 1780s. Gaslighting owed much to philosophical fireworks, as did Congreve's war rockets, but these links were obscured by marketplace competitions for patents and profits and by historical accounts that similarly stressed the origins of science and technology in the minds of singular experimenters and inventors, rather than in the shows of palaces, pleasure gardens, and theaters.

Meanwhile, the key statement in the history of spectacle in the nineteenth century was Claude-Fortuné Ruggieri's *Précis historique sur les fêtes, les spectacles et les réjouissances publiques*, published in 1829, which became, and remains, a significant source for the history of festivals and particularly the history of fireworks.[14] For different reasons, Ruggieri also erased most of the intimate interactions of pyrotechny and natural philosophy from this influential account. Written as a companion to his *Elémens de pyrotechnie* and *Pyrotechnie militaire*, the *Précis* was intended as a work demonstrating the forms and practices that public rejoicings and festivals had taken in the past, thus providing a model for future fêtes. Ruggieri presented a discussion of the origins of festival in ancient times, a taxonomy of the various forms of courtly, popular, urban, and provincial spectacles, and accounts of the most prominent fêtes in French history. Ruggieri's choice of what to include and what to exclude needs to be understood in the context of his practice during the late 1820s. Claude and his brother, Michel, had ridden out the turbulent politics of the Revolution and Napoleonic era by producing fireworks spectacles suited to the current regime, and, after the restoration of the monarchy and the coronation of Charles X in 1824, the Ruggieri performed fireworks glorifying the new monarch. But disaster struck in 1825, when an accident at fireworks staged by the brothers on the place Louis XV for Charles led to several deaths and numerous injuries. For the first time, relatives of the dead sued the brothers, who proceded to accuse one another of the accident during a prolonged trial, their fraternal attacks providing scandalous news for the Paris journals. The outcome was a substantial fine and a brief term of imprisonment for both brothers, whose rivalry would endure for many years.[15]

The *Précis* was written in the wake of these events, which caused serious damage to Claude Ruggieri's reputation and his standing with the king.[16] Certainly, the *Précis* read as an apology for Claude's skills and as a eulogy of Charles X since the historical part of the text concluded with a long description of festivals during Charles's reign, which were made to appear more sumptuous and splendid than all those of the past. Furthemore, in attempting to advertise his accomplishments to the king, Claude erased most of the pyrotechnic innovations of his predecessors in favor of emphasizing his own. A description of the Paris pleasure gardens thus identified volcanic and pyric pantomimes with the Jardin Ruggieri while excluding them from a brief account of Torré's gardens. Claude suggested that in the eighteenth century fireworks had been performed with much "slowness and difficulty" compared to his own.[17] The historical part of the *Précis* focused only on royal spectacles, at the expense of lesser establishments such as the *cabinets de physique*, and, while Claude detailed fireworks used in festivals of the past, he did so mostly by following official descriptions, which gave few details of pyrotechnics. In the end, the only philosophical fireworks to remain in the book were Claude's own: notably his colored and green fires. Discussion of these colored fires was delayed until the descriptions of Charles X's festivals, making it seem that the new fireworks had been used only in Charles's reign.[18] The first and most extensive work to recount the history of fireworks thus turns out to have erased much of the recent history of fireworks as Claude sought to appease the king and restore his reputation.

That Claude's was one of only a very few works recounting past festivals in the nineteenth century, and unique in its emphasis on fireworks, points to another reason why the history of fireworks has disappeared from the history of science. Outside the literature of pyrotechnic manuals, the memory of fireworks has existed almost exclusively in the realm of ephemeral literature and imagery—in newspaper advertisements, periodical articles, handbills, pamphlets, trade cards, engravings, and popular prints. Only a handful of books or academic works examining fireworks have been published since Ruggieri's *Précis*, and these have tended to focus on pyrotechnic manuals or official printed descriptions of displays. This book has been written at a fortunate moment when the ease with which ephemeral literature may be identified or explored has been rapidly increasing thanks to new digital technologies. Fireworks thus return to visibility with a change in media, and, for similar reasons, they may have disppeared in the past, in times when most pyrotechnic literature was still *ephemeral*, soon discarded and forgotten about.

Times have also changed with a renewed appreciation of geography in the making of knowledge. Before the advent of histories such as Priestley's and Smith's, the relations of fireworks and the sciences were more routinely made visible by artisans and the learned, who entertained a more spatialized sense of knowledge than their historical successors, often viewing knowledge as a map, a territory, or a domain. Furttenbach, it will be recalled, incorporated the map of mechanical knowledge within the decorations of his fireworks displays. In the early twenty-first century, as the era of histories of scientific progress wanes, historians are restoring this sense of space to history, revealing intersections in culture that have long been invisible. This book has returned to the scenes of performance, the places, practices, and skills of natural philosophy, art, and fireworks, in order to reveal their changing interrelations and shifting geographies.

Such geographies have provided new ways to think about the history of science and art in European history and warrant further efforts to show their interconnected and mutually constitutive histories, which are too often told separately and so obscured. Indeed, even *history of science* is something of a misnomer, erasing the multiple and often unexpected traditions of knowledge and practice that have constituted it.[19] Many of these traditions, local and global, no doubt remain hidden from view and deserve recovery.

ACKNOWLEDGMENTS

This book began as a study of spectacle in Russia and evolved into its current form through immensely enjoyable years at the University of Cambridge, the Max Planck Institute for the History of Science in Berlin, and the University of Washington in Seattle. Many colleagues and friends contributed essential criticisms, ideas, and support through this time, and I am very grateful to them all.

I am extremely fortunate to have worked with Simon Schaffer in Cambridge. Ever since I began my graduate career, his remarkable insight, energy, and encouragement have been an inspiration, for which I owe an immense debt. I also thank Daniel Alexandrov, whose enthusiasm and support always made doing research in Russia such a pleasurable experience. Special thanks also to Otto Sibum, in whose capable hands the "experimental knowledge" group at the Max Planck Institute offered ideal conditions for pursuing research. Otto, together with my regular colleagues, David Aubin, Charlotte Bigg, Arianna Borrelli, Frédéric Graber, Karl Hall, Annik Pietsch, Suman Seth, and Richard Staley, read many parts of the manuscript, and I appreciate their excellent advice and criticisms. Thanks also to Trevor Pinch, Andrew Pickering, and Andrew Warwick for their helpful comments, suggestions, and conversations at the Max Planck Institute. Lorraine Daston, Michael Hagner, William Clark, Ursula Klein, Julia Voss, Margarethe Vöhringer, Gabor Zempelen, Mari Dumett, and Horst Bredekamp also provided many valuable contributions and much support in Berlin.

At the University of Washington, Bruce Hevly, Tom Hankins, Karl Hufbauer, Glennys Young, Benjamin Schmidt, and Douglas Smith have all offered thoughtful critiques of the manuscript that have been enormously helpful. I thank my colleagues in the Department of History for their support and members of the department's History Research Group, who provided valuable comments on several chapters. Thanks also to Monica Azzolini, Alison Wylie, and Kathy Woodward for their support. Beyond the University of Washington, I am grateful to Tatiana Artemeeva, Eric Ash, John Bennett, Beryl Curt, Dieter Daniels, Nicholas Dew, Graham Burnett, Patricia Fara, Jan Golinski, Michael Gordin, Oliver Hochadel, Natasha Lee, Michael Lynn, Mikhail Mikeshin, Henry Nicholls, Christelle Rabier, Kapil Raj, Kara Reilly, Lissa Roberts, Olga Roussinova, Pamela Smith, and Jeremy Vetter for their help and critical feedback. Thanks also to Heather Ewing, Rajeev Raizada, Tanzeem Choudhury, Philip Howard, John Tresch, Rebekah Wray-Rogers, and Tanja Weingärtner for their friendship and many stimulating conversations.

It has been a great pleasure to work with my editor, Karen Merikangas Darling, at the University of Chicago Press, and I thank her for her patience, professionalism, and enthusiasm. I am also indebted to two anonymous readers for the press, who provided valuable critiques of the manuscript. Thanks also to my copyeditor, Joseph Brown, and to Abby Collier and Erik Carlson at the University of Chicago Press.

Furthermore, I could not have completed this book without the generous assistance of librarians and archivists at the Library and Archive of the St. Petersburg Academy of Sciences, St. Petersburg; the Bibliothèque nationale de France and the Cabinet des arts graphiques at the Musée Carnavalet, Paris; the Musée Lambinet at Versailles; the British Library, the Wellcome Library, the National Gallery, the Department of Prints and Drawings at the British Museum, and the Public Record Office in London; the National Museums of Scotland, Edinburgh; the New York Public Library; the Beinecke Rare Book and Manuscript Library, Yale University; the University of Chicago Library and Special Collections Center; the University of Pennsylvania Library; the William Andrews Clark Memorial Library, Los Angeles; the Rijksmuseum, Amsterdam; the Royal Library, Copenhagen; the Stadt- und Universitätsbibliothek, Frankfurt am Main; the Germanisches Nationalmuseum, Nuremberg; and the Sächsische Landesbibliothek, Staats- und Universitätsbibliothek, Dresden. Special thanks to the staff of the Hagley Museum and Library, Delaware; the Getty Research Institute in Los Angeles; Art Resource, the Bridgman Art Library, and Image Works in New York for their kind assistance with archives and illustrations. This project benefited from generous grants from the Darwin Trust of Edinburgh; the Getty Grant Program; the Hanauer and Keller funds of the Department of History at the University of Washington; the Simpson Center for the Humanities at the University of Washington; and the Hagley Museum and Library, Delaware, for which I am very grateful.

My family has also been a great source of support throughout this project. This book is dedicated to Anna Märker. Anna patiently read through the manuscript many times, always offering incisive editorial and intellectual guidance, and I cannot say adequately how much her friendship and kindness have meant to me while writing this book.

Unless otherwise specified, all translations into English are my own.

# NOTES

INTRODUCTION

1. Black, *G. F. Müller and the Imperial Russian Academy*, 100–101.
2. Stählin, *Izobrazhenie feierverka i illuminatsii kotoryia v novyi 1748. god*, 5.
3. Frazer, *The Golden Bough*, 731–89; Bachelard, *La flamme d'une chandelle*.
4. Raggio, "The Myth of Prometheus."
5. Livingstone, *Putting Science in Its Place*; Ophir, Shapin, and Schaffer, eds., "The Place of Knowledge."
6. Kaufmann, *Toward a Geography of Art*.
7. Withers, *Placing the Enlightenment*; Ogborn and Withers, *Georgian Geographies*.
8. Principal works have included Zilsel, *The Social Origins of Modern Science*; and Rossi, *Philosophy, Technology, and the Arts*. For an overview, see Cohen, *The Scientific Revolution*, 322–51.
9. Zilsel, *The Social Origins of Modern Science*, 82.
10. Smith, *The Body of the Artisan*. See also Smith, "Art, Science, and Visual Culture."
11. For example, Bennett, "The 'Mechanics' Philosophy and the Mechanical Philosophy"; Sibum, "Reworking the Mechanical Equivalent of Heat"; and Schaffer, "Experimenters' Techniques."
12. Findlen, "Jokes of Nature," and "Between Carnival and Lent"; Biagioli, *Galileo, Courtier*, 103–59.
13. Shakespeare, *As You Like It*, act 2, scene 7.
14. Schaffer, "Natural Philosophy and Public Spectacle." See also Bensaude-Vincent and Blondel, eds., *Science and Spectacle in the European Enlightenment*, 7–8.
15. On changing meanings of art and science, see Dear, *The Intelligibility of Nature*; and Werrett, "The Techniques of Innovation."
16. For the early history of Chinese fireworks, see Needham, *Military Technology*; Partington, *A History of Greek Fire and Gunpowder*, 237–97; and Ling, "Gunpowder and Firearms in China." On early fireworks in India, see Mitra, *Fire Works and Fire Festivals in Ancient India*; and Gode, *The History of Fireworks in India*.

CHAPTER ONE

1. On the lowly and disreputable status of gunners, see Liebe, "Die soziale Wertung der Artillerie"; and Walton, "The Art of Gunnery," 280.

2. On Suenck (also known as Christoffer Schwenke), see Wade, *Triumphus Nuptialis Danicus.*
3. See the title page of Babington, *Pyrotechnia.*
4. Francis Malthus, *A Treatise of Artificial Fire-Works,* dedicatory verses by the mathematician William Bastian. Malthus described himself as "Gentil-homme Anglois, Commissaire general des Feux & Artifices de l'Artillerie de France, Capitaine general des Sappes & Mines d'icelle, & Ingenieur [d]és Armées du Roy" (Malthe [Malthus], *Practiqve de la gverre,* title page).
5. On the emergence of other genres of political spectacle in the Renaissance, see, e.g., Orgel, *The Illusion of Power;* Strong, *Art and Power;* and Mulryne, Watanabe-O'Kelly, and Shewring, eds., *Europa Triumphans.*
6. "Description des artifices et magnificence faictes à Bordeaux avec le combat et les feux artificiels de sieur Morel et Jumeau, representez sur la Garonne, en la presence de leurs Majestez" (Larroque, ed., *Une fête bordelaise,* 3).
7. As the artisan-author John Bate noted of crackers in 1635: "It is well knowne that every boy can make these" (*The Mysteries of Nature and Art,* 122). For one of the earliest European references, see Bacon, *Opus Maius,* 2:629–30. On Chinese fireworks, see n. 16 of the introduction.
8. On the earliest incendiaries and the introduction of gunpowder weapons into Europe, see Partington, *History of Greek Fire and Gunpowder,* 174 (on the Chioggia bombardment); Hall, *Weapons and Warfare in Renaissance Europe,* 41–66; and Buchanan, ed., *Gunpowder.*
9. See, e.g., the images of the sixteenth-century Ranuzzi pyrotechnic manuscript in Wells, "A Flourish of Pyres."
10. The *Annales* of C. Pulce quoted in "A Spectacle at Pentecost, Vicenza, 1379," 432. See also Schaub, "Pleasure Fires," 39.
11. Ogden, *The Staging of Drama in the Medieval Church* 104–7, 221 n. 69; Muir, *The Biblical Drama of Medieval Europe;* Butterworth, *Theatre of Fire.*
12. Butterworth, *Theatre of Fire,* 42, quoting Abramo, the Russian bishop of Suzdal, discussing the Florence Annunciation of 1439.
13. Biringuccio, *Pirotechnia,* 440–42. See also Schaub, "Pleasure Fires," 46–52.
14. This view follows Brock, *A History of Fireworks,* 39–41, and informs, e.g., Schaub's "Pleasure Fires."
15. Marcigliano, "The Development of the Fireworks Display," 1–2.
16. See, e.g., Fagiolo dell'Arco, *L'effimero barocco,* 1:95.
17. Tassoni, *L'Isola Beata,* 9–10, quoted (in translation) in Marcigliano, "The Development of the Fireworks Display," 8.
18. M. Savonarola quoted (in translation) in Marcigliano, "The Development of the Fireworks Display," 7.
19. Lotz, *Das Feuerwerk,* 12, 22; Boorsch, *Fireworks!* 15.
20. On early German fireworks, see Fähler, *Feuerwerke des Barock.*
21. Schaub, "Pleasure Fires," 40–41.
22. The print, by Hans Sebald Beham of Nuremberg, is discussed in *Carlos V,* 400–402.
23. Fähler, *Feuerwerke des Barock,* 48; Lotz, *Das Feuerwerk,* 21, 22, 29, 33, 43; Schaub, "Pleasure Fires," 60.
24. One of the most celebrated was a naumachia on the Rhine at Düsseldorf in June 1585 for the marriage of Duke Johann Wilhelm of Berg. See Graminäus, *Fvrstliche Hochzeit;* Fähler, *Feuerwerke des Barock,* 88–93; and Sievernich and Budde, *Das Buch der Feuerwerkskunst,* 38–40.

25. A detailed account of these German guilds remains to be written and is beyond the scope of this book, but see Fähler, *Feuerwerke des Barock*, 72–75, 209; Lotz, *Das Feuerwerk*, 29, 44; and Schaub, "Pleasure Fires," 44–45.
26. On the history of the Girandola, see Cavazzi, *"Fochi d'allegrezza" a Roma*, 83–97. For descriptions of the displays, see, e.g., Schiavo, *Il diario romano*, 58–59; and Biringuccio, *Pirotechnia*, 442–43.
27. Biringuccio, *Pirotechnia*, 442.
28. Ibid., 443.
29. Brambilla quoted (in translation) in Salatino, *Incendiary Art*, 78.
30. "Immediately after the tribulation of those days shall the sun be darkened, and the moon shall not give her light, and the stars shall fall from heaven, and the powers of the heavens shall be shaken: And then shall appear the sign of the Son of man in heaven" (Matt. 24:29; see also Mark 13:24–25). On the portentousness of such imagery, see Schechner Genuth, *The Birth of Modern Cosmology*, 38–45.
31. In 1594, Cardinal Pietro Aldobrandini founded the gunners' guild or Compagnia dei Bombardieri di Castel S. Angelo, which staged subsequent displays.
32. Biringuccio, *Pirotechnia*, 446 (see 444–46 generally).
33. On early fireworks in France, see Christout, "Les feux d'artifices en France"; Weigert, "Les feux d'artifice"; and Lotz, *Das Feuerwerk*, 56–78.
34. On Morel, see Lignereux, *Lyon et le roi*, 434; and McGowan, *L'art du ballet de cour*, 119 n. 19, 126–29, 138. Other gunners staging fireworks at this time were Jumeau, *artificier du roi* and *armurier*. See Fournier, ed., *Variétés historiques et littéraires*, 6:13–14; and Beaurepaire, *Louis XIII et l'assemblée des notables à Rouen*, 22, 32, 54–57. Bagot was "ordinary artillerist to the King [*Artilier ordinaire du Roy*]" (*Discovrs sur les triomphes*, 8). Denis Caresme was *artillier et artificier de la ville de Paris* (Canova-Green, "Fireworks and Bonfires," 146).
35. *Réception de tres-chrestien, tres-iuste, et tres-victorieux monarque Louys XIII*, quoted (in translation) in Boorsch, *Fireworks!* 16.
36. [Morel], *Sviet dv fev d'artifice*.
37. Christout, "Les feux d'artifices en France," 248–52.
38. Nichols, *The Progresses . . . of King James the First*, 2:84, 90–91, 162.
39. Ibid., 90–91.
40. Ibid., 93.
41. *The Magnificent Marriage of the Two Great Princes*, n.p.; Taylor, *Heauens Blessing, and Earths Ioy*, n.p.
42. Lucar, "A Treatise Named Lucar Appendix," 83. See also Biringuccio, *Pirotechnia*, 443.
43. For star recipes, see, e.g., Malthus, *A Treatise of Artificial Fire-Works*, 83–85; Hanzelet, *La pyrotechnie*, 256–57, and *Mathematicall Recreations*, 268–69; and Babington, *Pyrotechnia*, 11–14.
44. Richard Wright, "Notes on Gunnery" (1563), Society of Antiquaries, London, MS 94, transcribed in Walton, "The Art of Gunnery," 357–88. The calendar appears in Walton on 386–87 (fol. 36).
45. Fulke, *Meteors*. Heninger (*A Handbook of Renaissance Meteorology*, chap. 3) describes "the phenomena of exhalations."
46. Malthus, *A Treatise of Artificial Fire-Works*, 82. For other rain recipes, see, e.g., Babington, *Pyrotechnia*, 14; and Bate, *The Mysteries of Nature and Art*, 112.
47. Hanzelet, *Mathematicall Recreations*, 269.
48. Hanzelet, *La pyrotechnie*, 255.

49. Smith, *The Body of the Artisan*, 140–47.
50. Smith, *The Body of the Artisan*, 143–44; Long, *Openness, Secrecy, Authorship*, 179–81; Newman, *Promethean Ambitions*, 127–32; Eamon, *Science and the Secrets of Nature*, 117–20.
51. For a reproduction of the German manuscript, see Hassenstein, ed., *Das Feuerwerkbuch von 1420*, 41–78. My translations follow the English version translated by Kramer, *The Firework Book*.
52. Hassenstein, ed., *Das Feuerwerkbuch von 1420*, 47; Kramer, *The Firework Book*, 26.
53. Franz Helm, *Buch von den probierten Künsten* (Cologne, 1535), Universitätsibliothek Heidelberg, MS cpg 128. The fireworks platform is shown on fol. 66v. For background, see Leng, *Ars belli*, 1:334–50, 2:154–55.
54. Hassenstein, ed., *Das Feuerwerkbuch von 1420*, 47; Kramer, *The Firework Book*, 26.
55. Hassenstein, ed., *Das Feuerwerkbuch von 1420*, 43; Kramer, *The Firework Book*, 22 (quote), 71 (the attribution to Geber).
56. Hassenstein, ed., *Das Feuerwerkbuch von 1420*, 47, 60; Kramer, *The Firework Book*, 21, 34–35.
57. Hassenstein, ed., *Das Feuerwerkbuch von 1420*, 71; Kramer, *The Firework Book*, 55.
58. Hassenstein, ed., *Das Feuerwerkbuch von 1420*, 45–46; Kramer, *The Firework Book*, 25. Another source described Berthold as "skillful in Gebers Cookery or Alchimy" (Camden, *Remaines*, 241). On the Berthold legend, see Tittmann, "Der Mythos vom 'Schwarzen Berthold.'"
59. This was certainly the strategy of seventeenth-century chemical laborers. See Smith, *The Business of Alchemy*.
60. Thus, John Dee, J. B. van Helmont, and George Starkey called their alchemical art *Pyrotechny*. The London Paracelsan John Hester sold fireworks—along with medical compositions, the philosopher's stone, and the elixir of life—at his premises by the Thames under the sign "Furnaces" and later "Stillatory" (Bennell, "Hester, John," 887).
61. The alchemical origins of laboratories are familiar, those of the fireworks laboratory less so. See Smith, "Laboratories"; and Shapin, "The House of Experiment," 377–78.
62. Zimmermann's dream, from which the quotations are taken, is reproduced in Romocki, *Geschichte der Explosivstoffe*, 257–63. For background, see Leng, *Ars belli*, 1:353–58, 2:29.
63. Smith, *The Body of the Artisan*, 114–27.
64. Malthus, *A Treatise of Artificial Fire-Works*, 1–2.
65. Hanzelet and Thybourel, *Recueil de plusieurs machines militaires*, bk. 4, p. 7. On Hanzelet and Thybourel's collaboration, see Hodgkin, "Halinitropyrobolia," 12–13.
66. Hanzelet and Thybourel, *Recueil de plusieurs machines militaires*, bk. 4, p. 8.
67. Norton, *The Gunner*, 149 (quote), 150 (on fire as the basis of generation).
68. Siemienowicz, *The Great Art of Artillery*, 99.
69. Biringuccio, *Pirotechnia*, 442. For other wheel recipes, see, e.g., Hanzelet and Thybourel, *Recueil de plusieurs machines militaires*, bk. 5, pp. 22–23; Malthus, *A Treatise of Artificial Fire-Works*, 110–12; Babington, *Pyrotechnia*, 19–25; Bate, *The Mysteries of Nature and Art*, 116–17; and Siemienowicz, *The Great Art of Artillery*, 319–22.
70. Malthus, *A Treatise of Artificial Fire-Works*, 79.
71. Babington, *Pyrotechnia*, 14. For serpent recipes, see, e.g., Hanzelet and Thybourel, *Recueil de plusieurs machines militaires*, bk. 5, p. 16; Norton, *The Gunner*, 151; and Malthus, *A Treatise of Artificial Fire-Works*, 78–80.
72. Smith, *The Body of the Artisan*, 117–22.
73. On early modern automata, see Bredekamp, *The Lure of Antiquity*.

74. Taylor, *Heauens Blessing, and Earths Ioy*, n.p.
75. Smith, *The Body of the Artisan*, 120. For a pyrotechnic salamander, see, e.g., the engraving in *Die triumphirende Liebe* (Hamburg, 1653), reproduced in Boorsch, *Fireworks!* 20.
76. See, e.g., the recipes for human and animal figures in Siemienowicz, *The Great Art of Artillery*, 375–77.
77. Norton, *The Gunner*, 151.
78. For *saucisson* recipes, see, e.g., Malthus, *A Treatise of Artificial Fire-Works*, 88–89; and Bate, *The Mysteries of Nature and Art*, 123–24.
79. For aquatic fireworks recipes, see, e.g., Della Porta, *Natural Magick*, 295–96; Babington, *Pyrotechnia*, 55–59; Hanzelet, *La pyrotechnie*, 259–60; and Siemienowicz, *The Great Art of Artillery*, 182–86.
80. Malthus, *A Treatise of Artificial Fire-Works*, 108.
81. Ibid., 105.
82. Laneham, *A Letter*, 24.
83. *The Mariage of Prince Fredericke, and the Kings Daughter*, fol. A2v.
84. Manzini, *Applausi festivi fatti in Roma*, 42, quoted (in translation) in Salatino, *Incendiary Art*, 54.
85. On the culture of wonders and curiosities, see Daston and Park, *Wonders and the Order of Nature*; and Daston, "Curiosity in Early Modern Science." On the *Kunstkammer*, see Bredekamp, *The Lure of Antiquity*; and Findlen, *Possessing Nature*. On artificial waterworks, see MacDougall and Miller, eds., *Fons sapientiae*; and Werrett, "Wonders Never Cease."
86. On the value of risk for courtly reputation, see Biagioli, *Galileo, Courtier*, esp. 167–69; and Elias, *The Court Society*.
87. Siemienowicz, *The Great Art of Artillery*, 380–81.
88. For honors to a wounded fireworker, see Gordon, *Passages from the Diary of General Patrick Gordon*, 174. On the distressed crowd as an amusement to the court, see Jordan, *London in Luster*, 16; and Waliszewski, *L'heritage de Pierre le Grand*, 274. For reprimands of fireworkers, see Stählin, "Kratkaia istoriia," 247, 253–55.
89. Kemp, "From Mimesis to Fantasia"; Kushner, "The Concept of Invention"; Lee, "*Ut pictura poesis*."
90. Castiglioni, *The Book of the Courtier*, 33. See also Biagioli, "Galileo the Emblem Maker," 236.
91. Dedicatory verses of Richard Rotheruppe, "souldier," to the gunner Thomas Smith's *The Art of Gvnnery*, n.p.
92. Thorndike, *A History of Magic and Experimental Science*, 4:150–82, esp. 172–75 (on Fontana); Alberti, *Ludi rerum mathematicarum*. Valturio's *De re militari* also included descriptions of cannon, siege towers, bridges, and battering rams.
93. Leonardo da Vinci, Codex Madrid I (Biblioteca Nacional de España), fol. 59v, fol. 81v, Codex Arundel (British Library, London), fol. 54r, and RL 12651 (Royal Library, Windsor Castle). See Dibner, *Leonardo da Vinci*, 17–20.
94. Cuomo, "Shooting by the Book"; Henninger-Voss, "The 'New Science' of Cannons," and "Working Machines and Noble Mechanics."
95. On the different forms and uses of sixteenth-century artisanal treatises, see Long, "Power, Patronage, and the Authorship of Ars."
96. Malthus, *A Treatise of Artificial Fire-Works*, 61. Among the many dedications of treatises, Hanzelet and Thybourel's *Recueil de plusieurs machines militaires* was dedicated to Charles de Haraucourt, Baron de Chamblay, artillery general of Lorraine; the works

of Robert Norton were dedicated to King Charles I and the Duke of Buckingham; and John Babington's *Pyrotechnia* was dedicated to the Earl of Newport, master general of the Ordnance.

97. Schmidlap von Schorndorff, *Künstliche und rechtschaffene Feuerwerck* (1560).
98. Siemienowicz, *Artis magnae artilleriae*. A French translation by Pierre Noizet followed as Siemienowicz, *Grand arte d'artillerie*. An English translation was made from the French by George Shelvocke the Younger: Siemienowicz, *The Great Art of Artillery*. For background on Siemienowicz, see Ivashkiavichius, *Kazimir Semenovich*.
99. For details of early printed fireworks treatises, see Philip, *A Bibliography of Firework Books*.
100. Fronsperger, *Von dem Lob dess Eigen Nutzen*. Fronsperger's pyrotechnic works included *Vonn geschütz vnnd Fewrwerck* and *Kriegs Ordnung und Regiment*. On his career, see Leng, *Ars belli*, 1:304–5.
101. On the troubles of credit and piracy in early modern print, see Johns, *The Nature of the Book*.
102. Ufano, *Tratado dela artilleria*.
103. Malthus, *A Treatise of Artificial Fire-Works*, preface.
104. Romano divides fireworks thus:

> First Part—Festive Artificial Fires called Aerial; Simple—Moveable, such as Rockets; Immovable, such as Squibs, (Roman) Candles.
>
> More complicated—Movable, such as Flying Dragons, Balls of Various Kinds, Clubs[.]
>
> Immovable, such as Shields, Towers.
>
> Second Part. Various Balls, burning in the Water. Constructed of Linen, Wood, Metal; Simple (charged with powder only); Loaded—with rockets; with smaller balls. (*Pyrotechnia*, 1–2, quoted [in translation] in Hodgkin, "Halinitropyrobolia," 10)

105. Tho. Stutevill, "In laudem Authoris Ioannis Babington Amici & in Arte Mathesios celeberrima Socii," in Babington, *Pyrotechnia*, n.p.
106. Norton, *The Gunner*, dedication.
107. Ibid., 3–16.
108. Ibid., 4.
109. ibid., 7.
110. Bate, *The Mysteries of Nature and Art*, 93–94.
111. Tartaglia, *Three Books of Colloquies*, dedication.
112. Bourne, *Inuentions and Deuices*, "To the Reader."
113. Biringuccio, *Pirotechnia*, 428.
114. Ibid., 434.
115. Ibid., 434.
116. Ibid., 428.
117. On variety in sixteenth- and seventeenth-century culture, see Hunt, *Garden and Grove*, 83–89; and Shearman, *Mannerism*, 140–51.
118. Biringuccio, *Pirotechnia*, 428–29.
119. Larroque, ed., *Une fête bordelaise*, 7.
120. "The Secrets of Gunmen," English manuscript, MS Ashmole, fols. 128r–139r, Bodleian Library, Oxford, fol. 128r.

121. John Chamberlain, commenting on fireworks on the Thames in 1613, quoted in Cressy, *Bonfires and Bells*, 88.
122. Barclay, *Barclay His Argenis*, bk. 3, chap. 24, p. 228.
123. Norton, *The Gunners Dialogue*, 34.
124. Hanzelet and Thybourel, *Recueil de plusieurs machines militaires*, bk. 5, p. 29 (see also pp. 23, 34).
125. Norton, *The Gunner*, 153, 154.
126. Malthus, *A Treatise of Artificial Fire-Works*, 88–89, 92, 108. It was also proposed that fireworks be arranged according to "whatsoever the fancie of an ingenious head may allude unto" (Hanzelet, *Mathematicall Recreations*, 276).
127. Vasari quoted (in translation) in Pallen, ed., *Vasari on Theater*, 67. See also Lorenzo, *Il teatro del fuoco*, 35–36.
128. *A Description of the Seuerall Fireworkes Inuented. and Wrought by His Maiesties Gonners and What Is. Intended to Be Performed (to the View) in Euerie of. Them, Five Set-Pieces* (ca. 1613), British Library, Department of Manuscripts, Royal 17 C XXXV. The plans were published as Taylor, *Heauens Blessing, and Earths Ioy*.
129. Yates, *Rosicrucian Enlightenment*, 6. For another interpretation of the allegory, see Limon, *The Masque of Stuart Culture*, 110–33. Limon notes (ibid., 110) that the author of the general scheme of the festivities may have been Francis Bacon.
130. [Morel], *Sviet dv fev d'artifice*, 3.
131. Walton, "The Art of Gunnery," 129.
132. Graves, *Thomas Norton*.
133. Babington, *Pyrotechnia*, "To the Reader." On gunnery and print, see also Hale, "Printing and Military Culture."
134. Butterworth, *Theatre of Fire*, 24; Orden, *Music, Discipline, and Arms*, 105 n. 80; Bayard, "Les faiseurs d'artifices."
135. Peacham, *The Compleat Gentleman*, 195.
136. Norton, *The Gunner*, dedicatory verses by Captain John Smith.
137. Camden, *Remaines*, 240–41; Boillot, *Artifices de feu*, 82–83. Christopher Marlowe invoked the image of Mephistopheles with fireworks soon after. See Marlowe, *The Tragicall History of Dr. Faustus*, fol. C2r.
138. Boillot, *Artifices de feu*, "Au Lecteur."
139. See Simms, "Archimedes and the Invention of Artillery and Gunpowder"; Laird, "Archimedes among the Humanists"; and Norton, *The Gunner*, 38.
140. Hogg, *The Royal Arsenal*; Rodzevich, *Istoricheskoe opisanie Sanktpeterburgskogo Arsenala*; Wackernagel, *Das Münchner Zeughaus*; Müller, *Das Berliner Zeughaus*.

CHAPTER TWO

1. Devon, ed., *Issues of the Exchequer*, 107, 114, 157, notes payments to Sir Roger Dallison, lieutenant of His Majesty's Ordnance, of £700, £600, and £1,000, respectively, for "the wages of the gunners, carpenters, and other artificers employed about the provision of fireworks" in 1610–12. The present value was calculated using the Economic History Services online converter, http://measuringworth.com/calculators/ppoweruk. In 1625, the French *artificier* Horace Morel received 1,355 livres for fireworks in a court ballet. See McGowan, *L'art du ballet de cour*, 140.
2. Hall, *Ballistics in the Seventeenth Century*, 164.
3. Pepys, *Diary and Correspondence*, 1:180 (April 24, 1661).
4. Baillet, *The Life of Monsieur Des Cartes*, 29.

5. On the history of natural magic, see Thorndike, *A History of Magic and Experimental Science*; Webster, *From Paracelsus to Newton*; Vickers, ed., *Occult and Scientific Mentalities*; and Eamon, *Science and the Secrets of Nature*.
6. Cardano, *De rerum varietate libri xvii*, 440. On earlier compositions, see, e.g., the works of Fontana, discussed in chapter 1 above.
7. See Della Porta, *Natural Magick*, 289–304 ("The Twelfth Book of Natural Magick: Of Artificial Fires").
8. Ibid., 289.
9. Ibid., 293. On Della Porta in Rome, see Clubb, *Giambattista Della Porta*, 19–21.
10. Della Porta, *Natural Magick*, 294.
11. Consider the range of reading in which the Cambridge scholar Gabriel Harvey sought out recipes on pyrotechnia: "Agrippa's pyrographia, or pyromachia, is unprecedented. Vanocchio's Pyrotechnia stands out, and several experiments on fire by Cardano, De Subtilitate; natural magic furnishes many others . . . Mr Digges Pyrotechny: & Fortification" ("Agrippae pyrographia, sive pyromachia, non est in praedicamentis. Vanoccii Pyrotechnia exstat: et nonnulla Cardani Experimenta pyria, De Subtilitate. Qualia etiam plura suggerit naturalis magia . . . Mr Digges Pyrotechny: & Fortification," marginalia to Harvey's copy of Peter Whitehorne's treatise on gunnery, *Certaine Wayes for the Ordering of Soldiours in Battelray*, quoted in Stern, "The Bibliotheca of Gabriel Harvey," 39).
12. Colie, "Cornelis Drebbel and Salomon de Caus," 249–50, quoting Cornelius van der Woude, *Kronyk der Stad Alkmaar* (Amsterdam, 1725), 7. See also Burton, *The Anatomy of Melancholy*, 2:93.
13. Kircher, *Ars magna lucis et umbrae*, pt. 2, pp. 826–27, and *Mundus subterraneus*, 2:467–80 (bk. 9, sec. 5, pt. 4, "Ars pyrabolica").
14. Schott, *Magia universalis*, 4:91–223 (bk. 2, "De magia pyrotechnica").
15. Findlen, "Inventing Nature."
16. For dragon recipes, see, e.g., Schmidlap von Schorndorff, *Künstliche und rechtschaffene Feuerwerck* (1591), fol. A4v; Ufano, *Vraye instruction de l'artillerie*, 152–54; Babington, *Pyrotechnia*, 36–41; Bate, *The Mysteries of Nature and Art*, 117–18; and Hanzelet and Thybourel, *Recueil de plusieurs machines militaires*, bk. 5, pp. 17–19.
17. Hill, *A Contemplation of Mysteries*, fols. 25r–26v, which explains how dragons emerged when vapors coagulated between cold and hot clouds. As the vapors were repelled by the cold and attracted by the heat, so they formed into thinner or thicker parts to create the shape of the dragon. See also Fulke, *Meteors*, 20–23; and Comenius, *Naturall Philosophie Reformed*, 132–33.
18. Houpreght, *Aurifontina chymica*, 148–49.
19. On the history of dragons in Europe, see Ingersoll, *Dragons and Dragon Lore*.
20. Quoted in Brock, *A History of Fireworks*, 32.
21. "After this, an artificiall fireworke with great wonder was seene flying in the ayre, like unto a dragon, against which another fiery vision appeared flaming like to Saint George on Horsebacke, brought in by a burning Inchantresse, betweene which was fought a most strange battell continuing a quarter of an hower or more; the dragons being vanquished, seemed to rore like thunder, and withal burst into pieces, and vanished" (*The Mariage of Prince Fredericke, and the Kings Daughter*, fol. A2v). See also, e.g., Davies, *Chesters Triumph*, fol. A3v (items 2–5).
22. See, e.g., the prints from the Brock Fireworks Collection of the Getty Research Institute, Los Angeles, reproduced in Salatino, *Incendiary Art*, 13, 16, 35, 43–44, 50.
23. *Lordonnance et ordre*, n.p. See also Hodgkin, "Halinitropyrobolia," iii.

24. In the following description, describing a civic rather than a courtly display, both the crowd and the nobility enjoy one another's distress: "The disorder'd People below in the Street was an excellent Scene of confusion to the Spectators above in the Belconies . . . and the Gallantry above were as pleasurable a sight to the Spectators below, where hundreds of various defensive postures were screw'd, for prevention of the fiery Serpents and Crackers that instantly assaulted the Perukes of the Gallants, and the Merkins of the Madams" (Jordan, *London in Luster*, 16).
25. Nichols, *The Progresses . . . of Queen Elizabeth*, 1:319–20, describing a display at Temple Fields for Elizabeth's visit to Ambrose Dudley, Earl of Warwick.
26. Canova-Green, "Fireworks and Bonfires," 150, quoting (in translation) the anonymous author of an account of the royal entry into La Rochelle. For similar contrasts, see Graminäus, *Fvrstliche Hochzeit*, 4.
27. See Biagioli, "Galileo the Emblem Maker," 236.
28. Laneham, *A Letter*, 16.
29. Hanzelet, *Mathematicall Recreations*, 283.
30. Bourne, *Inuentions and Deuices*, 99.
31. Della Porta, *Natural Magick*, 4. See also Eamon, *Science and the Secrets of Nature*, 221–29; and Daston, "Preternatural Philosophy," esp. 31–32.
32. Wecker, *De secretis*, 690–92, 936–38; Della Porta, *Natural Magick*, 409; Schwenter, *Deliciae physico-mathematicae*, vol. 3; Harsdörffer, *Delitiae Philosophicae et Mathematicae*, 54–55; Kircher, *Mundus subterraneus*, 2:479; Kestler, *Physiologia Kircheriana*, 118–19, 246–47.
33. Kestler, *Physiologia Kircheriana*, 118.
34. Ibid.
35. Ibid.
36. See the many examples in Flemming, *Geschichte des Jesuitentheaters*, 157–67.
37. On church fireworks in Lima, Peru, see Rodríguez-Camilloni, "Utopia Realized in the New World," 44–45. On Jesuit fireworks in India, see Camps, *Jerome Xavier, S.J.*, 190, 229, 231–33, 235.
38. Le Jeune, *Relation de . . . Nouvelle France*, 19. See also Thwaites, ed., *The Jesuit Relations*, 11:66–71, 22:147; and Faillon, *Histoire de la colonie française*, 1:291–92.
39. Le Jeune, *Relation de . . . Nouvelle France*, 18.
40. Harriot, *A Brief and True Report*, E4. On Harriot's work in gunnery, see Walton, "The Art of Gunnery," 15–65.
41. Della Porta, *Natural Magick*, 409.
42. Ibid.
43. Burattini quoted in Targosz, "'Le dragon volant,'" 71. See also Mersenne, *Correspondance*, 15:561, 581; and Hart, *The Prehistory of Flight*, 135–45. Thanks to Nick Wilding for drawing my attention to Burattini's dragon.
44. Cyrano de Bergerac, *Histoire comique*, 25, trans. in Cyrano de Bergerac, *Selenarhia*, n.p.
45. Ibid., 24. The story is widely taken as the first suggestion of using rockets to reach the moon. See, e.g., Bleeker and Huber, eds., *The Century of Space Science*, 1:27. I would argue that dragons, not rockets, were the firework most closely associated with space flight prior to the end of the eighteenth century.
46. Bate, *The Mysteries of Nature and Art*, 117–20.
47. Babington, *Pyrotechnia*, 44–46. See also Hart, *Kites*, 76.
48. William Stukeley, *Memoirs of Sir Isaac Newton's Life, Being Some Account of His Family; & Chiefly of the Junior Part of His Life* (1752), Royal Society, London, MS 142, fol. 38r.

49. Ibid., fol. 38r; John Conduitt's Life of Newton, King's College, Cambridge, Keynes MS 130.2, fols. 22–23.
50. Cyrano de Bergerac, *Histoire comique*, 27, trans. in Cyrano de Bergerac, *Selenarhia*, n.p.
51. Hanzelet, *Recreation mathematicqve*.
52. Heeffer, "*Récréations mathématiques*." For discussion and an alternative attribution to Van Etten, see Hall, *Old Conjuring Books*, 83–119; and Eamon, *Science and the Secrets of Nature*, 307–8.
53. Hanzelet, *Mathematicall Recreations*, "The Epistle Dedicatory."
54. Ibid., "By Way of Advertisement."
55. Ibid.
56. Ibid., "The Epistle to the Reader." The attribution to Malthus has been based on the last paragraph of the book, which states: "The making of . . . rare and excellent Fireworkes, and other practises, not only for recreation, but also for service: you may finde in a booke intituled *Artificiall Fire-Workes*, made by Mr. *Malthus* (a master of this knowledge) and are to be sold by *Rich. Hawkins* at his shop in *Chancery Lane* neare *Sarjants Inne*" (ibid., 285–86). For the section "Artificiall Fire-Workes," see ibid., 265–86.
57. Ibid., "The Epistle to the Reader."
58. Guerlac, "The Poet's Nitre"; Debus, "The Paracelsian Aerial Niter"; Jankovic, "Meteors under Scrutiny," 41–45.
59. Aristotle, *Meteorologica*; Jankovic, "Meteors under Scrutiny," 17–32.
60. Paracelsus, *Opera Bücher und Schrifften*, 2:32, 82–83. On Paracelsus and the arts, see Smith, *The Body of the Artisan*, 82–93.
61. Wallis, "A Letter of Dr. Wallis to Dr. Sloane," 317. See also Locke, "Elements of Natural Philosophy," 199.
62. Newton, *The Mathematical Principles of Natural Philosophy*, 2:202.
63. Biringuccio, *Pirotechnia*, 411. Siemienowicz proposed that the gunner should be someone "of an Intrepidity not to be shaken by the horrid Bellowing of Cannon, nor dismayed at the Tempestuous Iron Hail projected from those merciless Thunderers" (Siemienowicz, *The Great Art of Artillery*, 173).
64. This image endured into the eighteenth century. See, e.g., Barlow, *Meteorological Essays*, 76–78.
65. Guerlac explains the delay by proposing that "weapons using gunpowder were largely experimental until after the middle of the 15th century" ("The Poet's Nitre," 246 n. 14), but this still leaves a century without a gunpowder theory of meteors.
66. Laneham, *A Letter*, 16.
67. Fulke, *Meteors*, 15.
68. Ibid., 19–20.
69. Leonardo da Vinci, *The Literary Works*, 1:321.
70. Della Porta, *De aeris transmutationibus*, 148–49. My thanks to Arianna Borrelli for this reference and a translation.
71. Kepler, *Gesammelte Werke*, 8:226; Boner, "Kepler on the Origins of Comets," esp. 38.
72. Fludd explains that a star "is sustained and elevated in its proper place, no otherwise then we see that the artificiall squib according unto the proportion of its artificiall and fading fire, with the ponderosity of its body, is, during the time of the gunpowders force, raised in the aire to a certain height, moving neither lower nor higher, then the form all vigor affordeth it vertue, and there remaineth untill the force of the corruptible and wasting fire be spent; and then it falleth down againe" (*Mosaicall Phi-*

*losophy*, 68). Because the fifth element is perfect, the star remains in its place, whereas a squib falls back to earth.

73. Castelli, *Incendio del Monte Vesuvio*, 34, quoted (in translation) in Cocco, "Vesuvius and Naples," 236 (see also 237–39).
74. Comenius, *Naturall Philosophie Reformed*, 133.
75. Descartes quoted (in translation) in Descartes, *The Philosophical Writings of Descartes*, 2:405.
76. Descartes, *Discourse on Method*, 263.
77. Descartes, *Discourse on Method*, 344. See also Werrett, "Wonders Never Cease," 144–45.
78. Descartes, *Discourse on Method*, 322–31.
79. Bernier, *Abrégé de la philosophie de M. Gassendi*, 5:269–70.
80. *Bureau d'addresse et rencontre*, 74 (conference 12). Théophraste Renaudot, who organized the Bureau d'addresse in Paris, knew of Horace Morel's fireworks, as indicated in his newspaper, *La gazette française*, March 29, 1636, Livret, 34: "Les esclairs, les feux, les flammes, avec les habits, les pas, et les figures du ballet des monstres estoyent de l'invention de Monsieur Morel, très habile aux feux d'artifice."
81. Biringuccio, *Pirotechnia*, 444–46; Werrett, "Das Feuer und die Höfe der Spätrenaissance."
82. Debus, "The Paracelsian Aerial Niter," 44–48.
83. Quersitanus, *The Practise of Chymicall, and Hermeticall Physicke*, fol. Y4v; Debus, "The Paracelsian Aerial Niter," 52.
84. Quersitanus, *Practise of Chymicall, and Hermeticall Physicke*, fols. P1–P4; Debus, "The Paracelsian Aerial Niter," 53.
85. [William Derham], "Theory of Storms" (n.d.), Royal Society, Cl. P./4i/53, quoted in Jankovic, "Meteors under Scrutiny," 231.
86. Browne, *Pseudodoxia epidemica*, 93. See also Merton, "Sir Thomas Browne's Theories."
87. On new scientific academies, see Biagioli, "Scientific Revolution, Social Bricolage, and Etiquette"; and McClellan, *Science Reorganized*.
88. Biagioli, "Galileo the Emblem Maker," and *Galileo, Courtier*.
89. Campanella, *La città del sole*, and *Advice to the King of Spain*. On the arts in *City of the Sun*, see Hall, "Appreciation of Technology"; Rossi, *Philosophy, Technology, and the Arts*, 100–102; and Bredekamp, *The Lure of Antiquity*, 58–60.
90. Campanella, *La città del sole*, 37.
91. Ibid., 87.
92. Campanella, *Advice to the King of Spain*, 183.
93. Campanella was well versed in military spaces, writing a treatise on fortifications. See Firpo, "Il Campanella scrittore di cose militari."
94. See, e.g., Rossi, *Philosophy, Technology, and the Arts*, 80–87, 117–21; Smith, *The Body of the Artisan*, 231–33; and Colie, "Cornelius Drebbel and Salomon de Caus."
95. Rossi, *Philosophy, Technology, and the Arts*, 174–86.
96. Francis Bacon, "The Refutation of Philosophies," quoted in Rossi, *Philosophy, Technology, and the Arts*, 85.
97. Coffey, "'As in a theatre.'"
98. Bacon, *New Atlantis*, 32.
99. Ibid., 27.
100. Bacon, *Sylua syluarum*, 203. Kircher (*Ars magna lucis*, 823–24) claimed that no fires could burn under water, a point that Constantin Huygens later disputed in correspondence with Marin Mersenne: "Contre ce qu'il dit que rien ne brusle soubs l'eau,

ce nous est chose familiere aux feux d'artifice de veoir jetter certaines bales enflammées au fonds d'un estang et dans bien 7 ou 8 minutes apres s'en reuenir de là tourjours ardentes" (Constantin Huygens to Marin Mersenne, November 26, 1646, in Mersenne, *Correspondance*, 14:636).

101. Bacon, *New Atlantis*, 33.
102. Bacon, *The Works of Francis Bacon*, 8:363.
103. Ibid., 6:262–63.
104. Ibid., 8:241.
105. Ibid., 6:268–69.
106. Ibid., 8:241.
107. Babington, *Pyrotechnia*, "The Epistle Dedicatory."
108. The best study of Furttenbach's career is Berthold, "Joseph Furttenbach."
109. Furttenbach's diary of his travels has been published as Furttenbach, *Newes Itinerarium Italiae*.
110. Berthold, "Joseph Furttenbach," 175.
111. His pyrotechnic works were *Halinitro-Pyrobolia*, *Architectura martialis*, and *Büchsenmeisterey-Schul*. See Fähler, *Feuerwerke des Barock*, 36–38.
112. Berthold, "Joseph Furttenbach," 167–68, 176–79.
113. Ibid., 175 n. 193.
114. Ibid., 175–76.
115. The oil painting is in the Germanisches Nationalmuseum, Nuremberg, inventory no. Gm 595. See also Fähler, *Feuerwerke des Barock*, 204; Berthold, "Joseph Furttenbach," 175; and Lotz, *Das Feuerwerk*, 30.
116. Furttenbach, *Mechanische ReißLaden*.
117. Leibniz, "Drole de pensée," 768, trans. in Jones, *The Good Life in the Scientific Revolution*, 180. For a partial English translation of Leibniz's essay, see Wiener, "Leibniz's Project."
118. Leibniz, "Drole de pensée," 759, trans. in Wiener, "Leibniz's Project," 235. On the Leibnizian theater of nature and art, see Bredekamp, "Leibniz's Theater der Natur und Kunst."
119. Leibniz, "Drole de pensée," 761, 763, trans. in Wiener, "Leibniz's Project," 237, 238.
120. Leibniz, "Drole de pensée," 764, trans. in Wiener, "Leibniz's Project," 239.
121. Van Peursen, "Ars Inveniendi bei Leibniz." On the history of the art of combinations, see Zielinski, *Deep Time of the Media*, 79–83, 118–19, 147–55.
122. Smith, *The Business of Alchemy*, 247–71. On Leibniz's interest in fireworks, see his correspondence with the Jesuit in China Claudio Filippo Grimaldi, in Leibniz, *Sämtliche Schriften und Briefe*, vol. 4, *Juli 1683–1690*, 410, 413 (see generally 410–16).
123. The "theaters of machines" of this period by Agostino Ramelli, Jacques Besson, Vittorio Zonca, and Georg Andreas Böckler also produced variations by multiplying possible designs from a series of components. See Séris, *Machine et communication*; and Ferguson, "Leupold's 'Theatrum Machinarum.'"
124. Leibniz, "Zwei Pläne zu Societäten," 135.
125. Leibniz, *Theodicy*, 65.

CHAPTER THREE

1. Hankins and Silverman, *Instruments and the Imagination*, 14–36; Findlen, "Jokes of Nature."
2. Iliffe, "Lying Wonders and Juggling Tricks," 187. See also Agnew, *Worlds Apart*; and Barish, *The Anti-Theatrical Prejudice*.

3. Dear, "Miracles, Experiments"; Findlen, "Between Carnival and Lent"; Shapin and Schaffer, *Leviathan and the Air Pump*; Golinski, "A Noble Spectacle."
4. I have made a similar argument in "Healing the Nation's Wounds."
5. Thompson, "Patrician Society, Plebeian Culture."
6. *Calendar of State Papers, Domestic Series, of the Reign of Elizabeth*, 446, 450.
7. Cressy, *Bonfires and Bells*, 141–55; Buchanan, Cannadine, et al., *Gunpowder Plots*; Sharpe, *Remember, Remember*.
8. Browne, *A Modell of the Fire-Workes*; Cressy, *Bonfires and Bells*, 162–64; Clifton, "The Popular Fear of Catholics."
9. *Mercurius Elencticus*, no. 2 (November 5–12, 1647), 1.
10. Starkey, *Pyrotechny Asserted*, esp. 4–5. On Boyle and Starkey's relationship, see Newman and Principe, *Alchemy Tried in the Fire*.
11. Webster, *Academiarum examen*, 70.
12. Ibid., 106.
13. Newcastle quoted in Slaughter, ed., *Ideology and Politics*, 45. See also Reedy, "Mystical Politics."
14. Edward Conway quoted in Harris, "'Venerating the honesty of a tinker,'" 203.
15. Hogg, *The Royal Arsenal*, 1:200. On Beckman, see Wauchope, "Beckman, Sir Martin"; Johnson, "Fireworks and Firemasters of England," 1065; and Jocelyn, "The Connection of the Ordnance Department."
16. See chapter 5 below.
17. On the Royal Society's tenure in Gresham College, see Chartres and Vermont, *A Brief History of Gresham College*, 30–41.
18. Sprat, *History of the Royal Society*, 58–59.
19. Birch, *The History of the Royal Society*, 1:37.
20. Brouncker, "Experiments of the Recoiling of Guns"; Hall, "Gunnery, Science and the Royal Society," 125.
21. Birch, *The History of the Royal Society*, 1:69, 281–85, 292–303, 310, 322, 425–28, 444–45, 450, 455.
22. On Boyle's and Hooke's experiments, see McKie, "Fire and the Flamma Vitalis."
23. Mendelsohn, "Alchemy and Politics in England," esp. 65–66. Mendelsohn argues for continuity between radical Helmontianism in the Interregnum and its manifestations in the early years of the Restoration. See also Cook, "The Society of Chemical Physicians."
24. Johnson, *Agyrto-mastix*, 41–42.
25. On anti-Catholicism in this period, see Jones, *Country and Court*, chap. 9; and Miller, *Popery and Politics in England*.
26. *Protestant (Domestick) Intelligence*, nos. 102 (March 4, 1681), 103 (March 8, 1681). See Harris, *London Crowds in the Reign of Charles II*, 111.
27. *Pyrotechnica Loyolana*, 124.
28. "A Narrative of the Pope's Late Fire-Works," in Bedloe, *A Narrative . . . of the Horrid Popish Plot*, 1–3. See also the techniques for burning London discussed by Bedloe in the "Epistle Dedicatory." E.C., *A Full and Final Proof*, "Preface to the Members of the Romish Church," noted Catholic "skill in making Fire-balls." See also Rolle, *Shilhavtiyah*, 89–128.
29. Evelyn, *Diary*, 2:232 (November 14, 1668); Hogg, *The Royal Arsenal*, 1:200.
30. *An Act for Preventing and Suppressing of Fires*, 9. Further edicts were issued by the court and the corporation of London in 1674, 1682, and 1683.
31. Sprat, *History of the Royal Society*, 85.

32. Ibid., 91.
33. Ibid., 437.
34. Ibid., 38.
35. Ibid., 214.
36. Ibid., 91.
37. On the notion of "matters of fact" and the material technology of experiment in Restoration sciences, see Shapin and Schaffer, *Leviathan and the Air Pump*, 22–79.
38. Sprat, *History of the Royal Society*, 62. On the Society's antirhetorical stance, see Dear, "Totius in Verba"; and Shapin and Schaffer, *Leviathan and the Air Pump*, 65–69.
39. Sprat, *History of the Royal Society*, 215.
40. For gunpowder experiments and Henshaw's history, see Sprat, *History of the Royal Society*, 215–17, 228–29, 233–39, 260–76, 277–83. On the History of Trades project, see Ochs, "History of Trades Programme"; and Eamon, *Science and the Secrets of Nature*, 341–50.
41. Likewise, the Society could be accused of being enthusiasts. See Heyd, "The New Experimental Philosophy."
42. Stubbe, *An Epistolary Discourse Concerning Phlebotomy*, n.p.
43. Stubbe, "Animadversions upon the History of Making SALT-PETRE," 35.
44. Stubbe, *Campanella Revived*, 8.
45. See Hunter, *Science and Society in Restoration England*, 45.
46. C. Hatton to Charles Hatton, November 22, 1677, in Thompson, ed., *Correspondence of the Family of Hatton*, 1:157. See also Furley, "Pope-Burning Processions"; and Williams, "Pope-Burning Processions."
47. Corporation of London, Lord Mayor, *Divers Rude and Disordered Young-Men*.
48. Hooke, *Diary*, 68 (November 5, 1673), 192 (November 5, 1675).
49. Boyle, *Tracts*. Partington ("The Life and Work of John Mayow," 409) claims that the *Tracts* were published in 1673 and presented to the Royal Society by Boyle in April of that year.
50. Boyle, *Tracts*, 71–79. Boyle (ibid., 74) gives the recipe for powder burned under water as three ounces of gunpowder, one dram of charcoal, two drams of flower of brimstone, and twelve drams of saltpeter. This is quite different than any gunners' recipes available at the time.
51. Ibid., 76–77.
52. Boyle, "Suspicions about Some Hidden Qualities in the Air," 82. See also Partington, "The Life and Work of John Mayow," 406–7.
53. Willis, *Five Treatises*, 40, 42, 129, *An Essay of the Pathology of the Brain and Nervous Stock*, 2–5, and "Of Convulsive Diseases," 59. Walter Charleton contested the idea on relevant grounds: "What dominion could the Soul have over the Muscles . . . if they were agitated every moment by Squibbs or Crackers breaking within them? certainly she could never moderate such violent and tumultuose explosions" (*Three Anatomic Lectures*, 99).
54. Schaffer, "Regeneration," 89, 100–105.
55. Mayow, *Medico-Physical Works*.
56. Ibid., 35.
57. Ibid., 109.
58. "An Accompt of Two Books," *Philosophical Transactions of the Royal Society* 9 (1674): 101–20, 106. (The "Accompt" is a summary of Mayow's *Tractatus*.) Newton's circulating ether was close to this model. See Newton, "Of Natures Obvious Laws." Hall

compares Newton and Mayow's views in more detail in "Isaac Newton and the Aerial Nitre."

59. Boyle, *Some Considerations*, 410.
60. Boyle, *About the Excellency and Grounds of the Mechanical Hypothesis*, 24–25.
61. Daston, "The Cold Light of Facts."
62. On the study of phosphorus in the seventeenth century, see Golinski, "A Noble Spectacle"; and Harvey, *A History of Luminescence*, 424–33.
63. Jobst Dietrich Brandshagen to Leibniz, January 1682, in Leibniz, *Sämtliche Schriften und Briefe*, vol. 3, *1680–Juni 1683*, 540 (see generally 539–43). See also Brandshagen to Leibniz, October 26 (November 5), 1682, in ibid., 730–33; and Kragh, *Phosphors and Phosphorus*, 46–49.
64. Boyle, "Short Memorial," 60.
65. Hall, "Frederick Slare," 29.
66. John Beale quoted in Golinski, "A Noble Spectacle," 27. See also Beale's comment on Shadwell in ibid., 38.
67. Boyle, *The Aerial Noctiluca*, 20–21.
68. "An Elegy on the Death of the Honourable Robert Boyle," 68.
69. *Heraclitus Ridens*, January 10, 1681, n.p.
70. *For the Preventing Tumultuous Disorders; At the Court at Whitehall.*
71. *Calendar of State Papers* entry quoted in Harris, "'Venerating the honesty of a tinker,'" 214.
72. "Rules, Orders and Instructions for the Future Government of the Office of the Ordnance, Drawn up by Command of King Charles II in 1683," British Library, King's 70, fol. 1–42. Reproduced as "Instructions for the Government of Our Office of Ordnance under the Master-General Thereof, Committed to Five Principal Officers," in Cleaveland, *Notes on the Early History*, 53–90. Matross was a rank corresponding to a gunner's assistant.
73. Hogg, *The Royal Arsenal*, 1:200; *Calendar of State Papers, Domestic Series, James II*, 249.
74. Baigent, "Borgard, Albert"; Cleaveland, *Notes on the Early History*, 141; Hogg, *The Royal Arsenal*, 2:1110; Lefroy, *War Services*, 6.
75. Hall, "Gunnery, Science and the Royal Society," 121–23.
76. Evelyn, *Diary*, 2:436 (November 15, 1684).
77. Lowman, *Exact Narrative and Description*, 2. For a less flattering account of this firework, see Aubrey, *Miscellanies*, 41.
78. Birch, *The History of the Royal Society*, 4:187.
79. Fred[erick] Slare, "An Account of Some Experiments Made at Several Meetings of the Royal Society by the Ingenious Fred. Slare M.D.," *Philosophical Transactions of the Royal Society* 13 (1683): 289–302, 291.
80. Evelyn, *Diary*, 3:11–12 (December 13, 1685).
81. Slare, "An Account of Some Experiments Made at Several Meetings of the Royal Society" (n. 79 above), 291.
82. Slare, "An Account of Some Experiments," 50.
83. Evelyn, *Diary*, 3:12 (December 13, 1685).
84. For Molyneaux, see Birch, *The History of the Royal Society*, 4:374–75. For Gale, see ibid., 296, 298. For Papin, see ibid., 467, 470–517, 522–23, 526–27, 531, 537, 551–52.
85. Carrel, Fox, and Lonsdale, *History of the Counter-Revolution in England*, 248. See also Harris, *Revolution*, 269–70.

86. Evelyn, *Diary*, 3:50–51 (July 17, 1688).
87. Wood, *Life and Times*, 3:281 (November 5, 1688).
88. Schwoerer, "The Glorious Revolution as Spectacle"; Claydon, *William III and the Godly Revolution;* Harris, *Revolution*, 355–56.
89. Beek, *The Triumphs of William III*, 9. See also Griffiths, *The Print in Stuart Britain*, 304–5; and Maccubbin and Hamilton-Phillips, *The Age of William III and Mary II*, 39.
90. 9 & 10 Will. 3, c. 7: "An Act to prevent the throwing or firing of squibs, serpents, and other fireworks . . . after the 25th day of March, 1698, it shall not be lawful for any person or persons, of what age, sex, degree, or quality soever, to make or cause to be made, or to sell or utter or offer to expose to sale, any squibs, rockets, serpents, or other fireworks, or any cases, moulds, or other implements for the making any such squibs [etc.] . . ." The act did not apply to the master of the Ordnance, royal militias, and members of the artillery. See *An Exact Abridgment of All the Statutes*, 406–7. William's own sympathizers in the provinces were now obliged to celebrate anti-Catholic sentiments by illuminating their windows with lanterns and candles, instead of using fireworks. See Hutton, *Stations of the Sun*, 397.
91. Jacob, "Boyle's Atomism."
92. Boyle, *Free Enquiry*, 213.
93. Boyle, "Fragments of . . . 'Essay on Spontaneous Generation,'" 288.
94. Ray, *The Wisdom of God*, 8–9.
95. Burnet, *The Theory of the Earth*, bk. 3, p. 57.
96. My thanks to Susanne Pickert for pointing this out.
97. "Pyrotechny," in Phillips, *The New World of Words* (5th ed., 1696), n.p. The first edition of Phillips in 1658 gave the definition as "*Pyrotechnie.* (Greek) any structure or machination made by fire-works."
98. Chambers, *Cyclopaedia*, 2:920.
99. Frederick Slare, "An Account of Some Experiments Relating to the Production of Fire and Flame, Together with an Explosion; Made by the Mixture of Two Liquors Actually Cold," *Philosophical Transactions of the Royal Society* 18 (1694): 201–18, 215.
100. Newton, *Opticks*, 354.
101. Ibid., 355.
102. On Crane Court, see Martin, "Former Homes of the Royal Society," 14–15.
103. Schaffer, "Natural Philosophy and Public Spectacle"; Stewart, *The Rise of Public Science;* Morton, ed., "Science Lecturing in the 18th Century." On the notion of the *public* in England, see Melton, *The Rise of the Public*, 19–44.
104. Boerhaave, *A New Method of Chemistry*, 1:173 n. r; Shaw, *Chemical Lectures*, 416.
105. Shaw, *Chemical Lectures*, 438.
106. For perspectives on the history of modeling, see de Chadarevian and Hopwood, eds., *Models*.
107. Boerhaave, *A New Method of Chemistry*, 1:173.
108. Slare, "Account of Some Experiments Made at Several Meetings of the Royal Society" (n. 79 above), 289–91; John Evelyn to Thomas Tennison, October 15, 1692, in Evelyn, *Diary*, 3:467–72. In this letter, Evelyn claimed that subterranean fires meeting beds of niter and sulfur caused volcanoes, earthquakes, and the rise of mountains. Perhaps they were also the location of hell ("& I pray God we may never know"), and perhaps "this vast terraqueous globe may not one day breake like a granado about our eares, & cast itselfe into another figure than the deluge did" (470).

109. Nieuwentyt, *The Religious Philosopher*, 1:226. For fireworks experiments, see also ibid., 2:314, 324.
110. [Joseph Wasse], "Two Letters on the Effects of Lightning, from the Reverend Mr. Jos. Wasse, Rector of Aynho in Northamptonshire, to Dr. Mead, by Jos. Wasse," *Philosophical Transactions of the Royal Society* 33 (1724): 366–70, 369–70.
111. Morton, "Concepts of Power"; Stewart, "A Meaning for Machines."
112. Wallace, *The Social Context of Innovation*, 39, quoting a parliamentary memorandum. On Vauxhall, see Thorpe, "The Marquis of Worcester and Vauxhall."
113. Desaguliers, *A Course of Experimental Philosophy*, 2:465; "Engine," in Harris, *Lexicon technicum*, vol. 1, n.p.; Schaffer, "The Show That Never Ends," 184–87.
114. On the production of the thanksgiving fireworks, see the letter from Ja. Pound to the Board of Ordnance, June 11, 1713, Public Record Office, London, SP 34/21; Duncan, *History of the Royal Regiment of Artillery*, 1:95; and [Boyer], *The Political State of Great Britain*, 6:31–38.
115. *The Spectator*, 8:330 (no. 616, September 5, 1714).
116. The fireworks are listed on an etching by Sir James Thornhill, *Exact Draught of the Firework That Was Performed on the River Thames, July 7th, 1713, Being the Thanksgiving Day for the Peace Obtain'd by the Best of Queens* (London, ca. 1713), British Museum Department of Prints, Crace Collection, portfolio 5, sheet 29. For bee-swarm recipes, see, e.g., Smith, *The Laboratory*, lxix, lxxii, and fig. 57.
117. *The Guardian*, 2:112 (no. 103, July 9, 1713). On Steele and Burnet, see Stewart, *The Rise of Public Science*, 36. On Steele and lectures, see Loftis, "Richard Steele's Censorium."
118. *The Guardian*, 2:112 (no. 103, July 9, 1713).
119. Stewart, *The Rise of Public Science*, 56; *The Guardian*, 2:506–10 (no. 175, October 1, 1713).
120. *The Guardian*, 2:115–16 (no. 103, July 9, 1713).
121. Ibid., 2:133–34 (no. 107, July 14, 1713). On Whiston, see Force, *William Whiston*; and Snobelen, "William Whiston."
122. Whiston, *A New Theory of the Earth*.
123. Duffy, "'Whiston's affair'"; Snobelen, "Caution, Conscience and the Newtonian Reformation."
124. Stewart, *The Rise of Public Science*, 91–92.
125. Harris, *The Evil and Mischief of a Fiery Spirit*, 3.
126. Ibid., 8. Harris's *Lexicon technicum* explained: "Sulphureous, Nitrous, and other light and active mineral Particles do form Meteors in the Air, and particularly are the Cause of Thunder and Lightning &c. and other fiery Compositions there" ("Meteors," in Harris, *Lexicon technicum*, vol. 2, n.p.).
127. Harris, *The Evil and Mischief of a Fiery Spirit*, 3, 14–15.
128. *The Guardian*, 2:134–36 (no. 107, July 14, 1713). Steele's *Englishman* (no. 29 [December 8, 1713]) also advertised the scheme.
129. Whiston and Ditton, *A New Method* (1714). The second edition (*A New Method*, 1715) expanded the project and included an introduction explaining the background to Whiston's project.
130. Whiston and Ditton, *A New Method* (1715), 30.
131. Ibid., 23.
132. Ibid., 41.
133. Ibid.

134. Ibid., 23–24.
135. [Boyer], *The Political State of Great Britain*, 9:76.
136. For the subsequent history of the prize, see Howse, *Greenwich Time*.
137. N.N., *Will-with-a-Wisp*, "Advertisement."
138. Ibid., 59, 61.
139. From Hogarth's *Rake's Progress* and Jonathan Swift's *Gulliver's Travels* to Dava Sobel's recent *Longitude* and the film based on the book, Whiston's scheme has been presented as a crackpot idea.
140. See Stewart, *The Rise of Public Science*, 213–54 and passim; Rowbottom, "John Theophilus Desaguliers"; and Wigelsworth, "Competing to Popularize Newtonian Philosophy."
141. Stewart, *The Rise of Public Science*, 122–30; Schaffer, "The Show That Never Ends."
142. Desaguliers, *A Course of Experimental Philosophy*, 1:402.
143. Ibid., 403.
144. Desaguliers's argument was later invoked by the Harvard philosopher John Winthrop to explain extraordinary earthquakes and commotions of the sea. See Winthrop, *A Lecture on Earthquakes*, 24–26.
145. [Boyer], *The Political State of Great Britain*, 15:549.
146. Mariotte, *The Motion of Water*; Rowbottom, "John Theophilus Desaguliers," 213–14.
147. Hobbes, *Seven Philosophical Problems*, 7:12.
148. "A rocket, whose construction I here suppose is known, is only a little light cannon, which by the impulse of the lighted matter it contains, rebounds upwards in the direction of its breech, and this with the same velocity, as that wherewith the combustible matter streams out at its mouth, which looks downwards in effect; this rebound is the rise of the rocket." La Hire, "Sur les effets du ressort de l'air," discusses rockets on 11–12. For an English translation, see La Hire, "On the Effects of the Elasticity of the Air," 1:316–17.
149. Mariotte, *The Motion of Water*, 116.
150. Ibid., 287 (see generally 286–88).
151. Ibid., 287. Versions of the Mariotte theory endured well into the twentieth century. See, e.g., MacCulloch, "Pyrotechny," 242–43; and "Topics of the Times," *New York Times*, January 13, 1920, 12.
152. Ackermann and Wess, "Between Antiquarianism and Experiment," 155.
153. [Boyer], *The Political State of Great Britain*, 56:554–55; Galt, *George the Third*, 1:86. At least one other lecturer showed fireworks. In Durham, in 1740, Messrs Dove and Booth advertised their "Mathematical lectures" as accompanied by "several Fireworks in Honour of the Birthday of the brave Admiral Vernon." See Wilson, "Empire, Trade and Popular Politics," 85.

CHAPTER FOUR

1. Algarotti, *Letters from Count Algarotti*, 1:2–3.
2. Ibid., 2.
3. Stephens, "Desaguliers, Thomas."
4. *Izobrazhenie onago feierverka . . . 9 iiulia 1739 goda*.
5. Mikhail Ivanov Sukhomlinov, ed., *Materialy dlia istorii Imperatorskoi Akademii nauk* (hereafter *MIAN*), 4:134–35, 6:488–89; Iushkevich and Kopelevich, *Kristian Goldbakh*.
6. On the foundation and history of the academy in the eighteenth century, see Kopelevich, *Osnovanie*; Ostrovitianov, ed., *Istoriia Akademii nauk*; Pekarskii, *Istoriia Imperatorskoi Akademii nauk*; and Werrett, "An Odd Sort of Exhibition."

7. Cahusac, "Feu d'artifice," 639.
8. On the history of Russia during this period, see, e.g., Hughes, *Russia in the Age of Peter the Great;* and Anisimov, *Rossiia bez Petra.*
9. Stählin, "Kratkaia istoriia," 243; Paul, "The Military Revolution in Russia."
10. Velimirovich, "Liturgical Drama," esp. 365; Karlinsky, *Russian Drama,* 4–7; Maggs, "Firework Art and Literature," 25–26.
11. Zelov, *Ofitsial'nye svetskie prazdniki,* 53–54.
12. "Gollandets Klenk v Moskovii."
13. Gordon, *Tagebuch des Generals Patrick Gordon,* 2:297, 385.
14. The fireworks are described in Vasil'iev, *Starinnye feierverki,* 17–19.
15. Rovinskii, *Opisanie feierverkov,* 186; Vasil'iev, *Starinnye feierverki,* 40–41.
16. See the catalog of Russian fireworks prints, Alekseeva, ed., *Feierverki i illiuminatsii,* 5–7.
17. Bogdanov, *Opisanie,* 134. Peter visited the Moscow laboratory on six occasions in January 1722, e.g., testing fire wheels before a display on the twenty-eighth of that month. See Rovinskii, *Opisanie feierverkov,* 195.
18. Rossiiskii gosudarstvennyi arkhiv drevnikh aktov, Moscow, kab. Petra I, sec. 1, bk. 4, fol. 75. On Petrine fireworks manuscripts, see Luk'ianov, *Istoriia khimicheskikh promyslov,* 5:85–89.
19. Notes of the Danish envoy Just Juel, January 12, 1710, quoted in Rovinskii, *Opisanie feierverkov,* 187.
20. Among the books Peter had translated into Russian were the Danzig engineer Ernst Braun's *Novissima fundamentum et praxis artilleriae* and vol. 2 of the Saxon artillerist Johann Sigmund Buchner's *Theoria et praxis artilleriae.* These were published in Russia in 1709 and 1711, respectively. Daniel de La Feuille's *Devises et emblèmes anciennes et modernes* was translated into Russian in 1705.
21. Notes of Just Juel, October 31, 1710, quoted in Rovinskii, *Opisanie feierverkov,* 188.
22. Staehlin [Stählin], *Original Anecdotes of Peter the Great,* 317–18.
23. Moran, "German Prince-Practitioners."
24. Johann Georg became interested in fireworks on seeing them at the wedding of Christian IV of Denmark in 1634. See *Verzeichnuss undt Ordnung der Feuerwergke, so itzige Churfl. Durchl. selbst angegeben und ferttigen lassen, auch theils selbst laboriren helffen und verbrannt,* Staatsbibliothek, Berlin, Handschriftenabteilung, MS Germ. Fol. 297. See also Fähler, *Feuerwerke des Barock,* 98–102; and Horn, *Der aufgeführte Staat,* 89–121.
25. Syndram, "Princely Diversion and Courtly Display"; Baur and Plaßmeyer, "Physikalische Apparate und mechanische Spielereien."
26. Guerrier, *Leibniz in seinen Beziehungen zu Russland;* Benz, *Leibniz und Peter der Große;* Ger'e, *Sbornik pisem . . . Leibnitsa.*
27. In 1722, the guards captain Lev Izmailov brought rockets from China that Peter tested himself. See Vasil'iev, *Starinnye feierverki,* 55.
28. Klopp, "Leibniz' Plan," 248.
29. This project is reproduced in Ostrovitianov, ed., *Istoriia Akademiia nauk,* 429–35.
30. Gordin, "Importation."
31. This was despite the fact that Wolff had as much interest in fireworks as Leibniz, including a whole book of recipes for fireworks in his compendious works. See Wolff, *Cours de mathématique,* 149–210.
32. On the academy buildings, see Safranovskii, "Les salles de l'Académie des sciences." On the Kunstkamera, see Staniukovich, *Kunstkamera;* and Buberl and Dückershoff, eds., *Palast des Wissens.*

33. Marker, *The Origins of Intellectual Life in Russia,* 44–45.
34. On these assemblies, see Sukhomlinov, ed., *MIAN,* 1:24, 6:103–4; Kopelevich, *Osnovanie,* 106–9; and Gordin, "Importation," 15–19.
35. The academy therefore partly resembled Italian academies of the seventeenth century, constituted and dissolved at the will of the prince. See Biagioli, "Scientific Revolution, Social Bricolage, and Etiquette."
36. On Menshikov, see Bespiatykh, *Aleksandr Danilovich Menshikov.*
37. Delisle and Bernoulli, *Discours;* Pekarskii, *Istoriia Imperatorskoi Akademii nauk,* 1:145–46.
38. Kantemir, "To My Mind," 158.
39. Haven, "Puteshestvie v Rossiiu," 380. See also G. F. Müller's comments in Marker, *The Origins of Intellectual Life in Russia,* 48.
40. Berk, "Putevye zametki," 186.
41. *Protokoly . . . Imperatorskoi Akademii,* 1:55; Sukhomlinov, ed., *MIAN,* 6:247–48. The last speeches were printed in *Sankt-Peterburgskie Vedemosti,* January 27, February 3, 1732. There were, however, occasional lectures to foreign visitors during the 1730s. See Sukhomlinov, ed., *MIAN,* 6:292, 294.
42. On the struggle for power, see Anisimov, *Rossiia bez Petra,* 127–98; and Dolgorukov, *Vremia Petra II.*
43. Pekarskii, *Istoriia Imperatorskoi Akademii nauk,* 1:xliii; Schulze, "The Russification of the . . . Academy," 310–11. Senators considered the academy "a useless and ill-considered affair, carrying no advantage for the country . . . they preferred to save the money that was being spent on it" (Vockerodt, "Rossiia pri Petre Velikom," 167).
44. J. D. Schumacher to L. Blumentrost, March 29, 1731, quoted in Pekarskii, *Istoriia Imperatorskoi Akademii nauk,* 1:xliv–xlv n. 3.
45. J. A. Korff to J. D. Schumacher, September 9, 1731, Archive of the St. Petersburg Academy of Sciences, f. 121, op. 2, no. 84, pp. 15–16.
46. On Münnich, see Büsching, "Lebensgeschichte Burchard Christophs von Münnich"; Vischer, *Münnich;* and Ley, *Le maréchal de Münnich.*
47. Stählin noted of nonacademicians: "It goes without saying that for [Russian] scholars this science was quite foreign, for not one of them understood anything in the invention of allegorical pictures" ("Kratkaia istoriia," 244).
48. The illuminations—centered on the figure of a triumphal arch, topped by a statue of Peter II in a crown and mantle with a scepter, with allegorical statues and emblems surrounding the base—were illustrated in engravings by the academic artisan Ottomar Elliger and included in a brochure explaining the allegorical figures in the display, published by the academy: *Relatsiia o illuminatsii.* See also Zelov, *Ofitsial'nye svetskie prazdniki,* 219–20; Sukhomlinov, ed., *MIAN,* 1:360–64; and Rovinskii, *Opisanie feierverkov,* 199. Beckenstein's role in designing fireworks is discussed in Stählin, "Kratkaia istoriia," 244–45.
49. Bogdanov, *Opisanie,* 155; Rodzevich, *Istoricheskoe opisanie Sanktpeterburgskogo Arsenala,* 65. On the history of the Chancellery from 1711 to 1762, see Skalon, ed., *Glavnoe artilleriiskoe upravlenie,* 27–225.
50. J. D. Schumacher to L. Blumentrost, March 4, 1728, quoted in Pekarskii, *Istoriia Imperatorskoi Akademii nauk,* 1:22.
51. Delisle, "Extrait d'une lettre."
52. On Anna's reign, see Solov'ev, *Empress Anna,* and *The Rule of Empress Anna;* and Curtiss, *A Forgotten Empress.*

53. On Anna's court, see Marsden, *Palmyra of the North*, 85–88; Semenova, *Byt naselenie Sankt-Peterburga*, 135–50; and Manstein, *Contemporary Memoirs of Russia*, 254–58. On Anna's relationship to the nobility, see Ransel, "The Government Crisis of 1730"; and Meehan-Waters, *Autocracy and Aristocracy*.
54. Claudius Rondeau to Lord Townshend, April 9, 1730, Public Record Office, London, SP 91/11, fol. 76. Again in 1755: "The most renowned nobles undertook . . . special festivities in their town palaces for the all-joyful reception of Her Imperial Majesty, Their Imperial Highnesses, and their retinues, and each tried to surpass the other with expenditures and splendors. Their private residences and palaces were wonderfully illuminated, and a few grandees built in conclusion special fireworks" (Stählin, "Kratkaia istoriia," 257).
55. On the fireworks theater, see Stählin, "Kratkaia istoriia," 245, 268 n. 9; Rovinskii, *Opisanie feierverkov*, 206; and Sukhomlinov, ed., *MIAN*, 2:72–73.
56. Zelov, *Ofitsial'nye svetskie prazdniki*, 208.
57. On Juncker, see Pekarskii, *Istoriia Imperatorskoi Akademii nauk*, 1:479–91.
58. G. F. Müller quoted in ibid., 481–82.
59. G. F. Müller quoted in ibid. A description of Juncker's illuminations was published as Iunker [Juncker], *Kratkoe opisanie toi illuminatsii*.
60. Pekarskii, *Istoriia Imperatorskoi Akademii nauk*, 1:484.
61. Speranskii, ed., *Polnoe sobranie zakonov Rossiskoi Imperii*, 9:243 (no. 6517).
62. Zelov, *Ofitsial'nye svetskie prazdniki*, 209.
63. Stählin, *Aus den Papieren Jacob von Stählins*, 20–28. On Stählin, see Stählin, *Zapiski Iakoba Shtelina*; Malinovskii, "Iakob fon Shtelin i ego zapiski"; Pekarskii, *Istoriia Imperatorskoi Akademii nauk*, 1:538–67; and Liechtenhan, "Jacob von Stählin."
64. Stählin, "Kratkaia istoriia," 247; Picinelli, *Mundus symbolicus*.
65. On Adudorov's designs, see Sukhomlinov, ed., *MIAN*, 3:51–52. On Goldbach's, see ibid., 3:15–17, 4:52, 135, 242, 296, 355, 754.
66. Much correspondence of the academy and chancellery concerning fireworks may be found in Sukhomlinov, ed., *MIAN*, passim.
67. Siemienowicz, *The Great Art of Artillery*, 359–64.
68. Ibid., 349–50.
69. Ibid., 352.
70. Ibid., 333–81. Among authors translated into Russian, Buchner (*Theoria et praxis artilleriae*, 48) also described invention and referred his readers to Siemienowicz.
71. Drawings and sketches of fireworks are included in the Getty collection of Russian fireworks prints, ID no. 92 F-291, fols. 41, 42, 46, 50, 52–54, 68–70. Other extant drawings and paintings are cataloged in Alekseeva, ed., *Feierverki i illuminatsii*. On the model, see Stählin, "Kratkaia istoriia," 255.
72. On the laboratory, see Rodzevich, *Istoricheskoe opisanie*, 63; Bogdanov, *Opisanie*, 149; Luk'ianov, *Istoriia khimicheskikh promyslov*, 107–11; Stählin, "Kratkaia istoriia," 245; and the autobiographical memoir written ca. 1771 by the Russian fireworker Mikhail Vasil'evich Danilov, "Zapiski," 31.
73. Danilov, "Zapiski," 46–47.
74. Frézier, *Traité des feux d'artifice* (1747), 376–77.
75. Ibid., 376.
76. Stählin, "Kratkaia istoriia," 245.
77. Mikhail Danilov et al., *Al'bom feierverkov*, Department of Manuscripts, Russian National Library, St. Petersburg, FX. III. 31.

78. Danilov, "Zapiski," 31.
79. Bennett, "Evolution of the Meanings of *Chin.*"
80. Bolotov, "Zapiski Andreia Timofeevicha Bolotova," 192, describing illuminations by Lomonosov on September 5, 1752.
81. Berk, "Putevye zametki," 159. See also the opinions of the foreign commentators Elizabeth Justice, Peder von Haven, and John Cook in Zelov, *Ofitsial'nye svetskie prazdniki,* 206.
82. Thomas Ward (writing from Moscow) to Lord Townshend, May 7, 1730, Public Record Office, London, SP 91/11, fol. 106.
83. Röhling, "Illustrated Publications on Fireworks and Illuminations"; Maggs, "Firework Art and Literature"; Zelov, *Ofitsial'nye svetskie prazdniki,* 208–9.
84. *Opisaniia* were printed in runs of between five hundred and one thousand copies. See Zelov, *Ofitsial'nye svetskie prazdniki,* 208.
85. The description of fireworks on January 28, 1733, noted, however: "The firework consists of a great number of rockets, squibs, battle fires [*streitfeierov*], crashing fires [*poltfeierov*], fountains, stars, fireballs of various kinds, and other things . . . the completion of the illumination required two thousand people working for ten weeks" (Rovinskii, *Opisanie feierverkov,* 207).
86. Stählin, "O pol'ze teatral'nykh deistve komedii," 340. See also Maggs, "Firework Art and Literature," 36–38.
87. Iunker [Juncker], *Kratkoe opisanie toi illuminatsii,* 9–10.
88. On the academy's fireworks theater, erected in 1749, see Sukhomlinov, ed., *MIAN,* 9:766; and Zelov, *Ofitsial'nye svetskie prazdniki,* 204–5. On the masquerade, see Chenekal and Topchiev, *Letopis' zhizni i torzhestva M. V. Lomonosova,* 243. Items from the Kunstkamera were often used for the court's amusement. See Sukhomlinov, ed., *MIAN,* 4:367–68; and Staniukovich, *Kunstkamera,* 75–76.
89. Berk, "Putevye zametki," 181.
90. McGowan, *Ideal Forms,* 9–50.
91. Erasmus quoted in ibid., 11.
92. Sukhomlinov, ed., *MIAN,* 2:282–84 (January 28, 1733). See also *Opisanie Velikoi illuminatsii v 28 Genvaria 1733 goda;* and Rovinskii, *Opisanie feierverkov,* 206–7. On cameralism in Russia, see Raeff, *The Well-Ordered Police State.*
93. Rovinskii, *Opisanie feierverkov,* 214, 219 (see also 203, 211).
94. Rovinskii, *Opisanie feierverkov,* 203, 209–10, 219, 221, 223. The image of Minerva was also used to depict Anna's successor, Elizabeth Petrovna, and, famously, Catherine II. See Wortman, *From Peter the Great to the Death of Nicholas I,* 110–41.
95. Rovinskii, *Opisanie feierverkov,* 209. See also Iunker [Juncker], *Kratkoe opisanie onago feierverka.*
96. Rovinskii, *Opisanie feierverkov,* 209–10.
97. Ibid., 210.
98. Lomonosov quoted in Baehr, *The Paradise Myth,* 39.
99. On gift exchange in the sciences, see Biagioli, *Galileo, Courtier,* 36–53; and Findlen, "The Economy of Scientific Exchange."
100. On Krafft and court astrology in Russia, see Pekarskii, *Istoriia Imperatorskoi Akademii nauk,* 1:461–62; Staehlin, *Original Anecdotes,* 394–99; and Pyliaev, *Zabytoe proshloe okrestnostei Peterburga,* 223–27.
101. Stählin quoted in Pekarskii, *Istoriia Imperatorskoi Akademii nauk,* 1:461–62.
102. *Kratkoe iz'iasnenie . . . feierverka . . . 28 Genvaria 1735,* in the collection of fireworks and illuminations descriptions of the Library of the Academy of Sciences, St. Pe-

tersburg, BAN-471 (1–10), 36–38, 36. The text is partly reproduced in Rovinskii, *Opisanie feierverkov*, 211.

103. *Kratkoe iz'iasnenie . . . feierverka . . . 28 Genvaria 1735*, 37.
104. Gordin, "Importation," 22–26.
105. *Kratkoe iz'iasnenie . . . feierverka . . . 28 Genvaria 1735*, 38.
106. Compare Rudolf II's 1577 entry to Vienna, celebrated with a triumphal arch depicting the previous emperor, Maximillian II, above a Ptolemaic celestial globe and Rudolf above a globe illustrating the Copernican system. See Kaufmann, *The Mastery of Nature*, 136–50. See also *Dessein du feu d'artifice*.
107. *Sanktpeterburgskie vedemosti*, March 3, 1735, quoted in Anisimov, *Anna Ivanovna*, 241–42. See also Pekarskii, *Istoriia Imperatorskoi Akademii nauk*, 1:462.
108. Desaguliers, *The Newtonian System of the World*, 34. Baillon ("Early Eighteenth-Century Newtonianism") interprets the verses as an attempt at minority assimilation in a foreign culture.
109. The coronation album was printed as *Opisanie koronatsii Eia Velichestva Imperatritsy*. On prints, see Sukhomlinov, ed., *MIAN*, 1:370, 461; and J. D. Schumacher to I. K. Kirilov, February 17, 1729, reproduced in Alekseeva, Vinogradov, Piatnitskii, and Levshin, eds., *Graviroval'naia palata*, 50–51. On other arts, see Sukhomlinov, ed., *MIAN*, 3:43 (triumphal arch by Nartov), 3:173, 184–85 (ivory portrait by Nartov), 3:842 (snuffbox by Isaac Bruckner), 4:760–61 (medal by Goldbach), 4:782–89 (catafalque by Stählin). On jewelry, see Pekarskii, *Istoriia Imperatorskoi Akademii nauk*, 1: xlvii; and Kopelevich, *Osnovanie*, 174.
110. For the complete personnel list, see Sukhomlinov, ed., *MIAN*, 4:552–61.
111. See Sukhomlinov, ed., *MIAN*, 6:246, 289, 316, 389; and Ostrovitianov, ed., *Istoriia Akademii nauk*, 435.
112. As G. F. Müller wrote in his history of the academy, fireworks composition "became a business of the Academy, to which each member gladly contributed" (quoted in Sukhomlinov, ed., *MIAN*, 6:145).
113. Quoted in Pekarskii, *Istoriia Imperatorskoi Akademii nauk*, 1:li.
114. Sukhomlinov, ed., *MIAN*, 2:367–77; Kopelevich, *Osnovanie*, 125–27. Anna awarded a single payment of 30,000 rubles in 1733; the Senate added 10,000 in 1736 and an extra 20,000 in 1737. Pekarskii, *Istoriia Imperatorskoi Akademii nauk*, 1:lii–liv. A petition explaining why arts required additional funds is in Sukhomlinov, ed., *MIAN*, 3:118–19. The academy's first charter followed another round of complaints over the costs of arts in 1745. See Alekseeva et al., eds., *Graviroval'naia palata*, 78–81.
115. Caresme remained the principal artificer in Paris until his death in 1680, when his son Denis Caresme replaced him. See Weigert, "Les feux d'artifices," 181, 194–97.
116. [Stefano della Bella], *Traicté des feux artificielz de joye & de recreation* (Paris, 1649), Bibliotheque nationale, Paris, MS fr. 1247. See also Dearborn Massar, "Stefano della Bella's Illustrations."
117. On the organization of fêtes in the ancien regime, see Weigert, "Les feux d'artifices," 178–79; Chennevières, *Les menus-plaisirs du roi*; Proudhomme and Crauzat, *Les menus-plaisirs du roi*; and Gruber, *Les grandes fêtes*.
118. See "Order Forbidding the Use of Guns and Fireworks during the Feste-Dieu, Denys Thierry," Public Record Office, London, SP 117/458; and "Order Forbidding the Use of Fireworks, Denys Thierry," Public Record Office, London, SP 117/489. French laws concerning fireworks and the use of gunpowder can be found in *Continuation du traité de la police*, 142–48.

119. Caresme performed fireworks in 1682 and Titon in 1698, according to Ruggieri, *Précis historique*, 230, 234. Morel, Drouilly, and Caboureux were other *artificiers* of this period. Their names, together with a detailed budget for fireworks at Versailles in 1697, can be found in "Estat de la despense se qui a esté faite pour le feu d'artifice qui a esté tiré a Versailles au bout de la piece des Suisses le 7. Decembre 1697," Archives Nationales, Paris, O/1/3263/1.
120. On Louis's ruling image, see Apostolidès, *Le roi-machine*; and Burke, *The Fabrication of Louis XIV*.
121. Menestrier's *Advis necessaires pour la conduite des feux d'artifice* is appended to *Les reiovissances de la paix*. My analysis and translations of this work follow Zanger, *Scenes from the Marriage of Louis XIV*, 100–111.
122. Menestrier, *Les reiovissances de la paix*, 15.
123. Ibid., 17.
124. Fireworks engravings from the reign of Louis XIV are collected in the Cabinet des arts graphiques, Musée Carnavalet, Paris, Histoire GC II (1661–1715).
125. Félibien, *Les divertissements de Versailles*, 80–81, quoted (in translation) in Marin, *Portrait of the King*, 202–3. See also Salatino, *Incendiary Art*, 10–14.
126. Chartier, *Forms and Meanings*, 51. Thus, Félibien wrote of the illuminations for the 1674 fête: "The whole court was surprised by the novelty and grandeur of the spectacle. . . . The King wanting to show beauties that had never yet been seen, seemed this time to have been served by Magic itself, so much did the eyes and mind find themselves surprised by the different marvels that charmed them" (Félibien, *Les divertissements de Versailles*, 85–86, quoted [in translation] in Marin, *Portrait of the King*, 204).
127. Of Catherine it was said: "In [her] deep knowledge are hidden the wondrous secrets of how to correct the capricious laws of crude nature and how to transform countries. . . . I am told that only a magic power . . . could create this transformation. . . . She says 'Let it be so,' and it is. . . . [She] utters a word and things are in a new form. Everything is subject to [her] words" ("Ode to the Planter of the Grape in the North" [in Russian], *Rastushchii vinograd*, December 1785, 1–7, trans. and quoted in Baehr, *The Paradise Myth*, 78–79).
128. Dear, "A Mechanical Microcosm"; Sutton, *Science for a Polite Society*.
129. Fontenelle, *Eloges des académiciens*, 1:337, trans. in Simon, *Chemistry, Pharmacy and Revolution*, 54 (see generally 53–61).
130. Lemery, "Sur les feux souterrains," 51–52. Lemery's experiment later became a firework. See Jones, *A New Treatise on Artificial Fireworks*, 50.
131. Fontenelle, *Eloges des académiciens*, 1:340, trans. in Simon, *Chemistry, Pharmacy and Revolution*, 54.
132. Lemery, *Le nouveau recueil de curiositez*, 1:240–47, trans. as *New Curiosities in Art and Nature*, 82–86.
133. Sutton, *Science for a Polite Society*, 203; Polinière, *Experiences de physique*, 2:9.
134. Dear, "Miracles, Experiments"; Daston, "Marvellous Facts."
135. Francis Bacon, "Gesta Grayorum," quoted in Spedding, *The Letters and the Life of Francis Bacon*, 1:335.
136. Algarotti quoted in Curtiss, *A Forgotten Empress*, 222.
137. Krafft and Weitbrecht, *Sermones*. Schumacher also attempted to revive lectures in the academy's university at this time, though with less success. See Ostrovitianov, ed., *Istoriia Akademii nauk*, 148.
138. Smagina, "Publichnye lektsii"; Brooks, "Public Lectures in Chemistry in Russia."

CHAPTER FIVE

1. Edward Wright quoted in Rykwert, *The First Moderns*, 399 n. 85.
2. Pinto, "Nicola Michetti," 295–300, 303–4. On the history of the Chinea, see Sassoli, *Apparati architettonici per fuochi d'artificio*; and Moore, "Prints, Salami, and Cheese," and "The Chinea."
3. Rykwert, *The First Moderns*, 358; Moore, "The Chinea," 25. For a detailed description of how the *macchina* was constructed and how the Chinea fireworks were organized, see Moore, "The Chinea," 229–71, 397–410. On other Italian *macchine* of this period, see Kiene, "L'image du Dieu vivant"; and Camerini, *Il magnifico apparato.*
4. On Roman-Parisian artistic links, see Smith, *Architectural Diplomacy.*
5. Rykwert, *The First Moderns*, 358–63; Pinto, "Nicola Michetti," 304–5.
6. On Servandoni, see Heybrock, *Jean-Nicholas Servandoni*; States, "Jean-Nicholas Servandoni"; and Olivier, "Jean-Nicolas Servandoni's Spectacles."
7. On the Paris theaters in the eighteenth century, see Howarth and Clarke, eds., *French Theatre in the Neo-Classical Era*; and Ravel, *The Contested Parterre.*
8. On Servandoni at the Opéra, see Olivier, "Jean-Nicolas Servandoni's Spectacles"; and Christout, "Décors d'opéra."
9. The performance is described in *Description de la feste et du feu d'artifice*; and "Description de la fête & du feu d'artifice."
10. [Boyer], *The Political State of Great Britain*, 39:193.
11. See *Mercure de France*, February 1730, 397–98. Instructions on making *jets de feu*, sometimes described as *fontaines* (fountains), are given in Frézier, *Traité des feux d'artifice* (1706), 204–7; Belidor, *Le bombardier françois*, 333–35; Frézier, *Traité des feux d'artifice* (1747), 305–8; and Alberti, *La pirotechnia*, 74–78.
12. On St. Sulpice, see Sturges, "Jacques-François Blondel," 18.
13. Blondel, *Description des festes.*
14. The fireworks are described in "Feste donnée par Mrs les Prévôs des Marchands & Echevins"; and Bracco and Lebovici, *Ruggieri*, 27–31.
15. Melton, *The Rise of the Public*, 176–77.
16. Bracco and Lebovici, *Ruggieri*, 22, 32. Ruggieri (*Précis historique*, 132) says that they arrived in 1739. Certainly, the Ruggieri's fireworks accompanied a play shown at the Comédie on July 2, 1743, entitled *Les petits maîtres*. See Clément and Laporte, *Anecdotes dramatiques*, 2:53–54.
17. On the Comédie, see Bernardin, *La comédie italienne en France.*
18. Melton (*The Rise of the Public*, 160–93) describes the commercial pressures on London and Paris theaters in the eighteenth century.
19. *Mercure de France*, vol. 17, *June–December 1729* (Geneva: Slatkine, 1968), 2259–66, 2265; Parfaict and Parfaict, *Dictionnaire des théâtres de Paris*, 2:566–73.
20. Frézier, *Traité des feux d'artifice* (1747), 439 (see generally 437–48).
21. Ruggieri, *Principles of Pyrotechnics*, 130; Chennevières, "Les Ruggieri," 136.
22. Barbier, *Chronique*, 184.
23. Translated from Bracco and Lebovici, *Ruggieri*, 31. See also Barbier, *Journal*, 2:241–42. The Dodemant (also spelled Dodemans and Dodumant) family executed fireworks for several decades in Paris, where they had a shop selling fireworks. See Ruggieri, *Précis historique*, 80–82, 271. For city contracts with Etienne-Michel Dodemant, see Archives Nationales, Paris, K1009, nos. 83, 103, 152.
24. *Brevet exclusif*; Lynn, "Sparks for Sale," 78.
25. Ruggieri, *Précis historique*, 271; "Feu d'artifice."

26. Horace Walpole to Sir Horace Mann, May 3, 1749, in Walpole, *Letters*, 2:152. For parallel displays, see Archives nationales, Paris, K1009, nos. 103–4, 168–69, 188, 189.
27. Frézier, *Traité des feux d'artifice* (1741), and *Traité des feux d'artifice* (1747). Two further editions of Frézier's revised *Traité des feux d'artifice* were published in 1747 by Charles Antoine Jombert and at the Quay des Augustins, Nyon. See also Perrinet d'Orval's *Essay*, his *Traité*, his *Manuel de l'artificier* (1755), and his *Manuel de l'artificier* (1757), 114. Some secondary sources incorrectly cite Frézier as the author of the *Manuel*.
28. Perrinet d'Orval, *Essay*.
29. Frézier, *Traité des feux d'artifice* (1747), 92. On Perrinet's life, see Kafker and Kafker, *The Encyclopedists as Individuals*, 296–97.
30. For the regulations on artificer's shops, see *Guide des corps des marchands*, 159–60 ("Artificiers"); and Robinet, *Dictionnaire universel*, 291–92 ("Artificier").
31. On French artillery schools, see Boblaye, *Esquisse historique*.
32. Bélidor, *Le Bombardier françois*, 312–66; *A Colbert: Traité des manoeuvres et de l'artif[i]ce de guerre et de joie pour l'artillerie, dédié á messieur[s] du Corps royal de l'artillerie, regiment de La Fère*, Guttmann Collection on Explosives, Firearms, and Military Science, 1450–1905, Hagley Museum and Library, Wilmington, Delaware, ref. 1277.
33. "Compositions differentes employé par ordre dans les Soleil tournant de neuf cartouches par le Sr. Colbert artificier, il ne faut rien changer à ce qu'il dit, parce que les feux seroient defigurés et perderoient leur beautés, ne faire rien" (*A Colbert: Traité des manoeuvres* [n. 32 above], fol. 113).
34. Notebooks by Gaetano and Petronio Ruggieri are in the Collection Ets Ruggieri S.A., Monteux. See also W.M.M., "Designs for Artificial Fire Works," English manuscript, after 1788, Hagley Museum and Library, Wilmington, Delaware, ref. 1986; and "Pirotechnie artificielle," French manuscript, after 1755, Guttmann Collection on Explosives, Firearms, and Military Science, 1450–1905, Hagley Museum and Library, Wilmington, Delaware, ref. 1277.
35. Perrinet d'Orval, *Essay*, 153.
36. Ibid., 153–80.
37. Ibid., iv.
38. Ibid., v.
39. *Nouvelle biographie générale*, 18:859–65 ("Frézier [Amédée-François]").
40. Frézier, *A Voyage to the South-Sea*.
41. See Kellman, "Discovery and Enlightenment at Sea," 264–85.
42. Frézier, *Traité des feux d'artifice* (1747), vi.
43. Ibid., xxiv.
44. Ibid., 392–93, 416–21, 475–78.
45. Ibid., 384.
46. Ibid., 393.
47. Ibid., 393.
48. Ibid., vii.
49. Ibid., 446 (see generally 437–48).
50. Bohun, "The Royal Fireworks and the Politics of Music."
51. On the Hanoverian court and public spectacle, see Smith, *Georgian Monarchy*, 232–43.
52. Voltaire, *Letters Concerning the English Nation*, 10.
53. See Joseph Addison's remarks on Italian opera in *The Spectator*, 1:27–32 (no. 5, March 6, 1711). See also Knif, *Gentlemen and Spectators*, 49–58. On attitudes toward con-

sumption, see Berg and Eger, eds., *Luxury in the Eighteenth Century*; and Berg, *Luxury and Pleasure*.

54. On theatrical and public venues, see Brewer, *The Pleasures of the English Imagination*; and Altick, *The Shows of London*.
55. Wroth, *London Pleasure Gardens*; Scott, *Green Retreats*; Southworth, *Vauxhall Gardens*, esp. 84–105; Sands, *Pleasure Gardens of Marylebone*, and *Invitation to Ranelagh*; Ogborn, *Spaces of Modernity*, 116–57.
56. Sands, *Pleasure Gardens of Marylebone*, 5.
57. "Rural Beauty, or Vaux-Hall-Gardens" (June 1739), in *Historical Collections Relative to Spring Gardens, Charing Cross, and to Vauxhall Gardens*, comp. J. H. Burns, British Library, CUP 401.k.7, fol. 124.
58. *The Evening Lessons; Being the First and Second Chapters of the Book of Entertainments* (London, 1742), in Burns, comp., *Historical Collections Relative to Spring Gardens* (n. 57 above), fols. 139–50, 149.
59. Stephens, "Desaguliers, Thomas"; Askwith and Kane, *List of Officers*, 228.
60. Horace Walpole to Henry Seymour Conway, October 6, 1748, in Walpole, *Letters*, 2:132. See also Aspinall-Oglander and Boscawen, *Admiral's Wife*, 123; and Ayres, *Classical Culture and the Idea of Rome*.
61. Several of these prints are in the Crace Collection, British Museum, portfolio 12, "St. James's Park," sheets 104–16. On prints and souvenirs, see O'Connell, Porter, Fox, and Hyde, *London, 1753*, 222–24; and *Gentleman's Magazine* 18 (February 1749): 55–56.
62. Horace Walpole to Sir Horace Mann, October 24, 1748, in Walpole, *Letters*, 2:134.
63. Johnson, "Letter on Fireworks," 549.
64. Ibid.
65. See the notice of January 10, 1748 [1749], in *Gentleman's Magazine* 18, suppl. (December 1748): 599.
66. Ruggieri, *A Description of the Machine for the Fireworks*, 3.
67. Ibid., 9.
68. Ibid., 4.
69. Despite an offer of £500 to George Vertue to complete twenty-eight plates of architectural drawings, no doubt on the model of Blondel's festival book of 1739, only one materialized, for which the Ordnance paid Vertue a measly twenty guineas. See O'Connell et al., *London, 1753*, 222–23; and Myrone, "Vertue, George."
70. Brewer, *The Pleasures of the English Imagination*, 25–29. On negotiations between Tyers, Handel, Montagu, and Frederick, see Hogwood, *Handel* (1984), 214–16. Thomas Gray described the pyrotechnics in *Atalanta*. See Thomas Gray to Horace Walpole, June 11, 1736, in Gray, *Correspondence*, 44–45 (esp. 44).
71. *Gentleman's Magazine* 19 (April 1749): 185.
72. *Gentleman's Magazine* 19 (April 1749): 186–87.
73. *Daily Advertiser*, April 29, 1749; *Gentleman's Magazine* 19 (April 1749): 187; John Byrom to Mrs. Byrom, February 11, 1748/9, in Byrom, *Private Journal*, 2, pt. 2:484–85.
74. *The Green-Park Folly*, 8.
75. *An Account of the Famous Sieur Rocquet*; *An Ode for the Thanksgiving-Day*; "A Pindaric Ode upon Oddities."
76. *Gentleman's Magazine* 19 (May 1749): 227.
77. Lady Jemima, Marchioness Grey, to Lady Mary Gregory, April 28, 1749, quoted in Hogwood, *Handel* (2005), 93.

78. Horace Walpole to Sir Horace Mann, May 3, 1749, in Walpole, *Letters*, 2:151–52. For another account, see "By Capt. Nealson, Arrived at York from Bristol, There Are the Following Advices. London, April 25," *Pennsylvania Gazette*, no. 1073 (July 6, 1749).
79. "From the Gazeteer, May 1. On the Fireworks," *Gentleman's Magazine* 19 (May 1749): 204–5, 204. See also the letter from "Anti-Pyrobolos," *Gentleman's Magazine* 19 (May 1749): 220–21.
80. *The Grand Whim for Posterity to Laugh At* (London, 1749), engraving, British Museum, Department of Prints and Drawings, Crace Collection.
81. Jones, *A New Treatise on Artificial Fireworks*, vii.
82. Ruggieri participated in fireworks for the birthday of Lord Granby at Woolwich in 1767. See Benjamin Allin to Captain Mollman, July 9, 1767, Public Record Office, London, Office of Ordnance, PRO Supp 5, 79; and Hogg, *The Royal Arsenal*, 1:449.
83. Four Clanfield trade cards may be found in the British Museum, Department of Drawings and Prints: Heal Collection of trade cards and shop bills, ref. 62.4 (n.d., ca. 1760s); Sarah Banks Collection of trade cards, 62.3 (n.d., after 1749); Sarah Banks Collection, ref. 62.4 (ca. 1780); Sarah Banks Collection, ref. 62.5 (ca. 1809). The first identifies "Samuel Clanfield, Original Engineer to Ranelagh Cupers and Marylebone Gardens, at the Royal Fireworks in Hosier Lane, West Smithfield, London."
84. Guildhall Library MS 14022/9, Union, Board of Directors Minutes, December 14, 1748. See also Pearson, *Insuring the Industrial Revolution*, 84.
85. Stewart, *The Rise of Public Science*, 137–38, 176–79, 269–70, 270–78.
86. Stephen Hales, "A Proposal for Checking in Some Degree the Progress of Fires," *Philosophical Transactions of the Royal Society* 45 (1748): 277–79; Stewart, *The Rise of Public Science*, 222–23, 231–34.
87. Cromwell Mortimer, "An Addition to Dr. Hales's Paper, p. 279," *Philosophical Transactions of the Royal Society* 45 (1748): 302.
88. On gunners' methods for fireproofing buildings, not the same as Hales's, see, e.g., Siemienowicz, *The Great Art of Artillery*, 111–15.
89. "A Geometrical Use Proposed from the Fire-Works," *Gentleman's Magazine* 18 (November 1748): 488.
90. Robins, *New Principles of Gunnery*; Steele, "Muskets and Pendulums."
91. *Proposal of William Gee to [Newcastle] for a System of Signalling by Rockets in the Evening to Be Instituted, to Be Used If the Army in the North Required Assistance against the Rebels from Other Parts of the Country or When an Enemy Invasion in Remote Parts Seemed Imminent. October 9th 1745*, Public Record Office, London, SP 36/71. For another scheme, see Captain Robert Hamblin, "Signal Lights" (1730), Patent no. 517, British Library.
92. "A Geometrical Use . . . from the Fire-Works" (n. 89 above), 488.
93. Philo Pyrotechnicus, "Rules for Computing the Height of Rockets, &c.," *Gentleman's Magazine* 18, suppl. (December 1748): 597; *Gentleman's Magazine* 19 (February 1749): 55–56. Similar ideas appeared in the Dutch press. See "Of Rockets, and What Useful Purposes They May Be Made to Serve: Taken from the Hague Gazette," *London Magazine; or, Gentleman's Monthly Intelligencer* 18 (1749): 212–14.
94. See the letter from Thomas Ap Cymra in Carmarthen, *Gentleman's Magazine* 19 (May 1749): 217–18.

95. Benjamin Robins, "Observations on the Height to Which Rockets Ascend," *Philosophical Transactions of the Royal Society* 46 (1749): 131–33.
96. Robins, "Observations" (n. 95 above), 132–33.
97. Ibid., 133.
98. John Ellicott, "An Account of Some Experiments, Made by Benjamin Robins Esq; F.R.S. Mr. Samuel Da Costa, and Several Other Gentlemen, in Order to Discover the Height to Which Rockets May Be Made to Ascend, and to What Distance Their Light May Be Seen," *Philosophical Transactions of the Royal Society* 46 (1749): 578–84. The trials were also noticed in Stukeley, Lukis, Gale, and Gale, *The Family Memoirs of the Rev. William Stukeley*, 2:374.
99. Ellicott, "An Account" (n. 98 above), 581.
100. Ibid., 583.
101. Ibid., 582.
102. John Canton, "A Letter to the Astronomer Royal, from John Canton, M.A. F.R.S. Containing His Observations of the Transit of Venus, June 3, 1769, and of the Eclipse of the Sun the Next Morning," *Philosophical Transactions of the Royal Society* 59 (1769): 192–94, esp. 193.
103. Francis Wollaston, "On a Method of Describing the Relative Positions and Magnitudes of the Fixed Stars; Together with Some Astronomical Observations," *Philosophical Transactions of the Royal Society* 74 (1784): 181–200.
104. "Asiatic Researches," *Monthly Review; or, Literary Journal, Enlarged: From May to August 1790 . . . Volume II* (London, 1790), 515–24, 516.
105. Stählin, "Kratkaia istoriia," 251. Recall that this is probably not the weight of the rockets but the weight of a shot fitting their molds.
106. Stählin ("Kratkaia istoriia," 259–61) briefly mentions Sarti, who should not be confused with the musician and *Kapellmeister* of the same name who later worked in Russia.
107. On Lomonosov, see Menshutkin, *Russia's Lomonosov*; and Leicester, ed., *Lomonosov on the Corpuscular Theory*.
108. Alekseeva et al., eds., *Graviroval'naia palata*, 18, 90–91; Malinovskii, "Iakob von Shtelin i ego zapiski," 174; Ostrovitianov, ed., *Istoriia Akademii nauk*, 184–86; Konopleva, *Teatral'nyi zhivopisets Dzhuzeppe Valeriani*. Peresinotti's and Gradizzi's names appear as draftsmen on numerous engravings of fireworks in Russia from the 1750s to the 1780s.
109. See Ritzarev, *Eighteenth-Century Russian Music*, 41–45.
110. Stählin, "Kratkaia istoriia," 247, 255.
111. Ibid., 259.
112. Danilov, "Zapiski," 37.
113. Rovinskii, *Opisanie feierverkov*, 252.
114. Ibid., 253.
115. Danilov, "Zapiski," 37–38.
116. "Green Fire," *Museum of Foreign Literature, Science, and Art* 1, no. 6 (1822): 568.
117. Brock (*A History of Fireworks*, 153) claimed that Hanzelet was exceptional and "far in advance of his time" in proposing verdigris as a means to color fire (Hanzelet, *La pyrotechnie*, 257), but the recipe was common. See, e.g., Wright, "Notes on Gunnery" (n. 44, chap. 1, above), fols. 20, 22, in Walton, "The Art of Gunnery," 165, 374–75; Siemienowicz, *The Great Art of Artillery*, 106; Venn, *The Compleat Gunner*, 22; Frézier, *Traité des feux d'artifice* (1747), 36; and Alberti, *La pirotechnia*, 31.

118. See de Sangro, *Lettera apologetica,* 215. De Sangro claimed that he could produce "diversa degradazione" of green, "com' è, il verde mare, il verde smeraldo, e 'l verde prato." See also "Extrait d'une lettre de Madame la Duchesse de S . . . contenant l'éloge & le dénombrement des découvertes dans les sçiences & dans les arts de D. Raimond de Sangro, prince de S. Severo, Napolitain," *Journal Oeconomique,* April 1757, 175–79, and June 1757, 137–46, 138; and Lorenzo, *Il teatro del fuoco,* 88–91. Thanks to Lucia Dacome for pointing out de Sangro's interest in fireworks.
119. Rovinskii, *Opisanie feierverkov,* 186–87; *Sankt-Peterburgskie vedemosti,* no. 9 (January 31, 1737): 70.
120. *Encyclopédie méthodique,* 119.
121. Baehr, *The Paradise Myth,* 65–89. On the meanings of green, see Gage, *Colour and Culture.*
122. Harley, *Artists' Pigments,* 80–83.
123. Mikhail Ivanovich Agentov composed the collection *Otkrytie sokrovennykh khudozhestv* from various German authors. For green paint recipes, see ibid., 3:249, 289.
124. Danilov, "Zapiski," 38.
125. Ibid. A "Columbus's egg" is something that appears difficult or impossible yet is easy once a way has been found to do it, following a story told of Columbus.
126. Bennett, "Evolution of the Meanings of *Chin.*"
127. Similarly, in seventeenth-century England, only gentlemanly status could authorize discoveries. See Shapin, "The Invisible Technician." On the protection of inventions in Russia, see Aer, *Patents in Imperial Russia.*
128. Danilov, "Zapiski," 38–39.
129. Ibid., 39.
130. Ibid., 39; Stählin, "Kratkaia istoriia," 269 n. 25.
131. On Lomonosov's pyrotechnic designs, see Pavlova, "Proekty illiuminatsii Lomonosova"; Lomonosov, *Polnoe sobranie sochineniia,* vol. 8, passim; and Pekarskii, *Istoriia Imperatorskoi Akademii nauk,* 2:910–16.
132. For Lomonosov's request to finish pyrotechnic work in March 1755, see Pekarskii, *Istoriia Imperatorskoi Akademii nauk,* 1:548. On Lomonosov's mosaics, see Leicester, "Mikhail Lomonosov"; and Menshutkin, *Russia's Lomonosov,* 93–105.
133. Lomonosov, *Polnoe sobranie sochinenii,* 8:552 (see also 528).
134. "Pis'mo o piricheskikh predstavleniiakh," *Ezhemesiachyia Sochineniia k pol'ze i uveseleniiu sluzhashchiia,* July 1755, 58–65.
135. Biliarskii, *Materialy dlia biografii Lomonosova,* 313.
136. Lomonosov, "Oration on the Origin of Light."
137. Rovinskii, *Opisanie feierverkov,* 246–47, 255, 263; *Opisanie allegoricheskago izobrazheniia . . . sentiabr 1762 goda,* n.p.
138. For Sarti's fireworks, see Rovinskii, *Opisanie feierverkov,* 250, 252, 257–58, 261, 264.
139. Lomonosov, "Oration on the Use of Chemistry," 198–99.
140. Lomonosov, "Oration on the Origin of Light," 269.
141. Ibid.
142. Other chemists drew on arts such as textile dyeing for new theories of light in the eighteenth century, though Lomonosov's reliance on mosaics and fireworks was unique. See Shapiro, *Fits, Passions and Paroxysms,* 242–67.
143. Lomonosov, "Oration on the Origin of Light," 264.
144. Ibid., 268.
145. Ibid., 269.

146. On Melissino, see Makovskaia, "Melissino I"; Stählin, "Kratkaia istoriia," 261, 263; and Danilov, "Zapiski," 39.
147. Stählin, "Kratkaia istoriia," 261.
148. For example, all three presented fireworks at New Year's 1760. See Sumarokov, *Opisanie ognennago predstavleniia*, 263–65.
149. Stählin, "Kratkaia istoriia," 269 n. 25.
150. There was "an agreeable vista of two rows of palm trees in green fire. At the center six rows of machines of an admirable and new invention were presented, changing their fires continuously" (*Opisanie allegoricheskago izobrazheniia . . . sentiabr 1762 goda*, 3).
151. Stählin, "Kratkaia istoriia," 264.
152. Jacob Stählin to Peter Stählin, December 1, 1770, Stählin Papers, Department of Manuscripts, Russian National Library, St. Petersburg, F. 871, no. 274, letter 44.

CHAPTER SIX

1. Langford, *A Polite and Commercial People*.
2. Boas, "The Arts in the *Encyclopédie*"; Ehrard, "La main du travailleur"; Sewell, "Visions of Labour"; Pannabecker, "Representing Mechanical Arts."
3. Schaffer, "Enlightened Automata"; Koepp, "The Alphabetical Order."
4. d'Alembert, *Preliminary Discourse to the Encyclopedia of Diderot*, 42 (see generally 3–59).
5. Siemienowicz, *The Great Art of Artillery*, 1–30.
6. Ibid., 2.
7. Frézier, *Traité des feux d'artifice* (1706), 62, and *Traité des feux d'artifice* (1747), 90. Hence, if a one-pound rocket's diameter ($d$) was one hundred parts and one sought the diameter ($d'$) of a two-pound rocket, this was

$$d = 100,$$
$$d^3 = 1{,}000{,}000,$$
$$d'^3 = \sqrt[3]{2{,}000{,}000},$$
$$d' = 125 \text{ parts.}$$

8. Perrinet d'Orval, *Essay*, 41. See also Frézier, *Traité des feux d'artifice* (1747), 89.
9. Perrinet's classification was as follows:

8 *lignes—petit partement;*
10 *lignes—partement;*
12 *lignes—la marquise;*
14 *lignes—double marquise;*
16 *lignes—fusées de trois douzaines;*
18 *lignes—fusées de quatre douzaines;*
21 *lignes—fusées de cinq douzaines.*

The names of the last three sizes referred to garniture, the number of serpents or stars that fitted inside the rocket's cap.

10. Perrinet d'Orval, *Essay*, 41.
11. Frézier, *Traité des feux d'artifice* (1747), 84–106.
12. Ibid., 90.
13. Ibid., 88.

14. Ibid., 89.
15. Ibid., 92.
16. Ibid., 93.
17. Perrinet d'Orval, *Traité*, preface.
18. Frézier, *Traité des feux d'artifice* (1747), 97–98.
19. Ibid., ix.
20. Langins, *Conserving the Enlightenment*, 234–37, and "Eighteenth-Century French Fortification Theory," 337.
21. Daston, "The Physicalist Tradition."
22. Kafker and Kafker, *The Encyclopedists as Individuals*, 296.
23. Cahusac, "Feu d'artifice," 639–40. On Cahusac, see Kafker and Kafker, *The Encyclopedists as Individuals*, 79–82.
24. Cahusac, "Feu d'artifice," 639 (my translation).
25. Cahusac, "Feu d'artifice," 639, trans. in Salatino, *Incendiary Art*, 23.
26. Cahusac, "Feu d'artifice," 639 (my translation).
27. Diderot, "Art." On Diderot and the arts, see Rossi, *Philosophy, Technology, and the Arts*, 134–36; and Furbank, *Diderot*, 38–40.
28. Diderot, "Art," 716–17.
29. Ibid., 717.
30. Cru, *Diderot as a Disciple of English Thought*, 245–55; Dieckmann, "The Influence of Francis Bacon."
31. Diderot, "Art," 717. Louis Jaucourt's article on invention in the *Encyclopédie* similarly demanded a critical role for the philosopher in artisanal innovation: "We need no longer fear discoveries will be lost from memory; the facts will be unveiled to the philosophers, and reflection will be able to simplify and clarify blind practice" (Jaucourt, "Invention," 849).
32. Diderot, "Art," 717.
33. Ibid., 715.
34. See, e.g., Pannabecker, "Representing Mechanical Arts"; Koepp, "The Alphabetical Order"; and Sewell, "Visions of Labour."
35. Diderot, "Art," 717.
36. Perrinet d'Orval, *Manuel de l'artificier* (1757), 114.
37. The distinction between concealment and technique is explored in Long, "Invention, 'Intellectual Property,' and the Origin of Patents," 860.
38. Ruggieri, *Principles of Pyrotechnics*, 24.
39. Parfaict and Parfaict, *Dictionnaire des théâtres de Paris*, 7:521–22; Gofflot, *Théâtre au collège*, 297–98. The Ruggieri may have shown the Chinese fire first in 1753, when a display celebrating the birth of the duc d'Aquitaine included rockets "of the Chinese composition" and a forty-foot-high cascade "also in Chinese fire," according to Ruggieri, *Précis historique*, 282–83. See also Luynes, *Mémoires*, 14:223.
40. See Spence, *The Memory Palace of Matteo Ricci*, 45; Ricci, *China in the Sixteenth Century*, 18, 77, 321.
41. Athanasius Kircher quoted in Nieuhof, *An Embassy from the East-India Company*, 426–27.
42. Honour, *Chinoiserie*; Jarry, *Chinoiserie*.
43. Izmailov quoted in Vasil'iev, *Starinnye feierverki*, 55.
44. Le Comte, *Memoirs and Observations*, 167.
45. Porte, *Le voyageur françois*, 5:117–18.

46. Only in the nineteenth century was it claimed that the search for green fire was an attempt to emulate the Chinese. See Cheleev, *Polnoe i podrobnoe nastavlenie*, 229.
47. Du Halde, *The General History of China*, 2:168.
48. Golikov, *Dopolneniia k Deianiiam Petra Velikogo*, 13:294.
49. Macknight, *The Life of Henry St. John, Viscount Bolingbroke*, 531.
50. Parfaict and Parfaict, *Dictionnaire des theatres de Paris*, 7:521. The artificer was presumably not Chinese.
51. These included accounts of Chinese horn lanterns, seed drills, varnishes, paper, and paints. See, e.g., Incarville, "Mémoire sur la manière singulière," and "Mémoire sur les vernis." See also Bernard, "Un correspondant de Bernard de Jussieu en Chine."
52. Perrinet d'Orval, "Feux d'artifice," 643, where Perrinet explained that he was "indebted" (*redevable*) to Incarville for this preparation, "which up to the present has been a secret."
53. The submission is noted in *Histoire de l'Académie royale des sciences*, 245. It was published as Incarville, "Maniére de faire les fleurs."
54. Incarville, "Maniére de faire les fleurs," 66–72.
55. Ibid., 74–87.
56. "Manner of Making Flowers in the Chinese Fire-Works, Illustrated with an Elegantly Engraved Copper-Plate.—From the Fourth Volume [just published] of the Memoirs Presented to the Academy of Sciences," *Universal Magazine* 34 (1764): 20–23. See also Jones, *A New Treatise on Artificial Fireworks*, 44, 104, 162, 193, 196.
57. Incarville, "Maniére de faire les fleurs," 83.
58. See, e.g., Heilbron, *Electricity*; Fara, *An Entertainment for Angels*.
59. Albrecht von Haller, "An Historical Account of the Wonderful Discoveries Made in Germany, &c. Concerning Electricity," *Gentleman's Magazine* 15 (1745): 193–97, 193.
60. Bohnenberger quoted in Hochadel, "Öffentliche Wissenschaft," 242. The book in question was Bohnenberger, *Beyträge zur theoretischen und praktischen Elektrizitätslehre*.
61. De Tott, *Memoirs*, 1, pt. 2:86.
62. Haller, "An Historical Account" (n. 59 above), 196.
63. Heilbron, *Electricity*, 312–16; Riskin, *Science in the Age of Sensibility*, 68–103.
64. Priestley, *The History and Present State of Electricity*, 327.
65. Ibid., 563.
66. William Watson, "Experiments and Observations, Tending to Illustrate the Nature and Properties of Electricity: By William Watson, Apothecary, F.R.S.," *Philosophical Transactions of the Royal Society* 43 (1745): 481–501, 483.
67. William Watson, "A Collection of the Electrical Experiments Communicated to the Royal Society by Wm. Watson, F.R.S. Read at Several Meetings between October 29. 1747. and Jan. 21. Following," *Philosophical Transactions of the Royal Society* 45 (1748): 49–120, 111.
68. Home, *The Effluvial Theory of Electricity*.
69. Dyche, *Manuel lexique*, 1:583.
70. Jallabert, *Expériences sur l'électricité*, 44.
71. "Le rècipient se remplit d'une grande quantité de jets de feu, qui se meuvent en serpentant avec une rapidité étonnante" (Nollet, *Lettres sur l'électricité* [1753], 75). See also Home, *The Effluvial Theory of Electricity*, 226–27. Nollet's discussion of rocket motion may be found in Nollet, *Lectures in Experimental Philosophy*, 270–71. Nollet served as a professor of physics at the artillery school of La Fère for several years. See Grandjean de Fouchy, "Eloge de J.-A. Nollet," 133.

72. Nollet, *Essai sur l'électricité*, 178–79.
73. Nollet, *Lettres sur l'électricité* (1767), 292.
74. Rabiqueau, *Le spectacle du feu élémentaire*, 43.
75. Benjamin Wilson, "Farther Experiments in Electricity," *Philosophical Transactions of the Royal Society* 51 (1760): 896–906, 906. Wilson refers to "table fireworks" discussed below.
76. Mills, *An Essay on the Weather*, 18–19.
77. Watson, "A Collection of the Electrical Experiments" (n. 67 above), 50.
78. Advertisement for Mulberry Gardens, *Daily Post*, no. 7165 (August 23, 1742).
79. "Une Homme éléctrisé touche-t-il de la main un tas de pierreries ou de vaisselle d'argent; on en voit sortir du feu de tous côtés. C'est le feu d'artifice le plus brillant qu'on ait jamais vu" (Massuet, *Suite de la science des personnes de cour*, 1:350).
80. See "Abstract of What Is Contained in a Book Concerning Electricity, Just Published at Leipzic, 1744. By John Henry Wintler, Greek and Latin Professor There; from Article 75 to Article 79," *Philosophical Transactions of the Royal Society* 43 (1744–45): 166–69; Winkler, *Elements of Natural Philosophy*, 1:291–92; Meya and Sibum, *Das fünfte Element*, 50–54.
81. See Kinnersley's advertisement in the *Boston Evening Post*, October 7, 1751, reproduced in Morse, "Lectures on Electricity in Colonial Times," 367–69. See also Jallabert, *Experiences sur l'électricité*, 52–54.
82. Walker, *Syllabus of a Course of Lectures*, 9. See also "Application of Electricity to Discharge Cannon."
83. On the techniques of illuminating letters with fireworks, see Bélidor, *Le Bombardier françois*, 360; and Smith, *The Laboratory*, 100–102.
84. Winkler, *Die Eigenschaften der electrischen Materie*, 66; Meya and Sibum, *Das fünfte Element*, 54.
85. Nollet, *Lettres sur l'électricité* (1767), 274–95.
86. Ibid., 275.
87. Ibid., 287, 290, 294. Joseph Priestley figured variation as critical to scientific epistemology, explaining how philosophical instruments made it possible to exhibit the "operations . . . of the God of nature himself, which are infinitely various. By the help of [natural philosophical instruments], we are able to put an endless variety of things into an endless variety of situations . . . hereby the laws of [Nature's] action are observed" (*The History and Present State of Electricity*, x). On the continuing value of variety in eighteenth-century aesthetics, see Hogarth, *The Analysis of Beauty*, 16–17; and Burke, *Philosophical Enquiry*, 299–301.
88. Nollet advocated nighttime performances, to make electrical effects more observable. See Nollet, *Lettres sur l'électricité* (1753), 75.
89. Ibid., 287–88.
90. Ibid., 292–94.
91. "The preceeding Part of the Week had been remarkably warm, and the Air very dry: than which nothing is more necessary towards the success of electrical Trials" (Watson, "Experiments and Observations" [n. 66 above], 481). The Boston lecturer Ebenezer Kinnersley noted air and wind conditions conducive to electrical experiments. See Kinnersley's advertisement (see n. 81 above) reproduced in Morse, "Lectures on Electricity in Colonial Times," 367–69.
92. Nollet, *Lettres sur l'électricité* (1767), 290–91.
93. For brevity's sake, I will not deal with electrical medicine here, but see Bertucci and Pancaldi, eds., *Electric Bodies*; and Morus, *Frankenstein's Children*, 125–52.

94. William Watson, "A Letter from Mr. William Watson, F.R.S. to the Royal Society, Declaring That He as Well as Many Others Have Not Been Able to Make Odours Pass thro' Glass by Means of Electricity; and Giving a Particular Account of Professor Bose at Wittemberg His Experiment of Beatification, or Causing a Glory to Appear Round a Man's Head by Electricity," *Philosophical Transactions of the Royal Society* 46 (1749): 348–56, 355.
95. Priestley, *The History and Present State of Electricity*, 630–31. Stukeley supported electrical models of earthquakes and attacked the pyrotechnic account. See Stukeley, *The Philosophy of Earthquakes*, 6–7; and Schaffer, "Experimenters' Techniques."
96. Kinnersley's advertisement (see n. 81 above) reproduced in Morse, "Lectures on Electricity in Colonial Times," 368.
97. Franklin, *Experiments and Observations*, 9–10.
98. Warner, "Lightning Rods and Thunder Houses."
99. Becket, *An Essay on Electricity*, 140–42. See also Briggs, "Explanations of the Aurora Borealis." The aurora borealis was regularly compared to fireworks. See, e.g., Regnault, *Philosophical Conversations*, 3:144. See also the discussion of Johann Georg Gmelin's account of the aurora borealis in Charles Blagden, "An Account of Some Late Fiery Meteors; with Observations. In a Letter from Charles Blagden, M.D. Physician to the Army, Sec. R.S. to Sir Joseph Banks, Bart. P.R.S.," *Philosophical Transactions of the Royal Society* 74 (1784): 201–32, 228.
100. Priestley, *The History and Present State of Electricity*, 523.
101. Nollet quoted in Torlais, *Un physicien au siècle des lumières*, 74.
102. Cavallo, *A Complete Treatise of Electricity*, 267–68.
103. Winkler, *Elements of Natural Philosophy*, 2:70.
104. Franklin, *Experiments and Observations*, 47–49.
105. Babington, *Pyrotechnia*, 45.
106. Anderson, *Institutes of Physics*, 122.
107. Captain William Gordon, "A Letter from Capt. William Gordon to Capt. Samuel Mead, F.R.S. Inclosing an Account of the Fire-Ball Seen Dec. 11. 1741," *Philosophical Transactions of the Royal Society* 42 (1742–43): 58–60.
108. Wilson, "Biographical Account of Alexander Wilson," 284–86.
109. Jankovic, "Meteors under Scrutiny," 276–84; Knowles Middleton, *A History of the Theories of Rain*, 111–15.
110. Priestley, *The History and Present State of Electricity*, 356, summarizing Stukeley.
111. Bertholon, *De l'électricité des météores*.
112. Bertholon, "Mémoire," 488.
113. Ibid. Bertholon explains how to imitate, and, thus, give "conviction" to the electrical explanation of, the aurora borealis in *De l'électricité des météores*, 2:76–82.
114. Ibid., 2:17.
115. Ibid., 1:341–42.
116. Ibid., 345.
117. Ibid., 2:14–18.
118. Winkler, *Elements of Natural Philosophy*, 2:68–75.
119. Erasmus Darwin, "Remarks on the Opinion of Henry Eeles, Esq; Concerning the Ascent of Vapour, Published in the Philosoph. Transact. Vol. xlix. Part i. 124," *Philosophical Transactions of the Royal Society* 50 (1757): 240–54.
120. "A Letter from Mr. Henry Eeles, to the Royal Society, Concerning the Cause of Thunder," *Philosophical Transactions of the Royal Society* 47 (1751–52): 524–29, 525. Eeles's views were developed in "Letters of Henry Eeles, Esq; Concerning the Cause of the

Ascent of Vapour and Exhalation, and Those of Winds; and of the General Phaenomena of the Weather and Barometer. To the Rev. Tho. Birch, D.," *Philosophical Transactions of the Royal Society* 49 (1755): 124–54. See also Eeles, *Philosophical Essays.*

121. Eeles, "A Letter from Mr. Henry Eeles" (n. 120 above), 527.
122. Darwin, "Remarks on . . . Henry Eeles" (n. 119 above), 241.
123. Ibid.
124. Ibid., 242.
125. Ibid., 247.
126. Hales, *A Treatise on Ventilators,* 304–11.
127. Ibid., 305.
128. Ibid., 306.
129. Ibid., 306–7.
130. Ibid., 307.
131. Beccaria, *Treatise,* 440, 450.
132. Lyon, *Experiments and Observations,* 214–40.
133. Blesson, "Observations on the Ignis Fatuus."
134. Cohen, *Benjamin Franklin's Experiments,* 212–36.
135. On the controversy, see Schaffer, "Fish and Ships," esp. 82–86; and Mitchell, "The Politics of Experiment."
136. Wilson, *An Account of Experiments Made at the Pantheon.* On the Pantheon, see Malcolm, *Anecdotes,* 2:276–78.
137. See the biographical account of Benjamin Wilson in Wilson, *Life of General Sir Robert Wilson,* 1:18–21 (see generally 3–43).

CHAPTER SEVEN

1. On the rise of the public, see Melton, *The Rise of the Public;* Goodman, "Public Sphere and Private Life"; Volkov, "The Forms of Public Life"; and Smith, *Working the Rough Stone,* 53–90.
2. Roche, *The Culture of Clothing.*
3. Bodleian Library, Oxford, MS Rawlinson D 862, fol. 83, quoted in Rogers, "Crowds and Political Festival," 239.
4. *Public Advertiser,* September 22, 1761.
5. Servandoni, *Description du Spectacle de Pandore,* quoted in Olivier, "Jean-Nicolas Servandoni's Spectacles," 37. See also Campardon, *Les spectacles de la foire,* 2:395.
6. *Annonces, affiches, et avis divers: Feuille périodique* (hereafter *Affiches de Paris*), no. 56 (July 20, 1747), no. 57 (July 24, 1747); Isherwood, *Farce and Fantasy,* 50.
7. Le Normand d'Étioles, *Memoirs,* 2:106.
8. Thus, Mrs. Benjamin Clitherow died when her home and workshop exploded in 1791. See Toone, *The Chronological Historian,* 386. Her death was made into a memento mori in verse. See *The True and Particular Account.* Mary Cooper's fate was recorded in "Judgment Mixed with Mercy, Exemplified in the Death of Mary Cooper," *Methodist Magazine* 22 (1799): 453–56. The table "Artificiers du roi de la ville et fauxbourgs de Paris" (*Essai sur l'almanach,* n.p.) identifies four widows as *artificiers.*
9. *Avant-coureur,* December 2, 1765, 750–51. See also Lynn, "Sparks for Sale," 86–87.
10. *Avant-coureur,* December 2, 1765, 750–51.
11. *Avant-coureur,* December 21, 1767, 803–4.
12. *Avant-coureur,* December 2, 1765, 750–51.
13. Höckely, *Anhang,* 143–80.

14. Ibid., 143. The small fireworks for a room advertised by Clitherow were evidently table fireworks.
15. Ibid., 142.
16. Saint-Julien, *L'art de composer et faire les fusées*; [Grignon], *La pyrotecnie pratique*; Morel, *Traité pratique des feux d'artifice*.
17. Danilov, *Dovol'noe i iasnoe pokazanie . . . delat' vsiakie feierverki*, preface.
18. Smith, *The Laboratory*, frontispiece.
19. Montucla, ed., *Récréations mathématiques*. On Montucla, see Sarton, "Montucla"; and Guyot, *Nouvelles récréations*.
20. Montucla, *Récréations mathématiques*, 4:358–61, 377–78.
21. Hutton, *Recreations*.
22. Hooper, *Rational Recreations*, 1:iv.
23. Daston and Park, *Wonders and the Order of Nature*, 329–63.
24. *The Green-Park Folly*, 10. The poet Edward Young expressed a similar contrast in natural theological terms: "Instead of Squibbs, & Crackers, I shall humbly content myself with Sun, Moon, & Stars. These glorious Fireworks of that Great King who in ye noblest sense is ye Author of Peace; & Lover of Concord" (Edward Young to Margaret Cavendish Bentinck, Duchess of Portland, February 9, 1749, in Young, *Correspondence*, 312).
25. Neufchâteau, *Lettre à un ami*, 11.
26. Rousseau, *Julie*, 491.
27. Hooper, *Rational Recreations*, 1:vi.
28. *London Chronicle*, no. 4927 (May 24, 1788), 509.
29. Daston and Vidal, eds., *The Moral Authority of Nature*.
30. Neufchâteau, *Lettre à un ami*, 6.
31. Sigaud de la Fond, *Description et usage d'un cabinet de physique expérimentale*, 2: 415–19.
32. Adams, *An Essay on Electricity*, 196–99.
33. Bennett, *Letters to a Young Lady*, 2:45.
34. Parfaict and Parfaict, *Dictionnaire des théâters de Paris*, 7:521–22.
35. *A View of the Grand Hydraulick Water Works (as Exhibited at Exeter-Change in the Strand) by James Bourier, Mechanist to the King of Poland* (mezzotint, late eighteenth century; a copy is available in the British Museum, Department of Prints and Drawings, British XVIIIc Mounted Roy). See also Altick, *The Shows of London*, 80.
36. *Mémoire sur les feux d'air inflammable par M. Diller*. For more on Diller's fireworks, see Werrett, "From the Grand Whim to the Gasworks."
37. "Philosophical Fire," *Scots Magazine* 50 (April 1788): 164.
38. Guyot, *Nouvelles récréations*, 2:269 (quote), 269–85, 333–34 (instructions). See also Lacombe, *Dictionnaire*, 524–28, 830–38; and Hooper, *Rational Recreations*, 4:156–71.
39. Hooper, *Rational Recreations*, 4:162.
40. Ibid., 2:189–91. On *vues d'optique*, see Stafford, Terpak, and Poggi, *Devices of Wonder*, 344–54.
41. Hooper, *Rational Recreations*, 4:163.
42. Ibid., 162. Similarly sizable optical fireworks were displayed by the Russian mechanic Ivan Kulibin for Catherine II and the imperial family in St. Petersburg. See Kulibin, *Rukopisnye materialy*, 451–52, 454; and Werrett, "Enlightenment in Russian Hands."

43. *Norfolk Chronicle,* June 2, 1781, 3. See also *Norfolk Chronicle,* June 15, 1782, 3. Another Italian working the pleasure gardens at this time was Antonio Invetto of Milan. See *Norfolk Chronicle,* June 15, 1782, 3; and *The Times,* no. 1247 (September 3, 1789): 2.
44. Trade card of Benjamin Clitherow, British Museum, London, Department of Prints and Drawings, Sarah Banks Collection, ref. 62.6.
45. Joseph Banks Papers: Series 06.028, microfilm ref. CY3003/126 and CY3003/127—Invoice and Receipt Received by Banks from Peter Caillott of Charlotte Street, March 13, 1772. The Basset family of Cornwall made one of the earliest recorded private purchases of fireworks from a commercial manufacturer, in this case in France. See Basset Family of Tehidy, Cornwall, Farm Accounts, Royal Institution of Cornwall Menwinnion Collection, MEN/9 (July 5, 1773), MEN/10 (August 8, 1773), MEN/11-1 (September 1773), MEN/11-2 (November 7, 1773).
46. *Mr. Flockton's Theatre.* On science in the pub, see Secord, "Science in the Pub."
47. See Turner, "Hengler's Fireworks"; McConnell, "Hengler, Sarah"; Hood, "Ode to Madame Hengler."
48. Brock, *A History of Fireworks,* 169, quoting an unspecified newspaper article. For a royal performance, see *The Times,* no. 12730 (August 12, 1825): 2.
49. *The Times,* no. 18051 (October 10, 1845): 8.
50. On the locations of public science in Paris in the eighteenth century, see Lynn, *Popular Science and Public Opinion,* 43–71.
51. Isherwood, *Farce and Fantasy,* 49.
52. *Avant-coureur,* no. 51 (December 23, 1765): 800–801.
53. *Journal de Paris,* no. 207 (July 26, 1778): 828. See also *Journal de Paris,* no. 193 (July 12, 1778): 772; and *Journal de Paris,* no. 214 (August 2, 1778): 856. On the artificer, see Delavarinière, *Réplique du sieur Delavarinière.*
54. For some of Stählin's later inventions, see Jacob Stählin to Peter Stählin, October 28, 1770, and Jacob Stählin to Peter Stählin, December 1, 1770, Stählin Papers, Department of Manuscripts, Russian National Library, St. Petersburg, F. 871, no. 274, letters 39 and 44; and Rovinskii, *Opisanie feierverkov,* 296–97, 299.
55. Rovinskii, *Opisanie feierverkov,* 294–95.
56. Petrov, "Palat' . . . Sheremeteva," 142; Smith, *Working the Rough Stone,* 73–74; Roosevelt, "Emerald Thrones and Living Statues."
57. Smith, *Working the Rough Stone,* 68–69, 74.
58. Burke, *Philosophical Enquiry,* 58–59.
59. Ibid., 60. These modifications included, as Kevin Salatino has noted, characteristics such as infinity, vastness, and magnificence. Thus, the disorderly infinity of stars in the heavens was exemplary of a magnificence that was sublime, an effect also produced, as Burke noted, by "a sort of fireworks . . . that in this way succeed well, and are truly grand" (ibid., 141). Salatino (*Incendiary Art,* 47–98) explores the relation of fireworks to the sublime.
60. At celebrations for the taking of Mahon in 1756, the fireworks alone cost 11,500 livres. See Bracco and Lebovici, *Ruggieri,* 48. The display, it should be noted, was performed by Torré. See Ruggieri, *Précis historique,* 286–87.
61. See the table "Artificiers du roi . . . de Paris" in *Essai sur l'almanach,* n.p. On competition between French and Italian artificers, see "Description du feu d'artifice"; Ruggieri, *Précis historique,* 290–95.
62. For Torré's biography, see Carver, "Torré, Giovanni Battista"; and the obituary "Sur Torré, artificier du roi," *L'esprit des journaux françois et étrangers,* December 1780, 191–204.

63. *Novelle letterarie pubblicate in Firenze* 13 (1752): 524; Cohen, "Two Hundredth Anniversary," 16–17.
64. "Sur Torré" (n. 62 above), 191.
65. Langlois, *Folies, tivolis, et attractions*, 71; Barrier, "Architectes entre Paris et Londres," 222.
66. *Affiches de Paris*, no. 15 (February 21, 1763): 130; *Affiches de Paris*, no. 43 (June 6, 1763): 386.
67. *Avant-coureur*, no. 32 (August 6th 1764): 505–7.
68. On Torré's Vauxhall, see Bachaumont, *Mémoires secrets*, passim in vols. 2–4, 6, 9–10, 12, 19; Ruggieri, *Précis historique*, 79–80; Isherwood, *Farce and Fantasy*, 202–5; Gruber, "Les 'Vauxhalls' Parisiens," esp. 126–29; and Goodman, "'Altar against altar.'"
69. Bachaumont, *Mémoires secrets*, 3:55. For the relevant passage in Virgil, see Virgil, *Aeneid*, 2:50–51. Burke thought Virgil's story "admirably sublime," that there was no more "grand and laboured passage" (Burke, *Philosophical Enquiry*, 328–29). *The Guardian*, 2:114–15 (no. 103, July 9, 1713), describes a firework with Vulcan and Cyclops very similar to Torré's, referring to verses on the Girandola by Famiano Strada. See Strada, *Prolusiones*, 320, trans. in *The Spectator*, 8:335–36 (no. 617, November 8, 1714).
70. Bachaumont, *Mémoires secrets*, 3:54.
71. The controversy is recounted in Sands, *Pleasure Gardens of Marylebone*, 86–102. See also "Epigram, on the Fire-Works Displayed in Marybone Gardens." Torré defended himself in the *Morning Chronicle and London Advertiser*, no. 943 (June 1, 1772).
72. On Torré's shows in London, see *Morning Chronicle and London Advertiser*, no. 937 (May 25, 1772); *General Evening Post*, no. 6037 (June 20, 1772); *Morning Chronicle and London Advertiser*, no. 1567 (June 1, 1774); Boswell, *The Life of Samuel Johnson*, 2:519; and Burney, *Evelina*, 233–34, 443.
73. "Marylebone Fireworks," *Morning Chronicle and London Advertiser*, no. 946 (June 4, 1772), 2.
74. Malcolm, *Anecdotes*, 2:276.
75. Jones, *A New Treatise on Artificial Fireworks*, 50; Fourcroy, *Elements of Chemistry*, 3:68; Chaptal, *Elements of Chemistry*, 2:103.
76. Wright quoted in Nicolson, *Joseph Wright of Derby*, 1:279. On the Vesuvius-Girandola pairs, see ibid., 9–10, 279–84; Egerton, ed., *Wright of Derby*, 145–49, 166–76; and Salatino, *Incendiary Art*, 49, 62–63, 66–67, 71.
77. *Morning Post*, no. 1725 (April 29, 1778), 2 (reviewing the Royal Academy exhibition of 1778), quoted in Egerton, ed., *Wright of Derby*, 175.
78. Disraeli, *Curiosities of Literature*, 2:418.
79. Jenkins and Sloan, *Vases and Volcanoes*.
80. Hamilton, *Observations*, 8.
81. Hamilton, *Campi Phlegraei*, and *Supplement to the Campi Phlegraei*; Moore, "Hamilton's Volcanology"; Knight, "Hamilton's *Campi Phlegraei*"; Wood, "Making and Circulating Knowledge."
82. Sorenson, "Hamilton's Vesuvian Apparatus."
83. Hamilton, *Observations*, 37–39. Some volcanoes, such as Vesuvius, do produce lightning and Saint Elmo's fire. Hamilton assumed that volcanic eruptions were caused by the release of pent-up subterraneous fires and exhalations, analogous to the "foul humours" exhausted from the human body. See ibid., 107–9.
84. Gregory, *The Economy of Nature*, 2:382.
85. Young, *Mind over Magma*, 3–15.

86. Martin Lyster [Lister], "Three Papers of Dr. Martin Lyster, the First of the Nature of Earth-Quakes; More Particularly of the Origine of the Matter of Them, from the Pyrites Alone," *Philosophical Transactions of the Royal Society* 14 (1684): 512–15, "The Second Paper of the Same Person Concerning the Spontaneous Firing of the Pyrites," *Philosophical Transactions of the Royal Society* 14 (1684): 515–17, and "The Third Paper of the Same Person, Concerning Thunder and Lightning Being from the Pyrites," *Philosophical Transactions of the Royal Society* 14 (1684): 517–19.
87. "An Abstract of a Letter from an English Gentleman at Naples to His Friend in London, Containing an Account of the Eruption of Mount Vesuvius, May 18. and the Following Days, 1737. N.S.," *Philosophical Transactions of the Royal Society* 41 (1739–41): 252–61, 254, 259. See also d'Holbach, "Volcans," 443.
88. Darwin, *The Botanic Garden*, 1:15.
89. See *Ranelagh*. John Aspinwall described a fifty-foot-high volcano at Ranelagh on June 5, 1795/6. See Aspinwall, *Travels in Britain*, 92; and Altick, *The Shows of London*, 96, 124.
90. On Torré, Loutherbourg, and Servandoni, see Sands, *Pleasure Gardens of Marylebone*, 87–88.
91. Cahusac had praised the "sublime Milton" in his article on fireworks for the *Encyclopédie*, and the scene of Pandaemonium was proposed as an ideal theme for fireworks earlier in the century. See Cahusac, "Feu d'artifice," 639; and *The Guardian*, 2:113–14 (no. 103, July 9, 1713).
92. Pyne, "The Eidophusikon," 1:298. See also Altick, *The Shows of London*, 117–27. Loutherbourg resided with Garrick's fireworks master, Domenico Angelo. See Angelo, *Reminiscences*, 1:12; and Brock, *Pyrotechnics*, 33.
93. Altick, *The Shows of London*, 126; Cheetham, "The Taste for Phenomena."
94. Umbach, "Visual Culture"; Salatino, *Incendiary Art*, 58, 75, 84–85.
95. On the Jardin Ruggieri, see Bachaumont, *Mémoires secrets*, 2:270 (September 17, 1765), 19:40 (January 20, 1769, on Ruggieri's initial plans), 45 (February 10, 1769), 54 (April 6, 1769), 71 (May 14, 1769, on Torré's reaction to the Ruggieri); Ruggieri, *Précis historique*, 77–79; Thiery, *Almanach du voyageur à Paris*, 7, pt. 2:220; Gruber, "Les 'Vauxhalls' Parisiens," 129–30; and Bracco and Lebovici, *Ruggieri*, 38–46. The Ruggieri opened a second, "Winter Waux-hall" in 1769. See Isherwood, *Farce and Fantasy*, 56–59, 153.
96. On Lavoisier's career at this time, see Poirier, *Lavoisier*, 22–46; and Donovan, *Antoine Lavoisier*, 25–44. On fireworks, see Poirier, *Lavoisier*, 35–36.
97. Lavoisier, "Remarques sur la composition."
98. Ibid., 111–12.
99. Ibid., 110–11.
100. Ibid., 111.
101. Ibid., 113 n. 1. His yellow composition is not recorded.
102. Ibid., 112.
103. Lavoisier applied his knowledge of artifice in the *Elémens de chimie*, where he noted how artificers used zinc to create a blue fire and compared the burning of phosphorus to the brilliant Chinese fire. See Lavoisier, *Elements of Chemistry*, 89–90, 437. For relations between Lavoisier's chemistry and gunpowder manufactures, see Mauskopf, "Gunpowder and the Chemical Revolution."
104. Bracco and Lebovici, *Ruggieri*, 44.
105. Ibid., 48–49.
106. Ruggieri, *Précis historique*, 295–300; Gruber, *Les grand fêtes*, 38–51; Lynn, "Sparks for Sale," 74.

107. Bachaumont, *Mémoires secrets,* 5:115–16. See also Lynn, "Sparks for Sale," 74–75; and Prosper-Hardy, *Mes loisirs,* 198–99.
108. Lavoisier, "Rapport sur une manière"; "Détail des illuminations."
109. Bachaumont, *Mémoires secrets,* 5:116–17; Bracco and Lebovici, *Ruggieri,* 48.
110. "[Some] people thought that this fire was a new genre of spectacle, which in effect presented a very beautiful coup d'oeil and lit up the square magnificently" (Bachaumont, *Mémoires secrets,* 5:117–18).
111. On the many accounts of the incident, see Tabournel, "La catastrophe de la rue Royale"; Prosper-Hardy, *Mes loisirs,* 200–206; *Mercure de France,* July 1770, 199–202; and Ruggieri, *Précis historique,* 302–3.
112. Trapp, *Proceedings,* 207. Note that the display took place on May 30, not April 19.
113. See, e.g., the 1771 performance by Torré, which was presented in a book of color paintings preserved in the Louvre: "Projet du feu d'artifice tiré à Versailles en presence de sa majesté, le 15 mai 1771 à l'occasion du mariage de Monsigneur le Comte de Provence, executée par les Srs Torré, Morel, & Séguin, artificiers de sa majesté," Louvre, Fonds des dessins et miniatures, Réserve des petits albums, RF 4190, 1.
114. Ruggieri, *Précis historique,* 308–11.
115. *Journal de Paris,* no. 193 (July 12, 1778): 772. See also Ruggieri, *Principles of Pyrotechnics,* 140–42. Hengler & Co., *New Fire-Works,* included the Salamander, with snakes "drawn by a natural Historic Painter": "This Piece is really moveable, and the Snakes will appear as if alive."
116. Campardon, *Les spectacles de la foire,* 2:341.
117. On fireworks imitated with air, see Lavoisier, "Rapport sur les procédés." On balloon ascents, see Ruggieri, *Précis historique,* 78; and Campardon, *Les spectacles de la foire,* 2:342. On flame-proof clothes, see Lynn, "Sparks for Sale," 89–90.
118. Bachaumont, *Mémoires secrets,* 23:158 (September 16, 1783). On the Colisée, see Ruggieri, *Précis historique,* 80–82; Goodman, "'Altar against altar'"; and Isherwood, *Farce and Fantasy,* 149–58.
119. Ruggieri, *Précis historique,* 79.
120. Siméon Prosper Hardy quoted in Luckett, "Hunting for Spies and Whores," 138–39.
121. Entry for Friday, August 29, 1788, in the journal of the Parisian bookseller Siméon-Prosper Hardy, Bibliothèque Nationale, Paris, Fonds Français 6687.
122. Carlyle, *The French Revolution,* 1:320.
123. Roberts, "The Death of the Sensuous Chemist."
124. Terrall, "The Gendering of Science."
125. Ashworth, "The Calculating Eye"; Bigg, "The Panorama."
126. Kim, "'Public' Science."
127. Hilaire-Pérez, "Invention and the State"; Gillispie, *Science and Polity in France,* 459–78.
128. On Bléton, see Lynn, "Divining the Enlightenment," and *Popular Science and Public Opinion,* 97–122. On Mesmer, see Darnton, *Mesmerism;* and Riskin, *Science in the Age of Sensibility,* 189–226.
129. *Avant-coureur,* no. 43 (October 22, 1764): 674–75.
130. *Mémoire sur les feux d'air inflammable par M. Diller.*
131. Ruggieri, *Elémens de pyrotechnie* (1st ed., 1801). I have also used an English translation of the third edition of 1821: Ruggieri, *Principles of Pyrotechnics.*
132. Ruggieri [?], *Précis pour Michel-Marie Ruggieri,* n.p., which notes that Michel-Marie was appointed *artificier du roi* on February 1, 1788. Michel was probably born in 1764. See Ruggieri, *Pyrotechnie militaire,* xvii.

133. Bracco and Lebovici, *Ruggieri*, 50. On revolutionary fêtes, see Ozouf, *Festivals and the French Revolution*. On revolutionary fireworks, see Ruggieri, *Précis historique*, 320–30.
134. Ruggieri, *Précis historique*, 320–22. On the Tivoli, see Ruggieri, *Précis historique*, 86–89.
135. On Garnerin, see Perrin, *La vie rocambolesque*. On Robertson, see Ruggieri, *Précis historique*, 97–98;
136. Napoleon, *Memoirs*, 1:334.
137. Ruggieri, *Principles of Pyrotechnics*, 28, 32, 219–20, 222, 309, and *Précis historique*, 332.
138. Ruggieri, *Elémens de pyrotechnie*, 1–4. On Chaptal's *Elémens*, see Bensaude-Vincent, "The Chemical Revolution."
139. Horn and Jacob, "The Cultural Roots of French Industrialization."
140. Brock, *A History of Fireworks*, 74.
141. On Napoleon's coronation fireworks, see Ruggieri, *Principles of Pyrotechnics*, 310–11; and Salatino, *Incendiary Art*, 92–94.
142. Ruggieri, *Principles of Pyrotechnics*, 316–18 and pl. 27. On Bénard, see Ruggieri, *Principles of Pyrotechnics*, 308, 316–17; and Leith, *Space and Revolution*, 190, 192–93.
143. Ruggieri, *Principles of Pyrotechnics*, 28.
144. Ruggieri, *Elémens de pyrotechnie*, xii–xiii. Ruggieri (*Pyrotechnie militaire*, 27) called the *Elémens* "un petit traité *chimico-pyrotechnique*."
145. Ruggieri, *Elémens de pyrotechnie*, 33–34.
146. Ruggieri, *Elémens de pyrotechnie*, 335–78, and *Principles of Pyrotechnics*, 269–305; Bensaude-Vincent and Abbri, eds., *Lavoisier in European Perspective*.
147. Ruggieri, *Principles of Pyrotechnics*, 36, where Ruggieri also invokes Chaptal to make this point.
148. Ruggieri, *Principles of Pyrotechnics*, 31.
149. Garnerin and Garnerin, *Détails des trois premiers voyages aériens*; Wairy, *Mémoires de Constant*, 2:133–34.
150. Ruggieri, *Principles of Pyrotechnics*, 32.
151. Cheleev, *Polnoe i podrobnoe nastavlenie*, 229–30.
152. Ibid., 230.
153. Ibid., 230–34.
154. Ibid., 231. Makoveev (also spelled Macaveyef) had executed fireworks designed by Melissino in 1796. See Melissino, *Opisanie feierverka . . . 1 sentiabria 1796 goda*.
155. Ruggieri, *Principles of Pyrotechnics*, 78–79, 110–12, 258, and pl. 26, fig. 1.
156. Ruggieri, *Principles of Pyrotechnics*, 318.
157. [Beyer], *Aux amateurs de physique*, 15–16.
158. On Berthollet, see Mauskopf, "Gunpowder and the Chemical Revolution"; Sadoun-Goupil, *Le chimiste Claude-Louis Berthollet*; and Brock, *A History of Fireworks*, 196.
159. "Green Fire" (n. 116, chap. 5, above), 568; Cutbush, "The Composition and Properties of the Chinese Fire," 133.
160. Ruggieri, *Principles of Pyrotechnics*, 33, 316–17.
161. "Green Fire" (n. 116, chap. 5, above), 568.
162. Cutbush, *A System of Pyrotechny*, 77.
163. Chertier, *Essai*; Tessier, *Chimie pyrotechnique*.
164. Thus, Brock, whose opinion is often repeated, wrote that Ruggieri was "the first writer to make use of metal salts in the production of coloured flame—apart, that is, from the isolated use by Hanzelet of verdigris. He also introduced sal-ammoniac (ammo-

nium chloride), which, by volatizing the metal, greatly assisted colour production. This was a great step forward" (*A History of Fireworks*, 155).

165. See, e.g., Richardson and Watts, *Chemical Technology*, 1, pt. 4:551–611 ("Pyrotechny").

CONCLUSION

1. On the history of Congreve rockets, see Winter, *The First Golden Age of Rocketry*.
2. Moore, *Treatise*. See also Johnson, "Comments and Commentary."
3. MacCulloch, "Pyrotechny," 267.
4. See, e.g., Say, *A Treatise on Political Economy*, 110; and Marcet, *Conversations on Political Economy*, 450.
5. *The Magic Lantern*, 54–55.
6. Brewster, *The Kaleidoscope*, 136.
7. *The Times*, no. 26151 (June 15, 1868): 6 (describing fireworks at the Crystal Palace). See also *The Times*, no. 22379 (May 28, 1856): 9.
8. Watanabe-O'Kelly, "Fireworks Displays."
9. Priestley, *The History and Present State of Electricity*, i.
10. Ibid., ii.
11. Smith, "The History of Astronomy," 20.
12. Ibid., 23.
13. Werrett, "From the Grand Whim to the Gasworks."
14. Ruggieri, *Précis historique*.
15. *The Times*, no. 12723 (August 4, 1825): 3; Lesur, *Annuaire historique universel*, app., 227–28.
16. Ruggieri, *Notes explicites*.
17. Ruggieri, *Précis historique*, 2.
18. Ibid., 5, 357, 361–62, 375.
19. See Dear, "What Is the History of Science the History *Of*?"

# BIBLIOGRAPHY

Full references to manuscript sources and short newspaper and magazine articles are given in the footnotes, as are full bibiographic details of articles from the *Philosophical Transactions of the Royal Society*. These are not repeated in the bibliography.


MANUSCRIPT COLLECTIONS

Archives nationales de France, Paris.
Archive of the St. Petersburg Academy of Sciences, St. Petersburg.
Bodleian Library, Oxford.
British Library, London.
Cabinet des arts graphiques, Musée Carnavalet, Paris.
Department of Manuscripts, Russian National Library, St. Petersburg.
Department of Prints and Drawings, British Museum, London.
Getty Research Institute Special Collections, Los Angeles.
Guildhall Library, London.
Hagley Museum and Library, Delaware.
Handschriftenabteilung, Staatsbibliothek, Berlin.
Public Record Office, Kew, London.
Rossiiskii gosudarstvennyi arkhiv drevnikh aktov, Moscow.
Society of Antiquaries, London.

JOURNALS

*Adventurer.*
*Affiches de Paris.* See *Annonces, affiches, et avis divers.*
*Annonces, affiches, et avis divers: Feuille périodique* (*Affiches de Paris*).
*Athenian Oracle.*
*Avant-coureur.*
*Boston Evening Post.*
*Daily Advertiser.*
*Daily Post.*
*Domestick Intelligence.*
*Englishman.*
*L'esprit des journaux françois et étrangers.*

*Ezhemesiachyia Sochineniia k pol'ze i uveseleniiu sluzhashchiia.*
*La gazette française.*
*General Evening Post.*
*Gentleman's Magazine.*
*Heraclitus Ridens.*
*Journal de Paris.*
*Journal oeconomique.*
*London Chronicle.*
*London Magazine; or, Gentleman's Monthly Intelligencer.*
*Mercure de France.*
*Mercurius Elencticus.*
*Methodist Magazine.*
*Monthly Review; or, Literary Journal.*
*Morning Chronicle and London Advertiser.*
*Museum of Foreign Literature, Science, and Art.*
*New England Quarterly.*
*New York Times.*
*Norfolk Chronicle.*
*Novelle letterarie pubblicate in Firenze.*
*Pennsylvania Gazette.*
*Protestant (Domestick) Intelligence; or, News Both from City and Country.*
*Public Advertiser.*
*Rastushchii vinograd.*
*Scots Magazine.*
*Sankt-Peterburgskie vedemosti.*
*The Times* (London).
*Universal Magazine.*

PRINTED SOURCES

*An Account of the Famous Sieur Rocquet, Surgeon; Just Arrived from Paris. Necessary for All Gentlemen and Ladies, That Attend the Fire-Works.* London, 1749.

Ackermann, Silke, and Jane Wess. "Between Antiquarianism and Experiment: Hans Sloane, George III and Collecting Science." In *The British Enlightenment: Discovering the World in the Eighteenth Century*, ed. Kim Sloan, 150–57. Washington, DC: Smithsonian Books, 2003.

*An Act for Preventing and Suppressing of Fires within the City of London, and Liberties Thereof.* London, 1668.

Adams, George. *An Essay on Electricity.* London, 1792.

Aer, Anneli. *Patents in Imperial Russia: A History of the Russian Institution of Invention Privileges under the Old Regime.* Helsinki: Suomalainen Tiedeakatemia, 1995.

Agentov, Mikhail Ivanovich. *Otkrytie sokrovennykh khudozhestv, sluzhashchee dlia fabrikantov, manufacturistov, khudozhnikov, masterovykh liudei i dlia ekonomii.* 3 vols. Moscow, 1768–71.

Agnew, Jean-Christophe. *Worlds Apart: The Market and the Theater in Anglo-American Thought, 1550–1750.* New York: Cambridge University Press, 1986.

Alberti, Giuseppe Francesco Antonio. *La pirotechnia o sia trattato dei fuochi d'artificio.* Venice, 1749.

Alberti, Leon Battista. *Ludi rerum mathematicarum.* In *Opere volgari* (3 vols.), ed. Cecil Grayson, 3:131–73. Bari: Laterza, 1960–73.

Alekseeva, M. A., ed. *Feierverki i illiuminatsii v grafike XVIII veka: Katalog vystavki*. Leningrad: Gosudarstvennyi Russkii Muzei, 1978.

Alekseeva, M. A., Yu. A. Vinogradov, Yu. A. Piatnitskii, and B. V. Levshin, eds. *Graviroval'naia palata Akademii nauk XVIII veka: Sbornik dokumentov*. Leningrad: Nauka, 1985.

Algarotti, Francesco. *Letters from Count Algarotti to Lord Hervey and the Marquis Scipio Maffei, Containing the State . . . of the Russian Empire*. 2 vols. London, 1769.

Altick, Richard D. *The Shows of London*. Cambridge, MA: Belknap/Harvard University Press, 1978.

Anderson, John. *Institutes of Physics*. Glasgow, 1777.

Angelo, Henry. *The Reminiscences of Henry Angelo*. 2 vols. New York: Benjamin Blom, 1972.

Anisimov, Evgenii V. *Rossiia bez Petra, 1725–1740*. St. Petersburg: Lenizdat, 1994.

———. *Anna Ivanovna*. Moscow: Molodaia gvardiia, 2002.

Apostolidès, Jean-Marie. *Le roi-machine: Spectacle et politique au temps de Louis XIV*. Paris: Editions de Minuit, 1981.

"Application of Electricity to Discharge Cannon." *Journal of Natural Philosophy, Chemistry, and the Arts* 17 (1807): 232.

Aristotle. *Meteorologica*. Translated by H. D. P. Lee. Cambridge, MA: Harvard University Press, 1952.

Ashworth, William J. "The Calculating Eye: Baily, Herschel, Babbage and the Business of Astronomy." *British Journal for the History of Science* 27 (1994): 409–41.

Askwith, W. H., and John Kane. *A List of Officers of the Royal Regiment of Artillery from the Year 1716 to the Year 1899*. London: 1900.

Aspinall-Oglander, Cecil Faber, and Frances Evelyn Glanvill Boscawen. *Admiral's Wife; Being the Life and Letters of the Hon. Mrs. Edward Boscawen from 1719 to 1761*. London: Longmans, Green, 1940.

Aspinwall, John. *Travels in Britain, 1794–95: The Diary of John Aspinwall, Great-Grandfather of Franklin Delano Roosevelt, with a Brief History of His Aspinwall Forebears*. Edited by Aileen Sutherland Collins. Virginia Beach, VA: Parsons, 1994).

*At the Court at Whitehall, this Seventh Day of November 1683. . . .* London, 1683.

Aubrey, John. *Miscellanies upon the Following Subjects Collected by J. Aubrey, Esq*. London, 1696.

Ayres, Philip. *Classical Culture and the Idea of Rome in Eighteenth-Century England*. Cambridge: Cambridge University Press, 1997.

Babington, John. *Pyrotechnia; or, A Discourse of Artificiall Fire-works*. London, 1635.

Bachaumont, Louis Petit de. *Mémoires secrets pour servir à l'histoire de la république des lettres en France depuis 1762 jusqu'à nos jours*. 36 vols. London: John Adamson, 1784–89.

Bachelard, Gaston. *La flamme d'une chandelle*. Paris: Presses Universitaires de France, 1962.

Bacon, Francis. *Sylua syluarum; or, A Naturall Historie in Ten Centuries*. London, 1627.

———. *New Atlantis: A Work Unfinished*. London, 1658.

———. *The Works of Francis Bacon: Baron of Verulam, Viscount St. Albans, and Lord High Chancellor of England*. Edited by James Spedding, Robert Leslie Ellis, and Douglas Denon Heath. 14 vols. Boston, 1861–64.

Bacon, Roger. *The Opus Maius of Roger Bacon*. Translated by Robert Belle Burke. 2 vols. Philadelphia: University of Pennsylvania Press, 1928; reprint, Bristol: Thoemmes, 2000.

Baehr, Stephen Lessing. *The Paradise Myth in Eighteenth-Century Russia: Utopian Patterns in Early Secular Russian Literature and Culture*. Stanford, CA: Stanford University Press, 1991.

Baigent, Elizabeth. "Borgard, Albert (1659–1751)." In *Oxford Dictionary of National Biography* (61 vols.), ed. H. C. G. Matthew and Brian Howard Harrison, 6:658–59. Oxford: Oxford University Press, 2004.

Baillet, Adrien. *The Life of Monsieur Des Cartes, Containing the History of His Philosophy and Works*. London, 1693.

Baillon, Jean-François. "Early Eighteenth-Century Newtonianism: The Huguenot Contribution." *Studies in History and Philosophy of Science* 35 (2004): 533–48.

Barbier, Edmond Jean François. *Journal historique et anecdotique du règne de Louis XV*. 4 vols. Paris, 1847–56.

———. *Chronique de la régence et du règne de Louis XV (1718–1763); ou, Journal de Barbier; troisième série (1733–1744)*. Paris, 1857.

Barclay, John. *Barclay His Argenis; or, The Loves of Poliarchus and Argenis*. London, 1625.

Barish, Jonas. *The Anti-Theatrical Prejudice*. Berkeley and Los Angeles: University of California Press, 1981.

Barlow, Edward. *Meteorological Essays, Concerning the Origin of Springs, Generation of Rain, and Production of Wind*. London, 1715.

Barrier, Janine. "Architectes entre Paris et Londres à l'époque de Louis XV." In *Paris, capitale des arts sous Louis XV: Peinture, sculpture, architecture, fêtes, iconographie* (Annales du Centre Ledoux, vol. 1), ed. Daniel Rabreau, 219–36. Paris: Centre Ledoux, Université de Paris-I Panthéon-Sorbonne, 1997.

Bate, John. *The Mysteries of Nature and Art in Foure Severall Parts. The First of Water Works. The Second of Fire Works. The Third of Drawing, Washing, Limming, Painting, and Engraving. The fourth of sundry experiments*. London, 1635.

Baur, Désirée, and Peter Plaßmeyer. "Physikalische Apparate und mechanische Spielereien—Peters I. Besuche in Dresden, in der kurfürstlichen Kunstkammer und in den Werkstätten." In *Palast des Wissens: Die Kunst- und Wunderkammer Zar Peters des Großen . . . im Auftrag des Museums für Kunst und Kulturgeschichte Dortmund und des Schloßmuseums Gotha* (2 vols.), ed. Brigitte Buberl and Michael Dückershoff, 2:105–15. Munich: Hirmer, 2003.

Bayard, Marc. "Les faiseurs d'artifices: Georges Buffequin et les artistes de l'ephémère à l'époque de Richelieu." *XVII[e] Siècle* 230 (2006): 151–64.

Beaurepaire, Charles de. *Louis XIII et l'assemblée des notables à Rouen, en 1617: Documents recueillis et annotés*. Rouen, 1883.

Beccaria, Giambattista. *A Treatise upon Artificial Electricity, in Which Are Given Solutions of a Number of Interesting Electric Phoenomena, Hitherto Unexplained*. London, 1776.

Becket, John Brice. *An Essay on Electricity, Containing a Series of Experiments Introductory to the Study of That Science*. Bristol, 1773.

Bedloe, William. *A Narrative and Impartial Discovery of the Horrid Popish Plot*. London, 1679.

Beek, J. *The Triumphs of William III. King of England, Scotland, France and Ireland*. London, 1702.

Bélidor, Bernard Forest de. *Le Bombardier françois; ou, Nouvelle methode de jetter les bombes avec précision*. Amsterdam, 1734.

Bennell, John. "Hester, John (*d.* 1592)." In *Oxford Dictionary of National Biography* (61 vols.), ed. H. C. G. Matthew and Brian Howard Harrison, 26:887–88. Oxford: Oxford University Press, 2004.

Bennett, Helju Aulik. "Evolution of the Meanings of *Chin*: An Introduction to the Russian Institution of Rank Ordering and Niche Assignment from the Time of Peter the

Great's Table of Ranks to the Bolshevik Revolution." *California Slavic Studies* 10 (1977) 1–43.

Bennett, J. A. "The 'Mechanics' Philosophy and the Mechanical Philosophy." *History of Science* 24 (1986): 1–28.

Bennett, John. *Letters to a Young Lady, on a Variety of Useful and Interesting Subjects: Calculated to Improve the Heart.* 2 vols. Hartford, 1798.

Bensaude-Vincent, Bernadette. "A View of the Chemical Revolution through Contemporary Textbooks: Lavoisier, Fourcroy and Chaptal." *British Journal for the History of Science* 23 (1990): 435–60.

Bensaude-Vincent, Bernadette, and Ferdinando Abbri, eds. *Lavoisier in European Perspective: Negotiating a New Language for Chemistry.* Canton, MA: Science History, 1995.

Bensaude-Vincent, Bernadette, and Christine Blondel, eds. *Science and Spectacle in the European Enlightenment.* Aldershot: Ashgate, 2008.

Benz, Ernst. *Leibniz und Peter der Große.* Berlin: De Gruyter, 1947.

Berg, Maxine. *Luxury and Pleasure in Eighteenth-Century Britain.* Oxford: Oxford University Press, 2005.

Berg, Maxine, and Elizabeth Eger, eds. *Luxury in the Eighteenth Century: Debates, Desires, and Delectable Goods.* Basingstoke: Palgrave, 2003.

Berk, Karl Reinkhol'd. "Putevye zametki o Rossii." In *Peterburg Anny Ioannovny v inostrannykh opisaniakh,* ed. Iu. N. Bespiatykh, 111–302. St. Petersburg: Russko-Baltiiskii informatsionnyi tsentr, 1997.

Bernard, Henri. "Un correspondant de Bernard de Jussieu en Chine: Le Père Le Cheron d'Incarville, missionnaire français de Pékin." *Archives internationales d'histoire des sciences* 6 (1949): 333–62; 7 (1949): 692–717.

Bernardin, Napoléon-Maurice. *La comédie italienne en France et les théâtres de la foire et du boulevard, 1570–1791.* Paris: Edition de la Revue bleue, 1902.

Bernier, François. *Abrégé de la philosophie de M. Gassendi.* 7 vols. in 6. Paris, 1684.

Berthold, Margot. "Joseph Furttenbach von Leutkirch, Architekt und Ratsherr in Ulm (1591–1667)." *Ulm und Oberschwaben* 33 (1953): 119–79.

Bertholon, Pierre. "Mémoire sur de nouvelles illuminations électriques; par M. Bertholon, prêtre de Saint-Lazare, &c." *Observations et mémoires sur la physique, sur l'histoire naturelle et sur les arts* 7 (1776): 488–501.

———. *De l'électricité des météores.* 2 vols. Paris, 1787.

Bertucci, Paola, and Giuliano Pancaldi, eds. *Electric Bodies: Episodes in the History of Medical Electricity.* Bologna Studies in the History of Sciences, vol. 9. Bologna: Università di Bologna, Dipartimento di Filosofia, 2001.

Bespiatykh, Iu. N. *Aleksandr Danilovich Menshikov: Mify i real'nost'.* St. Petersburg: Istoricheskaia illustratsiia, 2005.

[Beyer]. *Aux amateurs de physique.* Paris, [after 1810].

Biagioli, Mario. "Galileo the Emblem Maker." *Isis* 81 (1990): 230–58.

———. "Scientific Revolution, Social Bricolage, and Etiquette." In *The Scientific Revolution in National Context,* ed. Roy Porter and Mikulas Teich, 11–54. Cambridge: Cambridge University Press, 1992.

———. *Galileo, Courtier: The Practice of Science in the Age of Absolutism.* Chicago: University of Chicago Press, 1993.

Bigg, Charlotte. "The Panorama; or, La nature a coup d'oeil." In *Observing Nature—Representing Experience: The Osmotic Dynamics of Romanticism, 1800–1850,* ed. Erna Fiorentini, 51–70. Berlin: Reimer, 2007.

Biliarskii, S. *Materialy dlia biografii Lomonosova*. St. Petersburg, 1865.

Birch, Thomas. *The History of the Royal Society of London for Improving of Natural Knowledge*. 4 vols. London, 1760.

Biringuccio, Vannoccio. *The Pirotechnia of Vannoccio Biringuccio: The Classic Sixteenth-Century Treatise on Metals and Metallurgy*. Translated and edited by Cyril Stanley Smith and Martha Teach Gnudi. Venice, 1540; reprint, New York: Dover, 1990.

Black, J. L. *G. F. Müller and the Imperial Russian Academy*. Montreal: McGill-Queen's University Press, 1986.

Bleeker, Johan, Johannes Geiss, and Martin C. E. Huber, eds. *The Century of Space Science*. 2 vols. Dodrecht: Kluwer, 2002.

Blesson, Major L. "Observations on the Ignis Fatuus; or, Will-with-the-Wisp, Falling Stars, and Thunder Storms." *Edinburgh New Philosophical Journal* 33 (1833): 90–94.

Blondel, Jean-François. *Description des festes, données par la ville de Paris, à l'occasion du mariage de Madame Louise-Elizabeth de France, & Dom Philippe, infant & grand amiral d'Espagne . . . le 29 & 30 Aout 1739*. Paris, 1740.

Blümel, Johann, Daniel. *Gründliche Anweisung zur Lust-Feuerwerkerey, besonders in denjenigen Stücken, die das Auge der Zuschauer am meisten erlustigen und in Verwunderung satzen. Mit einem Anhange vermehret, in welchem von wohlriechenden Tafel-Feuerwerk, wie auch von besonders artigen Kunststücken, in Tabacksdosen, in Bäumen &c. abzubrennen; Alles dieses ausgefertiget von Hr. Höckely*. Strasburg, Amand Koenig Buchhändler, 1771.

Boas, George. "The Arts in the *Encyclopédie*." *Journal of Aesthetics and Art Criticism* 23 (1964): 97–107.

Boblaye, Théodore Le Puillon de. *Esquisse historique sur les écoles d'artillerie*. Metz: Paris, 1858.

Boerhaave, Herman. *A New Method of Chemistry . . . Translated from Dr. Boerhaave's Elementa Chemiae*. Translated by Peter Shaw. 2 vols. London, 1753.

Bogdanov, A. I. *Opisanie Sanktpeterburga, 1749–1751*. St. Petersburg, 1776; reprint, St. Petersburg: TOO Katriona, 1997.

Bohnenberger, Gottlieb Christoph. *Beyträge zur theoretischen und praktischen Elektrizitätslehre*. 4 vols. Stuttgart, 1793–95.

Bohun, James. "The Royal Fireworks and the Politics of Music in Mid-Hanoverian Britain." Ph.D. thesis, University of Alberta, 1993.

Boillot, Joseph. *Artifices de feu, & diuerses instrumens de guerre*. Strasburg, 1603.

Bolotov, Andrei Timofeevich. "Zapiski Andreia Timofeevicha Bolotova: Tom pervyi, chast' I–VII." *Russkaia Starina* (St. Petersburg), 1870, 1–1017.

Boner, Patrick J. "Kepler on the Origins of Comets: Applying Earthly Knowledge to Celestial Events." *Nuncius* 21 (2006): 31–48.

Boorsch, Suzanne. *Fireworks! Four Centuries of Pyrotechnics in Prints and Drawings*. New York: Metropolitan Museum of Art, 2000.

Boswell, James. *The Life of Samuel Johnson, LL.D. Comprehending an Account of His Studies and Numerous Works*. 2 vols. London, 1791.

Bourne, William. *Inuentions and Deuices. Very Necessary for All Generalles and Captaines, or Leaders of Men, as Wel by Sea as by Land*. London, 1578.

[Boyer, Abel]. *The Political State of Great Britain*. 60 vols. London, 1711–40.

Boyle, Robert. *Some Considerations Touching the Vsefulnesse of Experimental Naturall Philosophy Proposd in Familiar Discourses to a Friend, by Way of Invitation to the Study of it*. London, 1663.

———. *Tracts Written by the Honourable Robert Boyle, Containing New Experiments, Touching the Relation betwixt Flame and Air. And about Explosions. . . .* London, 1672.

———. *About the Excellency and Grounds of the Mechanical Hypothesis*. Pt. 2 of *The Excellency of Theology Compard with Natural Philosophy (As Both Are Objects of Mens Study) Discoursd of in a Letter to a Friend by T.H.R.B.E . . . ; to Which Are Annexd Some Occasional Thoughts about the Excellency and Grounds of the Mechanical Hypothesis*. London, 1674.

———. "A Short Memorial of an Artificial Substance That Shines without Any Precedent Illustration." In *Lectures and Collections Made by Robert Hooke, Secretary of the Royal Society*, 57–66. London, 1678.

———. *The Aerial Noctiluca; or, Some New Phenomena, and a Proces of A Factitious Self-Shining Substance, Imparted in a Letter to a Friend, Living in the Country*. London, 1680.

———. *A Free Enquiry into the Vulgarly Receivd Notion of Nature Made in an Essay Addressd to a Friend*. London, 1686.

———. "Suspicions about Some Hidden Qualities in the Air." In *The Philosophical Works of the Honourable Robert Boyle Esq.* (3 vols.), ed. Peter Shaw, 3:76–98. London, 1725.

———. "Fragments of Boyle's 'Essay on Spontaneous Generation' (c. 1670s–80s)." In *The Works of Robert Boyle* (14 vols.), ed. Michael Hunter and Edward B. Davis, 13:275–88. London: Pickering & Chatto, 2000.

Bracco, Patrick, and Elisabeth Lebovici. *Ruggieri: 250 ans de feux d'artifice*. Paris: Denoël, 1988.

Braun, Ernst. *Novissima fundamentum et praxis artilleriae*. Danzig, 1682.

Bredekamp, Horst. *The Lure of Antiquity and the Cult of the Machine*. Princeton, NJ: Markus Wiener, 1995.

———. "Leibniz's Theater der Natur und Kunst." In *Theater der Natur und Kunst/Theatrum naturae et artis: Wunderkammern des Wissens* (2 vols.), ed. Horst Bredekamp, Jochen Brüning, and Cornelia Weber, 1:12–19. Berlin: Henscel, 2000.

*Brevet exclusif qui permet aux Sieurs Guérin, Testard, & autres artificiers, de faire & exécuter, pendant 12 ans, un feu d'artifice sur la rivière de Seine, la veille de la Fête de S. Louis*. Paris: Le Breton, 1741.

Brewer, John. *The Pleasures of the English Imagination: English Culture in the Eighteenth Century*. Chicago: University of Chicago Press, 1997.

Brewster, Sir David. *The Kaleidoscope: Its History, Theory, and Construction*. London, 1858.

Briggs, J. Morton, Jr. "Aurora and Enlightenment Eighteenth-Century Explanations of the Aurora Borealis." *Isis* 58 (1967): 491–503.

*The Britannic Magazine; or, Entertaining Repository of Heroic Achievements*. 12 vols. London, 1798–1807.

Brock, Alan St. Hill. *Pyrotechnics: The History and Art of Firework Making*. London: Daniel O'Connor, 1922.

———. *A History of Fireworks*. London: George G. Harrap, 1949.

Brooks, Nathan M. "Public Lectures in Chemistry in Russia, 1750–1870." *Ambix* 44 (1997): 1–10.

Brouncker, William. "Experiments of the Recoiling of Guns." In *History of the Royal Society of London, for the Improving of Natural Knowledge*, by Thomas Sprat, 233–39. London, 1667.

Browne, George. *A Modell of the Fire-Workes to Be Presented in Lincolnes-Inne Fields, on the 5th of Novemb. 1647. Before the Lords and Commons of Parliament, and the Militia of London, in Commemoration of Gods Great Mercy in Delivering This Kingdome from the Hellish Plots of Papists, Acted in the Damnable Gunpowder Treason*. London, 1647.

Browne, Thomas. *Pseudodoxia epidemica; or, Enquiries into Very Many Received Tenents and Commonly Presumed Truths Together with the Religio medici . . . the Sixth and Last Edition*. London, 1672.

Buberl, Brigitte, and Michael Dückershoff, eds. *Palast des Wissens: Die Kunst- und Wunderkammer Zar Peters des Großen . . . im Auftrag des Museums für Kunst und Kulturgeschichte Dortmund und des Schloßmuseums Gotha*. 2 vols. Munich: Hirmer, 2003.

Buchanan, Brenda J., ed. *Gunpowder: The History of an International Technology*. Bath: Bath University Press, 1996.

Buchanan, Brenda J., David Cannadine, et al. *Gunpowder Plots: A Celebration of 400 Years of British Carelessness with Explosives*. London: Allen Lane, 2005.

Buchner, Johann Sigmund. *Theoria et praxis artilleriae. Oder: Deutliche Beschreibung Der bey itziger Zeit bräuchlichen Artillerie*. 3 vols. Nuremberg, 1682–85.

*Bureau d'addresse et rencontre: A General Collection of Discourses of the Virtuosi of France, upon Questions of All Sorts of Philosophy, and Other Natural Knowledg Made in the Assembly of the Beaux Esprits at Paris, by the Most Ingenious Persons of That Nation, Renderd into English by G. Havers, Gent*. London, 1664.

Burke, Edmund. *A Philosophical Enquiry into the Origin of Our Ideas of the Sublime and Beautiful*. 4th ed. London, 1764.

Burke, Peter. *The Fabrication of Louis XIV*. New Haven, CT: Yale University Press, 1992.

Burnet, Thomas. *The Theory of the Earth Containing an Account of the Original of the Earth, and of All the General Changes Which It Hath Already Undergone, or Is to Undergo till the Consummation of All Things*. London, 1697.

Burney, Frances. *Evelina; or, The History of a Young Ladys Entrance into the World*. London, 1778; reprint, Oxford: Oxford Classics, 2002.

Burton, Robert. *The Anatomy of Melancholy*. Edited by Thomas C. Faulkner, Nicolas K. Kiessling, and Rhonda L. Blair. 3 vols. Oxford: Oxford University Press, 1989–2000.

Büsching, Anton F. "Lebensgeschichte Burchard Christophs von Münnich." *Magazin für die neue Historie und Geographie* 3 (1769): 387–536.

Butterworth, Philip. *Theatre of Fire: Special Effects in Early English and Scottish Theatre*. London: Society for Theatre Research, 1998.

Byrom, John. *The Private Journal and Literary Remains of John Byrom*. Edited by Richard Parkinson. 2 vols. in 4. London, 1854–57.

Cahusac, Louis de. "Feu d'artifice." In *Encyclopédie; ou, Dictionnaire raisonné des science, des arts et des métiers* (35 vols.), ed. Denis Diderot and Jean d'Alembert, 6:639–40. Geneva, Paris, and Neufchastel, 1751–80.

*Calendar of State Papers, Domestic Series, James II, vol. III, June 1687–February 1689*. Edited by E. K. Timings. London: Her Majesty's Stationery Office, 1972.

*Calendar of State Papers, Domestic Series, of the Reign of Elizabeth, 1591–1594*. Edited by Mary Ann Everett Green. London, 1867.

Camden, William. *Remaines Concerning Britaine but Especially England, and the Inhabitants Thereof*. London, 1614.

Camerini, Silvia. *Il magnifico apparato: Pubbliche funzioni, feste e giochi bolognesi nel Settecento*. Bologna: Clueb, 1982.

Campanella, Tomaso. *Thomas Campanella, an Italian Friar and Second Machiavel, His Advice to the King of Spain for Attaining the Universal Monarchy of the World*. London, 1660.

———. *La città del sole: Dialogo poetico/The City of the Sun: A Poetical Dialogue*. Translated and edited by Daniel J. Donno. Berkeley and Los Angeles: University of California Press, 1981.

Campardon, Emile. *Les spectacles de la foire: Théâtres, acteurs, sauteurs et danseurs de corde, monstres, géants, nains, animaux curieux ou savants, marionnettes, automates, figures de cire et jeux mécaniques des foires Saint-Germain et Saint-Laurent, des boulevards et du Palais-Royal, depuis 1595 jusquà 1791*. 2 vols. Paris, 1877.

Camps, Arnulf. *Jerome Xavier, S.J., and the Muslims of the Mogul Empire: Controversial Works and Missionary Activity*. Schöneck: Nouvelle revue de science missionaire Suisse, 1957.

Canova-Green, Maria-Claude. "Fireworks and Bonfires in Paris and La Rochelle." In *Europa Triumphans: Court and Civic Festivals in Early Modern Europe* (2 vols.), ed. J. R. Mulryne, Helen Watanabe-O'Kelly, and Margaret Shewring, 2:145–53. Aldershot: Ashgate, 2004.

Cardano, Girolamo. *De rerum varietate libri xvii*. Basel, 1557.

*Carlos V: Las armas y las letras: 14 de abril–25 de junio, 2000, Hospital Real, Granada*. Granada: Sociedad Estatal para la Conmemoración de los Centenarios de Felipe II y Carlos V, 2000. Exhibition catalog.

Carlyle, Thomas. *The French Revolution: A History in Three Volumes*. 3 vols. in 2. Boston, 1838.

Carrel, Armand, Charles James Fox, and John Lowther Lonsdale. *History of the Counter-Revolution in England, for the Establishment of Popery, under Charles II. and James II*. London: H. G. Bohn, 1857.

Carver, Gavin. "Torré, Giovanni Battista (*fl.* 1753–1776)." In *Oxford Dictionary of National Biography* (61 vols.), ed. H. C. G. Matthew and Brian Howard Harrison, 55:50–51. Oxford: Oxford University Press, 2004.

Castiglioni, Baldassare. *The Book of the Courtier*. Translated by Sir Thomas Hoby. London: David Nutt, 1900.

Cavallo, Tiberius. *A Complete Treatise of Electricity in Theory and Practice; with Original Experiments*. London, 1777.

Cavazzi, Lucia. *"Fochi d'allegrezza" a Roma dal Cinquecento all'Ottocento*. Rome: Quasar, 1982.

Chambers, Ephraim. *Cyclopaedia; or, An Universal Dictionary of the Arts and Sciences*. 2 vols. London, 1728.

Chaptal, Jean-Antoine-Claude. *Elements of Chemistry by M. I. A. Chaptal . . . Translated from the French*. 3rd ed. 3 vols. London, 1800.

Charleton, Walter. *Three Anatomic Lectures*. London, 1683.

Chartier, Roger. *Forms and Meanings: Texts, Performances, and Audiences from Codex to Computer*. Philadelphia, 1995.

Chartres, Richard, and David Vermont. *A Brief History of Gresham College, 1597–1997*. London: Gresham College, 1998.

Cheetham, Mark A. "The Taste for Phenomena: Mount Vesuvius and Transformations in Late 18th-Century European Landscape Depiction." *Wallraf-Richartz Jahrbuch* 45 (1984): 131–44.

Cheleev, Lt. Fedor. *Polnoe i podrobnoe nastavlenie o sostavlenii uveselitel'nykh ognei, feierverkami imenuemykh; s prisovokupleniem priugotovleniia voennykh ognestrel'nykh i zazhigatel'nykh veshchei v pol'zu artilleriistov i liubitelei sego uprazhneniia*. St. Petersburg, 1824.

Chenekal, V. L., and A. V. Topchiev. *Letopis' zhizni i torzhestva M. V. Lomonosova*. Moscow-Leningrad: Izdatel'stvo Akademii nauk SSSR, 1961.

Chennevières, Henry de. *Les menus-plaisirs du roi et leurs artistes*. Paris, 1882.

———. "Les Ruggieri: Artificiers, 1730–1885." *Gazette des beaux-arts* 36 (August 1887): 132–40.

Chertier, F. M. *Essai sur les compositions qui donnent les plus belles couleurs dans les feux d'artifice*. Paris, 1836.

Christout, Marie-Françoise. "Décors d'opéra et pièces à machine au XVIII$^{e}$ siècle: Servandoni, illusioniste." *Médécine de France* 109 (1960): 17–31.

———. "Les feux d'artifices en France de 1606 à 1628: Esquisse historique et ésthetique." In *Les fêtes de la Renaissance* (2nd ed., 3 vols.), ed. Jean Jacquot, 1:247–58. Paris: Editions du Centre National de la Recherche Scientifique, 1973.

Clark, William, Jan Golinski, and Simon Schaffer, eds. *The Sciences in Enlightened Europe.* Chicago: Chicago University Press, 1999.

Claydon, Tony. *William III and the Godly Revolution.* Cambridge: Cambridge University Press, 1996.

Cleaveland, Frederick Darby. *Notes on the Early History of the Royal Artillery.* Woolwich, 1892.

Clément, J. M. B., and J. de Laporte. *Anecdotes dramatiques.* 3 vols. Paris, 1775.

Clifton, Robin. "The Popular Fear of Catholics during the English Revolution." In *Rebellion, Popular Protest and Social Order in Early Modern England,* ed. Paul Slack, 129–61. Cambridge: Cambridge University Press, 1984.

Clubb, Louise George. *Giambattista Della Porta: Dramatist.* Princeton, NJ: Princeton University Press, 1965.

Cocco, Sean. "Vesuvius and Naples: Nature and the City, 1500–1700." Ph.D. thesis, University of Washington, 2004.

Coffey, Donna. "'As in a theatre': Scientific Spectacle in Bacon's *New Atlantis.*" *Science as Culture* 13 (2004): 259–90.

Cohen, H. Floris. *The Scientific Revolution: A Historiographical Inquiry.* Chicago: University of Chicago Press, 1994.

Cohen, I. Bernard. *Benjamin Franklin's Experiments.* Cambridge, MA: Harvard University Press, 1941.

———. "The Two Hundredth Anniversary of Benjamin Franklin's Two Lightning Experiments and the Introduction of the Lightning Rod." *Proceedings of the American Philosophical Society* 96 (1952): 331–66.

Colie, Rosalie L. "Cornelis Drebbel and Salomon de Caus: Two Jacobean Models for Salomon's House." *Huntington Library Quarterly* 18 (1954): 245–69.

Colley, Linda. *Britons: Forging the Nation.* New Haven, CT: Yale University Press, 1992.

Comenius, Johann Amos. *Naturall Philosophie Reformed by Divine Light; or, A Synopsis of Physicks.* London, 1651.

*Continuation du traité de la police, contenant l'histoire de son établissement, les fonctions & les prérogatives de ses magistrats; toutes les loix & les réglemens qui la concernent. Tome quatrième: De la Voirier.* Paris, 1738.

Cook, Harold J. "The Society of Chemical Physicians, the New Philosophy, and the Restoration Court." *Bulletin of the History of Medicine* 61 (1987): 61–77.

Corporation of London. Lord Mayor. *By the Maior by Corporation of London. Lord Mayor . . . Whereas Divers Rude and Disordered Young-Men, Apprentices and Others, Do Now . . . Throw about Squibs and Fireworks in the Streets.* London, 1674.

Cressy, David. *Bonfires and Bells: National Memory and the Protestant Calendar in Elizabethan and Stuart England.* London: Weidenfeld & Nicolson, 1989.

Cru, R. Loyalty. *Diderot as a Disciple of English Thought.* New York: Columbia University Press, 1913.

Cuomo, Serafina. "Shooting by the Book: Notes on Niccolò Tartaglia's 'Nova scientia.'" *History of Science* 35 (1997): 155–88.

Curtiss, Mina. *A Forgotten Empress: Anna Ivanovna and Her Era, 1730–1740.* New York: Ungar, 1974.

Cutbush, James. "Remarks on the Composition and Properties of the Chinese Fire, and on the So Called Brilliant Fires." *American Journal of Science* 7 (1824): 118–41.

———. *A System of Pyrotechny*. Philadelphia, 1825.

Cyrano de Bergerac, Savinien. *Histoire comique; contenant les éstats et empires de la lune*. Paris, 1657.

———. *Selenarhia; or, The Government of the World in the Moon, a Comical History*. London, 1659.

d'Alembert, Jean. *Preliminary Discourse to the Encyclopedia of Diderot*. Translated by Richard N. Schwab. Indianapolis: Bobbs-Merrill, 1963.

Danilov, Mikhail Vasil'evich. *Dovol'noe i iasnoe pokazanie po kotoromu Vsiakoi sam soboiu liuzhet prigotovliat' i delat' vsiakie feierverki i raznyia illuminatsii*. Moscow, 1777; 3rd ed., Moscow, 1785.

———. "Zapiski M. V. Danilova." Edited by Pavel Stroev. *Russkii Arkhiv* 21, pt. 3 (1883): 1–67.

Darnton, Robert. *Mesmerism and the End of the Enlightenment in France*. Cambridge, MA: Harvard University Press, 1968.

Darwin, Erasmus. *The Botanic Garden: A Poem, in Two Parts. Part I. Containing the Economy of Vegetation. Part II. The Loves of the Plants. With Philosophical Notes. . . .* 2nd ed. 2 vols. London, 1791.

Daston, Lorraine. "The Physicalist Tradition in Early Ninenteenth Century French Geometry." *Studies in the History and Philosophy of Science* 17 (1986): 269–95.

———. "Curiosity in Early Modern Science." *Word and Image* 11 (1995): 391–404.

———. "The Cold Light of Facts and the Facts of Cold Light: Luminescence and the Transformation of the Scientific Fact, 1600–1750." In *Signs of the Early Modern 2: 17th Century and Beyond* (EMF: Studies in Early Modern France, vol. 3), ed. David Lee Rubin, 17–44. Charlottesville, VA: Rookwood, 1997.

———. "Marvellous Facts and Miraculous Evidence in Early Modern Europe." In *Wonders, Marvels, and Monsters in Early Modern Culture*, ed. Peter G. Platt, 105–32. Newark: University of Delaware Press; London: Associated University Presses, 1999.

———. "Preternatural Philosophy." In *Biographies of Scientific Objects*, ed. Lorraine Daston, 15–41. Chicago: University of Chicago Press, 2000.

Daston, Lorraine, and Katharine Park. *Wonders and the Order of Nature, 1150–1750*. New York: Zone; London: MIT Press, 1998.

Daston, Lorraine, and Fernando Vidal, eds. *The Moral Authority of Nature*. Chicago: University of Chicago Press, 2004.

Davies, Richard. *Chesters Triumph in Honor of Her Price as It Was Performed vpon S. Georges Day 1610. in the Foresaid Citie*. London, 1610.

Dear, Peter. "Totius in Verba: Rhetoric and Authority in the Early Royal Society." *Isis* 76 (1985): 145–61.

———. "Miracles, Experiments, and the Ordinary Course of Nature." *Isis* 81 (1990): 663–83.

———. "A Mechanical Microcosm: Bodily Passions, Good Manners, and Cartesian Mechanism." In *Science Incarnate: Historical Embodiments of Natural Knowledge*, ed. Christopher Lawrence and Steven Shapin, 51–82. Chicago: University of Chicago Press, 1998.

———. "What Is the History of Science the History *Of*? Early Modern Roots of the Ideology of Modern Science." *Isis* 96 (2005): 390–406.

———. *The Intelligibility of Nature: How Science Makes Sense of the World*. Chicago: University of Chicago Press, 2006.

Dearborn Massar, Phyllis. "Stefano della Bella's Illustrations for a Fireworks Treatise." *Master Drawings* 7 (1969): 294–302.

Debus, Allen G. "The Paracelsian Aerial Niter." *Isis* 55 (1964): 43–61.

de Chadarevian, Soraya, and Nick Hopwood, eds. *Models: The Third Dimension of Science*. Stanford, CA: Stanford University Press, 2004.

de La Feuille, Daniel. *Devises et emblèmes anciennes et modernes*. Amsterdam, 1691.

Delavarinière, Sieur. *Réplique du sieur Delavarinière artificier aux défenses des sieurs Prévôt des marchands et échevins de la ville de Paris*. Paris, 1783.

Delisle, Joseph. "Extrait d'une lettre de M. Delisle, écrite de Petersbourg le 3. Janvier 1730." In *Mercure de France* (141 vols.), 18:378–83. Geneva: Slatkine, 1968–74.

Delisle, Joseph, and Daniel Bernoulli. *Discours lu dans l'assemblée de l'Académie des sciences le 2 Mars 1728, par Mr. de L'Isle avec la réponse de Mr. Bernoulli*. St. Petersburg, 1728.

Della Porta, Giambattista. *De aeris transmutationibus*. 1610. Edited by A. Paolelle. Naples: Scientifiche Steliane, 2000.

———. *Natural Magick by John Baptista Porta, a Neapolitane; in Twenty Books . . . Wherein Are Set Forth All the Riches and Delights of the Natural Sciences*. London, 1658.

Desaguliers, John Theophilus. *The Newtonian System of the World, the Best Model of Government: An Allegorical Poem*. London, 1728.

———. *A Course of Experimental Philosophy*. 2 vols. London, 1734–44.

de Sangro, Raimondo, Prince of San Severo. *Lettera apologetica dell'Esercitato accademico della Crusca [pseud.] contenente la difesa del libro intitolato Lettere d'una peruana, per rispetto alla supposizione de'quipu scritta alla duchessa di S**** e dalla medesima fatta pubblicare*. Naples, 1750.

Descartes, René. *Discourse on Method, Optics, Geometry, and Meteorology*. 1637. Translated by Paul J. Olscamp. Indianapolis: Hackett, 2001.

———. *The Philosophical Writings of Descartes*. Translated and edited by John Cottingham, Robert Stoothoff, Dugald Murdoch, and Anthony Kenny. 3 vols. Cambridge: Cambridge University Press, 1985–91.

*Description de la feste et du feu d'artifice: Qui doit être tiré sur la rivière, au sujet de la naissance de Monseigneur le Dauphin par ordre de Sa Majesté catholique Philippe V / et par les soins de leurs excellences MM. le Marquis de Santa-Cruz et de Barrenechea, ambassadeurs extraordinaires et plenipotentiaires du Roi d'Espagne*. Paris, 1730.

"Description de la fête & du feu d'artifice tiré sur la riviere. . . ." In *Mercure de France* (141 vols.), 18:390–403. Geneva: Slatkine, 1968–74.

"Description du feu d'artifice ordonné par la ville pour être tiré le 22 juin 1763 à l'occasion de l'inauguration de la statue du roi & de la publication de la paix." In *Mercure de France* (141 vols.), 85:153–59. Geneva: Slatkine, 1968–74.

*Dessein du feu d'artifice dressé sur la rivière de Saône par les ordres de Messieurs les Prevost des Marchands & Echevins de la ville de Lyon. Pour l'heureuse arrivée de Monseigneur le Duc de Bourgogne, & de Monseigneur le Duc de Berry*. Lyon, 1701.

"Détail des illuminations faites dans les jardins du château de Versailles le samedi 19 Mai 1770, pour le marriage de Mgr le Dauphin." In *Mercure de France* (141 vols.), 99:191–99. Geneva: Slatkine, 1968–74.

De Tott, François, Baron. *Memoirs of Baron de Tott: Containing the State of the Turkish Empire and the Crimea, during the Late War with Russia*. 2 vols. London, 1785.

Devon, Frederick, ed. *Issues of the Exchequer; Being Payments Made out of His Majesty's Revenue during the Reign of King James I*. London, 1836.

d'Holbach, Paul-Henri-Thiry, Baron. "Volcans (Histoire naturelle)." In *Encyclopédie; ou, Dictionnaire raisonné des science, des arts et des métiers* (35 vols.), ed. Denis Diderot and Jean d'Alembert, 17:443–46. Geneva, Paris, and Neufchastel, 1751–80.

Dibner, Bern. *Leonardo da Vinci: Machines and Weaponry*. Norwalk, CT: Burndy Library, 1974.

Diderot, Denis. "Art." In *Encyclopédie; ou, Dictionnaire raisonné des science, des arts et des métiers* (35 vols.), ed. Denis Diderot and Jean d'Alembert, 1:713–17. Geneva, Paris, and Neufchastel, 1751–80.

Diderot, Denis, and Jean d'Alembert, eds. *Encyclopédie; ou, Dictionnaire raisonné des science, des arts et des métiers*. 35 vols. Geneva, Paris, and Neufchastel, 1751–80.

Dieckmann, Herbert. "The Influence of Francis Bacon on Diderot's *Interprétation de la nature*." *Romanic Review* 34 (1943): 303–30.

*Discovrs sur les triomphes qui ont esté faicts le 25. 26. [et] 27. aoust 1613. dans la ville de Paris. A l'honneur & loüange de la feste S. Louys, & de Louys XIII par la grace de Dieu roy de France & de Nauarre. Ensemble les particularitez des feux artificiels décrites selon la disposition des sieurs Bagot, Iumeau & Morel, auteurs desdits artifices*. Lyon, 1613.

Disraeli, Isaac. *Curiosities of Literature; Consisting of Anecdotes, Characters, Sketches, and Observations, Literary, Critical, and Historical*. 4th ed. 2 vols. London, 1798.

Dolgorukov, V. *Vremia Petra II i imperatritsy Anny Ioannovny*. Moscow, 1909.

Donovan, Arthur. *Antoine Lavoisier: Science, Administration and Revolution*. Cambridge: Cambridge University Press, 1993.

Duffy, E. "'Whiston's affair': The Trials of a Primitive Christian, 1709–1714." *Journal of Ecclesiastical History* 27 (1976): 129–51.

Du Halde, Jean-Baptiste. *The General History of China; Containing a Geographical, Historical, Chronological, Political and Physical Description of the Empire of China*. 1735. 4 vols. London, 1741.

Duncan, Major Francis. *History of the Royal Regiment of Artillery*. 3rd ed. 2 vols. London, 1879.

Dyche, Thomas. *Manuel lexique; ou, Dictionnaire portatif des mots françois*. Edited by Abbé Prévost. 2 vols. Paris, 1767.

Eamon, William. *Science and the Secrets of Nature: Books of Secrets in Medieval and Early Modern Culture*. Princeton, NJ: Princeton University Press, 1994.

E.C. *A Full and Final Proof of the Plot from the Revelations*. London, 1680.

Eeles, Henry. *Philosophical Essays: In Several Letters to the Royal Society, Containing a Discovery of the Cause of Thunder*. London, 1771.

Egerton, Judy, ed. *Wright of Derby*. London: Tate Gallery, 1990.

Ehrard, Jean. "La main du travailleur, la plume du philosophe." *Milieux* 19/20 (1984–85): 47–53.

"An Elegy on the Death of the Honourable Robert Boyle, Esq; Fellow of the Royal Society." In *The Athenian Oracle* (3 vols., 3rd ed.), 1:65–69. London, 1706–16.

Elias, Norbert. *The Court Society*. Oxford: Blackwell, 1983.

*Encyclopédie méthodique: Arts et métiérs méchaniques*. Paris, 1782.

"Epigram, on the Fire-Works Displayed in Marybone Gardens, by Monsieur Torree, of Which Mrs. F—— complained to the Magistrates, That They Might Be Supressed." In *Nauticks; or, Sailor's Verses* (2 vols.), 2:44. London, 1783.

*Essai sur l'almanach général d'indication d'adresse personnelle et domicile fixe, des six corps, arts et métiers*. Paris, 1769.

Evelyn, John. *The Diary of John Evelyn, Esq. F. R. S.* Edited by William Bray. 4 vols. London, 1906.

*An Exact Abridgment of All the Statutes of King William and Queen Mary, and of King William III. And Queen Anne, in Force and Use*. London, 1704.

Fagiolo dell'Arco, Maurizio. *L'effimero barocco: Strutture della festa nella Roma del '600*. 2 vols. Rome: Bulzoni, 1977.

Fähler, Eberhard. *Feuerwerke des Barock: Studien zum öffentlichen Fest und seiner literarischen Deutung vom 16. bis 18. Jahrhundert*. Stuttgart: Metzler, 1974.

Faillon, Etienne Michel. *Histoire de la colonie française en Canada*. 3 vols. Villemarie, 1865–66.

Fara, Patricia. *An Entertainment for Angels*. New York: Totem, 2002.

Félibien, André. *Les divertissements de Versailles donnés par le roy à toute la cour au retour de la conquête de la Franche-Compté en l'année 1674*. Paris, 1674.

Ferguson, Eugene S. "Leupold's 'Theatrum Machinarum': A Need and an Opportunity." *Technology and Culture* 12 (1971): 64–68.

"Feste donnée par Mrs les Prévôs des Marchands & Echevins de la ville de Paris, pour le mariage de Madame de France, avec l'Infant Don Philipe, executée le Samedi 29. Août 1739." In *Mercure de France* (141 vols.), 37:2267–91. Geneva: Slatkine, 1968–74.

"Feu d'artifice." In *Mercure de France* (141 vols.), 41:2111–14. Geneva: Slatkine, 1968–74.

Findlen, Paula. "Jokes of Nature and Jokes of Knowledge: The Playfulness of Scientific Discourse in Early Modern Europe." *Renaissance Quarterly* 43 (1990): 292–331.

———. "The Economy of Scientific Exchange in Early Modern Italy." In *Patronage and Institutions: Science, Technology and Medicine at the European Court, 1500–1750*, ed. Bruce T. Moran, 5–24. Rochester, NY: Boydell, 1991.

———. *Possessing Nature: Museums, Collecting, and Scientific Culture in Early Modern Italy*. Berkeley and Los Angeles: University of California Press, 1994.

———. "Scientific Spectacle in Baroque Rome: Athanasius Kircher and the Roman College Museum." *Roma moderna e contemporanea*, 3 (1995): 625–65.

———. "Between Carnival and Lent: The Scientific Revolution at the Margins of Culture." *Configurations* 6 (1998): 243–67.

———. "Inventing Nature: Commerce, Art and Science in the Early Cabinet of Curiosities." In *Merchants and Marvels: Commerce, Science and Art in Early Modern Europe*, ed. Pamela H. Smith and Paula Findlen, 297–323. London: Routledge, 2001.

Firpo, Luigi. "Il Campanella scrittore di cose militari e un inedito discorso giovanile." *Giornale critico della filosofia italiana* 20 (1930): 472–80.

Flemming, Willi. *Geschichte des Jesuitentheaters in den Landen Deutscher Zunge*. Berlin: Selbstverlag der Gesellschaft für Theatergeschichte, 1923.

Fludd, Robert. *Mosaicall Philosophy Grounded upon the Essentiall Truth, or Eternal Sapience*. Goudae, 1638. 1st English ed. London, 1659.

Neufchâteau, Nicolas Louis François de. *Lettre à un ami sur le spectacle pyrique et hydraulique*. Paris, 1768.

Fontenelle, Bernard Le Bovier de. *Oeuvres diverses de M. de Fontenelle, de l'académie françoise: Nouvelle édition, augmentée et enrichie de figures gravées par Bernard Picart*. 3 vols. The Hague, 1728–29.

———. *Eloges des académiciens: Avec l'histoire de l'Académie royale des sciences en M.DC.XCIX; avec un discours préliminaire sur l'utilité des mathématiques*. 2 vols. The Hague, 1740.

*For the Preventing Tumultuous Disorders Which May Happen Hereafter upon Pretence of Assembling to Make Bonfires or Publick Fireworks. . . .* London, 1682.

Force, James E. *William Whiston: Honest Newtonian*. Cambridge: Cambridge University Press, 1985.

Fourcroy, Antoine-François de. *Elements of Chemistry, and Natural History: To Which Is Prefixed the Philosophy of Chemistry*. 4 vols. London, 1796.

Fournier, Eduoard, ed. *Variétés historiques et littéraires*. 10 vols. Paris, 1855–63.

Franklin, Benjamin. *Experiments and Observations on Electricity, Made at Philadelphia in America by Mr. Benjamin Franklin*. 5th ed. London, 1774.

Frazer, Sir James George. *The Golden Bough: A Study in Magic and Religion*. London: Penguin Classics, 1996.

Frézier, Amédée-François. *Traité des feux d'artifice pour le spectacle*. Paris, 1706.

———. *A Voyage to the South-Sea, and along the Coasts of Chili and Peru, in the Years 1712, 1713, and 1714 Particularly Describing the Genius and Constitution of the Inhabitants, as Well Indians as Diviards: Their Customs and Manners; Their Natural History, Mines, Commodities, Traffick with Evrope, &c. . . . with Postscript by Dr. Edmund Halley*. Paris, 1716. 1st English ed., London, 1717.

———. *Traité des feux d'artifice pour le spectacle*. The Hague, 1741.

———. *Traité des feux d'artifice pour le spectacle. Nouvelle édition. Toute changée, & considerablement augmentée*. 2nd ed. Paris, 1747.

Fronsperger, Leonard. *Vonn geschütz vnnd Fewrwerck, wie dasselb zuwerffen vnd schiessen. . . .* Frankfurt am Main, 1557.

———. *Kriegs Ordnung und Regiment*. Frankfurt am Main, 1564.

———. *Von dem Lob dess Eigen Nutzen*. Frankfurt am Main, 1564.

Fulke, William. *Meteors; or, A Plain Description of All Kind of Meteors as Well Fiery and Ayrie, as Watry and Earthy, Briefly Manifesting the Causes of All Blazing-Stars, Shooting Stars, Flames in the Aire, Thunder, Lightning, Earthquakes, Rain, Dew, Snow, Clouds, Sprigs, Stones, and Metalls*. 1563. London, 1655.

Furbank, Philip Nicholas. *Diderot: A Critical Biography*. London: Secker & Warburg, 1992.

Furley, O. W. "The Pope-Burning Processions of the Late Seventeenth Century." *History* 44 (1959): 16–23.

Furttenbach, Joseph. *Halinitro-Pyrobolia. Beschreibung Einer newen Büchsenmeisterey, nemlichen: Gruendlicher Bericht, wie der Salpeter, Schwefel, Kohlen, unnd das Pulfer zu praepariren, zu probieren, auch langwirzig gut zu behalten: Das Fewrwerck zur Kurtzweil und Ernst zu laboriren*. Ulm, 1627.

———. *Architectura martialis: Das ist, Ausführliche Bedencken, über das, zu dem Geschütz und Waffen gehörige Gebäw. . . .* Ulm, 1630.

———. *Architectura universalis*. Ulm, 1635.

———. *Büchsenmeisterey-Schul, Darinnen die dem angehende Büchsenmeister und Feuerwercker. . . .* Augsburg, 1643.

———. *Mechanische ReißLaden, Das ist, Ein gar geschmeidige, bey sich verborgen tragende Laden, daß .. alle fünffzehen Recreationen, als da seynd Arithmetica, Geometria, Planimetria . . . Prospectiva, Mechanica, Grottenwerck, Wasserlaitungen, Fewrwerck, Büchsenmeisterey, Architectura Militaris . . . könden exercirt werden*. Augsburg, 1644.

———. *Newes Itinerarium Italiae*. Hildesheim: Georg Olms, 1971.

Gage, John. *Colour and Culture: Practice and Meaning from Antiquity to Abstraction*. London: Thames & Hudson, 1997.

Galt, John. *George the Third: His Court and Family*. 2 vols. London, 1821.

Garnerin, André-Jacques. "Letter from M. Garnerin to the Editors of the Journal de Paris." *Select Reviews, and Spirit of the Foreign Magazines* (Philadelphia) 1 (1809): 124–26.

Garnerin, André-Jacques, and Elisa Garnerin. *Détails des trois premiers voyages aériens que M. Garnerin a fait en Russe le 20 juin et le 18 juillet à Saint-Petersbourg, le 20 septembre à Moscou 1803*. Moscow, 1803.

Ger'e, Vladimir. *Sbornik pisem i memorialov Leibnitsa otnosiashchikhsia k Rossii i Petru Velikomu*. St. Petersburg, 1873.

Gillispie, Charles Coulston. *Science and Polity in France at the End of the Old Regime*. Princeton, NJ: Princeton University Press, 1981.

Gode, P. K. *The History of Fireworks in India between A.D. 1400 and 1900*. Transactions of the Indian Institute of Culture, vol. 17. Bangalore: Indian Institute of Culture, 1953.

Gofflot, L. V. *Théâtre au collège du moyen age à nos jours*. 1907; reprint, New York: Franklin, 1964.

Golikov, Ivan Ivanovich. *Dopolneniia k Deianiiam Petra Velikogo*. 18 vols. Moscow, 1790–97.

Golinski, J. V. "A Noble Spectacle: Phosphorus and the Public Cultures of Science in the Early Royal Society." *Isis* 80 (1989): 11–39.

"Gollandets Klenk v Moskovii." *Istoricheskii vestnik* 57 (September 1894): 770.

Goodman, Dena. "Public Sphere and Private Life: Towards a Synthesis of Recent Historiographical Approaches to the Old Regime." *History and Theory* 31 (1992): 1–20.

Goodman, John. "'Altar against altar': The Colisée, Vauxhall Utopianism and the Symbolic Politics in Paris (1769–1777)." *Art History* 15 (1992): 434–69.

Gordin, Michael. "The Importation of Being Earnest: The Early St. Petersburg Academy of Sciences." *Isis* 91 (2000): 1–31.

Gordon, Patrick. *Tagebuch des Generals Patrick Gordon, während seiner Kriegsdienste unter den Schweden und Polen vom Jahre 1655 bis 1661 und seines Aufenthaltes in Russland vom Jahre 1661 bis 1699, zum ersten Male vollständig veröffentlicht durch Fürst M.A.* 3 vols. St. Petersburg, 1849–52.

———. *Passages from the Diary of General Patrick Gordon of Auchleuchries in the Years 1635–1699*. London: Cass, 1968.

Graminäus, Diederich. *Fvrstliche Hochzeit So der Durchluchtig hochgeb. Furst und Herr, herr Wilhelm Hertzog zu Gulich Cleue vnd Berg . . . in . . . Dusseldorff gehalttenn. . . . 1585*. Cologne, 1587.

Grandjean de Fouchy, Jean Paul. "Eloge de J.-A. Nollet." In *Histoire de l'Academie royale des sciences*, 121–36. Paris, 1773.

Graves, Michael A. R. *Thomas Norton: The Parliament Man*. Oxford: Blackwell, 1994.

Gray, Thomas. *Correspondence of Thomas Gray*. Vol. 1, *1734–1755*. Edited by Paget Toynbee and Leonard Whibley. Oxford: Oxford University Press, 1935.

*The Green-Park Folly; or, The Fireworks Blown Up; a Satire*. London, 1749.

Gregory, George. *The Economy of Nature Explained and Illustrated on the Principles of Modern Philosophy*. 3 vols. London, 1796.

Griffiths, Antony. *The Print in Stuart Britain, 1603–1689*. London: British Museum Press, 1998.

[Grignon]. *La pyrotecnie pratique; ou, Dialogues entre un amateur des feux d'artifice, pour le spectacle, et un jeune homme curieux de s'en instruire*. Paris, 1780.

Gruber, Alain-Charles. "Les 'Vauxhalls' Parisiens au XVIIIe siècle." *Bulletin de la Société de l'histoire de l'art français*, 1971, 125–43.

———. *Les grandes fêtes et leurs décors a l'époque de Louis XVI*. Paris: Droz, 1972.

*The Guardian*. 2 vols. London, 1714.

Guerlac, Henry. "The Poet's Nitre." *Isis* 45 (1954): 243–55.

Guerrier, Woldemar. *Leibniz in seinen Beziehungen zu Russland und Peter dem Großen: eine geschichtliche Darstellung dieses Verhältnißes nebst den darauf bezüglichen Briefen und Denkschriften*. St. Petersburg and Leipzig, 1839; reprint, Hildesheim: Gerstenberg, 1975.

*Guide des corps des marchands et des communautés des arts et métiers tant de la ville & fauxbourgs de Paris, que du royaume*. Paris, 1766.

Guyot, Edmé-Gilles. *Nouvelles récréations physiques et mathématiques, contenant ce qui a été imaginé de plus précieux dans ce genre et qui se découvre journellement; auxqelles on a joint les causes, leurs effets, la manière de les construire, et l'amusement qu'on en peut tirer pour étonner et surprendre agréablement.* 3 vols. Paris, 1799.

Hale, J. R. "Printing and Military Culture of Renaissance Venice." *Medievalia et Humanistica* 8 (1977): 21–62.

Hales, Stephen. *A Treatise on Ventilators.* London, 1758.

Hall, A. R. *Ballistics in the Seventeenth Century: A Study in the Relations of Science and War, with Reference Principally to England.* Cambridge: Cambridge University Press, 1952.

———. "Gunnery, Science and the Royal Society." In *The Uses of Science in the Age of Newton,* ed. John G. Burke, 111–41. Berkeley and Los Angeles: University of California Press, 1983.

———. "Isaac Newton and the Aerial Nitre." *Notes and Records of the Royal Society of London* 52 (1998): 51–61.

Hall, Bert S. *Weapons and Warfare in Renaissance Europe: Gunpowder, Technology, and Tactics.* Baltimore: Johns Hopkins University Press, 2002.

Hall, Marie Boas. "Frederick Slare, F.R.S. (1648–1727)." *Notes and Records of the Royal Society of London* 46 (1992): 23–41.

Hall, Phyllis A. "The Appreciation of Technology in Campanella's 'The City of the Sun.'" *Technology and Culture* 34 (1993): 613–28.

Hall, Trevor H. *Old Conjuring Books: A Bibliographical and Historical Study with a Supplementary Check-List.* London: Duckworth, 1972.

Hamilton, William. *Observations on Mount Vesuvius, Mount Etna, and Other Volcanoes, in a Series of Letters, Addressed to the Royal Society.* London, 1772.

———. *Campi Phlegraei, Observations on the Volcanos of the Two Sicilies as They Have Been Communicated to the Royal Society of London.* Naples, 1776.

———. *Supplement to the Campi Phlegraei.* Naples, 1779.

Hankins, Thomas, and Robert Silverman. *Instruments and the Imagination.* Princeton, NJ: Princeton University Press, 1995.

Hanzelet, Jean Appier, dit. *Recreation mathematicqve, composee de plusieurs problemes plaisants et facetievx, en faict d'arithmeticque, geometrie, mechanicque, opticque, et autres parties de ces belles sciences.* Pont-à-Mousson, 1624.

———. *La pyrotechnie.* Pont-à-Mousson, 1630.

———. *Mathematicall Recreations; or, A Collection of Sundrie Problemes, Extracted out of the Ancient and Moderne Philosophers, as Secrets in Nature, and Experiments . . . Most of Which Were Written First in Greeke and Latine, Lately Compiled in French, by Henry Van Etten Gent.* Translated by Francis Malthus. London, 1633.

Hanzelet, Jean Appier, dit, and François Thybourel. *Recueil de plusieurs machines militaires, et feux artificiels pour la guerre, & recreation.* Pont-à-Mousson, 1620.

Harley, R. D. *Artists' Pigments, c. 1600–1835: A Study in English Documentary Sources.* London: Archteype, 2001.

Harriot, Thomas. *A Brief and True Report of the New Found Land of Virginia: Of the Commodities There Found and to Be Raysed, as Well Marchantable, as Others for Victuall, Building and Other Necessarie Uses for Those That Are and Shalbe the Planters There; and of the Nature and Manners of the Natural Inhabitants Discovered by the English Colony There Seated by Sir Richard Greinvile Knight in the Year 1585.* London, 1588.

Harris, John. *The Evil and Mischief of a Fiery Spirit: A Sermon preach'd in the Parish Churches of St. Mildred Bread-Street and St. Matthew Fryday-Street, on February 27, 1709/10.* London, 1710.

———. *Lexicon technicum, or, An Universal English Dictionary of Arts and Sciences. . . .* 5th ed. 2 vols. London, 1736.

Harris, Tim. *London Crowds in the Reign of Charles II: Propaganda and Politics from the Restoration until the Exclusion Crisis*. Cambridge: Cambridge University Press, 1990.

———. "'Venerating the honesty of a tinker': The King's Friends and the Battle for the Allegiance of the Common People in Restoration England." In *The Politics of the Excluded, c. 1500–1850*, ed. Tim Harris, 195–232. Basingstoke: Palgrave, 2001.

———. *Revolution: The Great Crisis of the British Monarchy, 1685–1720*. London: Penguin, 2007.

Harsdörffer, Georg Philipp. *Delitiae Philosophicae et Mathematicae Der Philosophischen und Mathematischen Erquickstunden: Dritter Theil*. Nürnberg, 1653.

Hart, Clive. *Kites: An Historical Survey*. New York: Praeger, 1967.

———. *The Prehistory of Flight*. Berkeley and Los Angeles: University of California Press, 1985.

Harvey, E. Newton. *A History of Luminescence from the Earliest Times until 1900*. Philadelphia: American Philosophical Society, 1957.

Hassenstein, Wilhelm, ed. *Das Feuerwerkbuch von 1420: 600 Jahre deutsche Pulverwaffen und Büchsenmeisterei*. Munich: Deutschen Technik, 1941.

Haven, Peder von. "Puteshestvie v Rossiiu." In *Peterburg Anny Ioannovny v inostrannykh opisaniakh*, ed. Iu. N. Bespiatykh, 303–84. St. Petersburg: Russko-Baltiiskii informatsionnyi tsentr, 1997.

Heeffer, Albrecht. "*Récréations mathématiques*: A Study on Its Authorship, Sources and Influence." *Gibecière* 1 (2006): 79–167.

Heilbron, J. L. *Electricity in the Seventeenth and Eighteenth Centuries: A Study of Early Modern Physics*. Berkeley and Los Angeles: University of California Press, 1979.

Hengler & Co. *New Fire-Works, at the Prospect Hotel, Hooper's Hill, Margate. Signior Hengler & Co. Artists in Fire-Works of Vauxhall, Inform the Nobility . . . That They Will Exhibit a Grand and Brilliant Display of Fire-works, on Thursday Evening Next, the 7th August, 1800*. Margate: Warren, Printer, 1800. Handbill.

Heninger, S. K., Jr. *A Handbook of Renaissance Meteorology*. Durham, NC: Duke University Press, 1960.

Henninger-Voss, Mary J. "Working Machines and Noble Mechanics: Guidobaldo del Monte and the Translation of Knowledge." *Isis* 91 (2000): 233–59.

———. "How the 'New Science' of Cannons Shook Up the Aristotelian Cosmos." *Journal of the History of Ideas* 63 (2002): 371–97.

Heybrock, C. *Jean-Nicholas Servandoni: Eine Untersuchung seiner Pariser Bühnenwerke*. Cologne, 1970.

Heyd, Michael. "The New Experimental Philosophy: A Manifestation of 'Enthusiasm' or an Antidote to It?" *Minerva* 25 (1987): 423–40.

Hilaire-Pérez, Liliane. "Invention and the State in 18th-Century France." *Technology and Culture* 32 (1991): 911–31.

Hill, Thomas. *A Contemplation of Mysteries Contayning the Rare Effectes and Significations of Certayne Comets, and a Briefe Rehersall of Sundrie Hystoricall Examples, as Well Diuine, as Prophane*. London, 1574.

*Histoire de l'Académie royale des sciences: Année MDCCLIX*. Paris, 1765.

Hobbes, Thomas. *Seven Philosophical Problems and Two Propositions of Geometry*. 1662. In *The English Works of Thomas Hobbes of Malmesbury* (10 vols.), ed. Sir William Molesworth, 7:1–68. London: Longman, Brown, Green & Longmans, 1839–45.

Hochadel, Oliver. "Öffentliche Wissenschaft: Elektrizität in der deutschen Aufklärung." Ph.D. thesis, University of Goettingen, 2003.

Höckely, Michael. *Anhang zu Johann Daniel Blümels Anweisung zur Lust=Feuerwerkerey verfertiget von Hr. Michael Höckely, Königl. Feuerwerker zu Auxonne in Burgund.* Supplement to Johann Daniel Blümel, *Gründliche Anweisung zur Lust=Feuerwerkerey, besonders in denjenigen Stücken, die das Auge der Zuschauer am meisten erlustingen und in Verwunderung satzen. Mit einem Anhange vermehret, in welchem von wohirlechenden Tafel-Feuerwerk, wie auch von besonders artigen Kunststücken, in Tabacksdosen, in Bäumen &c. abzubrennen; Alles dieses ausgefertiget von Hr. Höckely*, 143–80. Strasburg: Amand Koenig, 1771.

Hodgkin, John Eliot. "Halinitropyrobolia." In *Rariora: Being Notes of Some of the Printed Books, Manuscripts, Historical Documents, Medals, Engravings, Pottery, etc., Collected (1858–1900) by John Eliot Hodgkin, F.S.A.* (3 vols.), 3:i–viii, 1–92. London: Low, Marston, 1902.

Hogarth, William. *The Analysis of Beauty: Written with a View of Fixing the Fluctuating Ideas of Taste*. London, 1753.

Hogg, O. F. G. *The Royal Arsenal: Its Background, Origin, and Subsequent History*. 2 vols. London: Oxford University Press, 1963.

Hogwood, Christopher. *Handel*. London: Thames & Hudson, 1984.

———. *Handel: Music for the Royal Fireworks*. Cambridge: Cambridge University Press, 2005.

Home, Roderick W. *The Effluvial Theory of Electricity*. New York: Arno, 1981.

Honour, Hugh. *Chinoiserie: The Vision of Cathay*. New York: Harper & Row, 1973.

Hood, Thomas. "Ode to Madame Hengler, Fire-Work Maker to Vauxhall." *Comic Annual* 1 (1830): 155–60.

Hooke, Robert. *The Diary of Robert Hooke F.R.S., 1672–1680: Transcribed from the Original in the Possession of the Corporation of the City of London*. Edited by H. W. Robinson and W. Adams. London: Taylor & Francis, 1935.

Hooper, William. *Rational Recreations, in which the Principles of Numbers and Natural Philosophy Are Clearly and Copiously Elucidated, by a Series of Easy, Entertaining, Interesting Experiments. . . .* 4th ed. 4 vols. London, 1794.

Horn, Christian. *Der aufgeführte Staat: Zur Theatralität höfischer Repräsentation unter Kurfürst Johann Georg II. von Sachsen*. Tübingen: Francke, 2004.

Horn, Jeff, and Margaret C. Jacob. "Jean-Antoine Chaptal and the Cultural Roots of French Industrialization." *Technology and Culture* 39 (1998): 671–98.

Houpreght, John Frederick. *Aurifontina chymica; or, A Collection of Fourteen Small Treatises Concerning the First Matter of Philosophers for the Discovery of Their (Hitherto So Much Concealed) Mercury. . . .* London, 1680.

Howarth, William Driver, and Jan Clarke, eds. *French Theatre in the Neo-Classical Era, 1550–1789*. Cambridge: Cambridge University Press, 1997.

Howse, Derek. *Greenwich Time and the Discovery of the Longitude*. Oxford: Oxford University Press, 1980.

Hughes, Lindsey. *Russia in the Age of Peter the Great*. New Haven, CT: Yale University Press, 1998.

Hunt, John Dixon. *Garden and Grove: The Italian Renaissance Garden in the English Imagination, 1600–1750*. London: Dent, 1986.

Hunter, Michael. *Science and Society in Restoration England*. Cambridge: Cambridge University Press, 1991.

Hutton, Charles. *Recreations in Mathematics and Natural Philosophy*. 4 vols. London, 1803.

Hutton, Ronald. *Stations of the Sun: A History of the Ritual Year in Britain*. Oxford: Oxford University Press, 1996.

Iliffe, Rob. "Lying Wonders and Juggling Tricks: Nature and Imposture in Early Modern England." In *"Everything Connects": In Conference with Richard H. Popkin: Essays in His Honor*, ed. J. Force and D. Katz, 183–210. Leiden: Brill, 1998.

Incarville, Pierre Nicholas le Chéron d'. "Mémoire sur la manière singulière dont les Chinois fondent la corne à lanterne." *Mémoires de l'Académie des sciences (savant-étrangés)*, 1755, 350–68.

———. "Mémoire sur les vernis de la Chine." *Mémoires de mathématique et de physique présentés à l'Académie royale des sciences* 3 (1760): 117–42.

———. "Maniére de faire les fleurs dans les feux d'artifice chinois." *Mémoires de mathématique et de physique présentés à l'Académie royale des sciences, par divers savans, & lus dans ses assemblées* 4 (1763): 66–94.

Ingersoll, Ernest. *Dragons and Dragon Lore*. New York: Payson & Clarke, 1928.

Isherwood, Robert M. *Farce and Fantasy: Popular Entertainment in Eighteenth-Century Paris*. Oxford: Oxford University Press, 1986.

Iunker [Juncker], Gottlob Fridrikh Vil'gel'm. *Kratkoe opisanie toi illuminatsii, kotoraia Aprelia 28 dnia, 1732 goda . . . pri feierverka v Sanktpeterburge predstavlena byla kupno s pozdravitel'nym vosklitsaniem k eia imperatorskamu velichestvu*. St. Petersburg, 1732.

———. *Kratkoe opisanie onago feierverka kotoroi Aprelia 28 dnia 1734 goda to est' v vysokotorzhestvennyi den' koronovaniia presvetleishiia derzhavneishiia velikiia gosudaryni anny ioannovny imperatritsy i samoderzhitsy vserossiiskiia i prochaia . . . velikoi illuminatsii v tsarstvuiushchem sanktpeterburge predstavlen byl*. St. Petersburg, 1734.

Iushkevich, A. P., and Iu. Kh. Kopelevich. *Kristian Goldbakh*. Moscow: Nauka, 1983.

Ivashkiavichius, A. *Kazimir Semenovich i ego kniga "Velikoe iskusstvo artillerii. Chast' pervaia."* Vilnius: Mintis, 1971.

*Izobrazhenie onago feierverka kotoroi po blagopoluchno sovershivshemcia brachnom sochetanii eia vysochestva gosudaryni printsessy anny . . . s svetleishim kniazem i gosudarem antonom ulrichom . . . v sanktpeterburge 9 iiulia 1739 goda predstavlen byl*. St. Petersburg, 1739.

Jacob, J. R. "Boyle's Atomism and the Restoration Assault on Pagan Naturalism." *Social Studies of Science* 8 (1978): 211–33.

Jallabert, Jean. *Expériences sur l'électricité, avec quelques conjectures sur la cause de ses effets*. Paris, 1749.

Jaucourt, Louis. "Invention." In *Encyclopédie; ou, Dictionnaire raisonné des science, des arts et des métiers* (35 vols.), ed. Denis Diderot and Jean d'Alembert, 8:848–49. Geneva, Paris, and Neufchastel, 1751–80.

Jankovic, Vladimir. "Meteors under Scrutiny: Private, Public, and Professional Weather in Britain, 1660–1800." Ph.D. thesis, University of Notre Dame, 1998.

Jarry, Madeleine. *Chinoiserie: Chinese Influence on European Decorative Art, 17th and 18th Centuries*. New York: Vendome, 1981.

Jenkins, Ian, and Kim Sloan. *Vases and Volcanoes: Sir William Hamilton and His Collection*. London: British Museum Press, 1996.

Jocelyn, Colonel J. R. J. "The Connection of the Ordnance Department with National & Royal Fire-Works, Including Some Account of Sir Martin Beckman, Colonel Henry John Hopkey, & Sir William Congreve (2nd Baronet)." *Journal of the Royal Artillery* 32 (1905–6): 481–503.

Johns, Adrian. *The Nature of the Book: Print and Knowledge in the Making*. Chicago: University of Chicago Press, 1998.

Johnson, Samuel. "Letter on Fireworks, From the Gentleman's Magazine, Jan. 1749." In *The Works of Samuel Johnson, LL.D.: With an Essay on His Life and Genius* (2 vols.), ed. Arthur Murphy, 2:549. New York, 1846.

Johnson, William. *Agyrto-mastix; or, Some Brief Animadversions upon Two Late Treatises One of Master George Thomsons, Entituled Galeno-Pale, the Other of Master Thomas O'Dowdes, Called the Poor Mans Physitian*. London, 1665.

Johnson, William. "Fireworks and Firemasters of England, 1662–1856." *International Journal of Mechanical Sciences* 36 (1994): 1061–67.

———. "Comments and Commentary on William Moore's *A Treatise on the Motion of Rockets and an Essay on Naval Gunnery*." *International Journal of Impact Engineering* 16 (1995): 449–521.

Jones, J. R. *Country and Court: England, 1658–1714*. London: Arnold, 1978.

Jones, Matthew L. *The Good Life in the Scientific Revolution: Descartes, Pascal, Leibniz, and the Cultivation of Virtue*. Chicago: University of Chicago Press.

Jones, Lt. Robert. *A New Treatise on Artificial Fireworks*. London, 1765.

Jordan, Thomas. *London in Luster: Projecting Many Bright Beams of Triumph: Disposed into Several Representations of Scenes and Pageants. Performed with Great Splendor on Wednesday, October XXIX. 1679. At the Initiation and Instalment of the Right Honourable Sir Robert Clayton, Knight, Lord Mayor of the City of London*. London, 1679.

Kafker, Frank A., and Serena L. Kafker. *The Encyclopedists as Individuals: A Biographical Dictionary of the Authors of the Encyclopédie*. Studies in Voltaire and the Eighteenth Century, vol. 257. Oxford: Voltaire Foundation at the Taylor Institution, Oxford University, 1988.

Kantemir, Antiokh. "To My Mind: On the Detractors of Learning." In *The Literature of Eighteenth Century Russia: A History and Anthology* (2 vols.), ed. and trans. Harold B. Segel, 1:151–63. New York: Dutton, 1967.

Karlinsky, Simon. *Russian Drama from Its Beginnings to the Age of Pushkin*. Berkeley and Los Angeles: University of California Press, 1985.

Kaufmann, Thomas DaCosta. *The Mastery of Nature: Aspects of Art, Science, and Humanism in the Renaissance*. Princeton, NJ: Princeton University Press, 1993.

———. *Toward a Geography of Art*. Chicago: University of Chicago Press, 2004.

Kellman, Jordan. "Discovery and Enlightenment at Sea: Maritime Exploration and Observation in the 18th Century French Scientific Community." Ph.D. thesis, Princeton University, 1998.

Kemp, Martin. "From Mimesis to Fantasia: The Quattrocento Vocabulary of Creation, Inspiration, and Genius in the Visual Arts." *Viator* 8 (1977): 347–98.

Kepler, Johannes. *Johannes Kepler Gesammelte Werke*. Edited by Max Caspar et al. 20 vols. to date. Munich: C. H. Beck, 1937–.

Kestler, Johann Stephan. *Physiologia Kircheriana experimentalis qua summa argumentorum multitudine & varietate naturalium rerum scientia per experimenta physica, mathematica, medica, chymica, musica, magnetica, mechanica comprobatur atque stabilitur*. Amsterdam, 1680.

Kiene, Michael. "L'image du Dieu vivant: Zum 'Aktionsbild' und zur Ikonographie des Festes am 30. November 1729 auf der Piazza Navona in Rom." *Zeitschrift für Kunstgeschichte* 54 (1991): 220–48.

Kim, Mi Gyung. "'Public' Science: Hydrogen Balloons and Lavoisier's Decomposition of Water." *Annals of Science* 63 (2006): 291–318.

Kircher, Athanasius. *Ars magna lucis et umbrae*. Rome, 1646.

———. *Mundus subterraneus in XII libros digestus*. 2 vols. in 1. Amsterdam, 1665.

Klopp, Onno. "Leibniz' Plan der Gründung einer Societät der Wissenschaften in Wien." *Archiv für österreichische Geschichte* 40 (1868): 157–255.

Knif, Henrik. *Gentlemen and Spectators: Studies in Journals, Opera and the Social Scene in Late Stuart London*. Helsinki: Finnish Historical Society, 1995.

Knight, Carlo. "Sir William Hamilton's *Campi Phlegraei* and the artistic contribution of Peter Fabris." In *Oxford, China and Italy: Writings in Honour of Sir Harold Acton on His Eightieth Birthday*, ed. Edward Chaney and Neil Ritchie, 192–208. London, 1984.

Knowles Middleton, W. E. *A History of the Theories of Rain and Other Forms of Precipitation*. New York: Franklin Watts, 1965.

Koepp, Cynthia J. "The Alphabetical Order: Work in Diderot's *Encyclopédie*." In *Work in France: Representations, Meaning, Organization, and Practice*, ed. Stephen Laurence Kaplan and Cynthia J. Koepp, 229–57. Ithaca, NY: Cornell University Press, 1986.

Konopleva, M. S. *Teatral'nyi zhivopisets Dzhuzeppe Valeriani: Materialy k biografii i istorii tvorchestva*. Leningrad: Gosudarstvennyi Ermitazh, 1948.

Kopelevich, Iu. Kh. *Osnovanie Peterburgskoi Akademii nauk*. Leningrad: Nauka, 1977.

Krafft, G. W., and J. Weitbrecht. *Sermones in solenni Academiae Scientiarum Imperialis conuentu die XXIX. Aprilis anni MDCCXLII. publice recitat*. St. Petersburg, 1742.

Kragh, Helge. *Phosphors and Phosphorus in Early Danish Natural Philosophy*. Copenhagen: Royal Danish Academy of Sciences and Letters, 2003.

Kramer, Gerhard W. *The Firework Book: Gunpowder in Medieval Germany*. *Journal of the Arms and Armour Society*, vol. 17, no. 1. London: Arms and Armour Society, 2001.

*Kratkoe iz'iasnenie izobrazheniia onago feierverka i illuminatsii kotorye v chest' eia imperatorskago velichestva samoderzhitsy vserossiiskiia v vysochaishii den' eia rozhdeniia 28 Genvaria 1735 goda pred imperatorskimi palatami v Sanktpeterburge zazhzheny byli*. St. Petersburg, 1735.

Kulibin, Ivan Petrovich. *Rukopisnye materialy I. P. Kulibina v Arkhive Akademii nauk SSSR: Nauchnoe opisanie s prilozheniem tekstov i chertezhei*. Edited by N. M. Raskin, B. A. Malkevich, and I. I. Artobolevskii (general ed.). Moscow; Leningrad: Izdatel'stvo Akademii nauk SSSR, 1953.

Kushner, Eva. "The Concept of Invention and Its Role in Renaissance Literary Theory." *Neohelicon: Acta comparationis litterarum* 8 (1980–81): 135–46.

Lacombe, Jacques. *Dictionnaire encyclopédique des amusemens des sciences*. Paris, 1792.

La Hire, Philippe de. "Sur les effets du ressort de l'air dans la poudre à canon, & dans le tonnere." In *Histoire de l'Académie royale des sciences, année MDCCII*, 9–14. Paris, 1702.

———. "On the Effects of the Elasticity of the Air in Gunpowder and Thunder, Translated by Mr. Chambers." In *The Philosophical History and Memoirs of the Royal Academy of Sciences at Paris* (5 vols.), trans. John Martyn and Ephraim Chambers, 1:313–19. London, 1742.

Laird, W. R. "Archimedes among the Humanists." *Isis* 82 (1991): 628–38.

Laneham, Robert. *A Letter Whearin Part of the Entertainment Vntoo the Queenz Maiesty at Killingwoorth Castl in Warwik Sheer in This Soomerz Progress 1575 Is Signified, from a Freend Officer Attendant in Coourt Vntoo Hiz Freend a Citizen and Merchaunt of London*. London, 1575.

Langford, Paul. *A Polite and Commercial People: England, 1727–1783*. Oxford: Oxford University Press, 1989.

Langins, Janis. *Conserving the Enlightenment: French Military Engineering from Vauban to the Revolution*. Cambridge, MA: MIT Press, 2004.

———. "Eighteenth-Century French Fortification Theory After Vauban: The Case of Montalembert." In *The Heirs of Archimedes: Science and the Art of War through the Age of Enlightenment*, ed. Brett D. Steele and Tamera Dorland, 333–60. Cambridge, MA: MIT Press, 2005.

Langlois, Gilles-Antoine. *Folies, tivolis, et attractions: Les premiers parcs de loisirs parisiens*. Paris: Délégation à l'action artistique de la ville de Paris, Difusion, Hachette, 1991.

Larroque, Tamizey de, ed. *Une fête bordelaise en 1615, relation contemporaine publiée avec un avertissement et des notes*. Bordeaux: A. Bellier, 1892.

Lavoisier, Antoine. "Remarques sur la composition de quelques feux d'artifice colorés en bleu et en jaune, dont je remets les recettes cachetées entre les mains de l'académie." 1766. In *Oeuvres de Lavoisier*, vol. 7, *Correspondence*, ed. René Fric, 109–13. Paris: Editions Albin Michel, 1955.

———. *Elements of Chemistry, in a New Systematic Order*. Translated by Robert Kerr. 4th ed. Edinburgh, 1799.

———. "Rapport sur les procédés d'artifice proposés par M. Ruggieri." In *Oeuvres de Lavoisier* (6 vols.), ed. Edouard Grimaux, 4:417–18. Paris, 1862–93.

———. "Rapport sur une manière d'allumer simultanément un grand nombre de lampions, du 4 février 1772." In *Oeuvres de Lavoisier* (6 vols.), ed. Edouard Grimaux, 4:106–8. Paris, 1862–93.

Le Comte, Louis. *Memoirs and Observations Typographical, Physical, Mathematical . . . Made in a Late Journey through the Empire of China*. London, 1697.

Lee, Rensselaer W. *"Ut pictura poesis": The Humanistic Theory of Painting*. New York, 1967.

Lefroy, General Sir J. H. *War Services of Lieutenant-General Albert Borgard*. Woolwich: Royal Artillery Institution, 1948.

Leibniz, Gottfried Wilhelm. "Zwei Pläne zu Societäten." In *Werke*, ed. Onno Klopp, 1:109–48. Hannover, 1864.

———. *Theodicy: Essays on the Goodness of God, the Freedom of Man, and the Origin of Evil*. Translated by E. M. Huggard. Edited by Austin Farrer. London: Routledge & Kegan Paul, 1952.

———. "Drole de pensée." *La Nouvelle Révue Française* 4 (1958): 758–68.

———. *Sämtliche Schriften und Briefe*. Ser. 3, *Mathematischer, naturwissenschaftlicher und technischer Briefwechsel*. Vol. 3, *1680–Juni 1683*. Berlin: Akademie, 1991.

———. *Sämtliche Schriften und Briefe*. Ser. 3, *Mathematischer, naturwissenschaftlicher und technischer Briefwechsel*. Vol. 4, *Juli 1683–1690*. Berlin: Akademie, 1995.

Leicester, Henry M. "Mikhail Lomonosov and the Manufacturing of Glass and Mosaics." *Journal of Chemical Education* 46 (1969): 295–98.

———, ed. *Lomonosov on the Corpuscular Theory*. Cambridge, MA: Harvard University Press, 1970.

Leith, James A. *Space and Revolution: Projects for Monuments, Squares, and Public Buildings in France, 1789–1799*. Montreal: McGill-Queen's University Press, 1991.

Le Jeune, Paul. *Relation de ce qui s'est passé en la Nouvelle France en l'année 1637: Envoyée au R. Père provincial de la Compagnie de Jésus en la province de France*. Rouen, 1638.

Lemery, Nicolas. *Le nouveau recueil de curiositez rares et nouvelles des plus admirables effets de la nature & de l'art . . . autres dont quel-que uns ont été tirez du cabinet de feu Monsieur le marquis de l'Hôpital*. 2 vols. Paris, 1685.

———. *New Curiosities in Art and Nature; or, A Collection of the Most Valuable Secrets in All Arts and Sciences . . . Translated into English from the Seventh Edition*. London, 1711.

———. "Sur les feux souterrains, les tremblemens de terre, & c. expliqués chimiquement." In *Histoire de l'Académie royale des sciences avec les mémoires de mathématique et de physique tirés des registres de cette académie, année 1700*, 51–52. Paris, 1761.

Leng, Rainer. *Ars belli: Deutsche taktische und kriegstechnische Bilderhandschriften und Traktate im 15. und 16. Jahrhundert*. 2 vols. Wiesbaden: Reichert Verlag, 2002.

Le Normand d'Étioles, Jeanne Antoinette, Marquise de Pompadour. *Memoirs of the Marchioness of Pompadour*. 2 vols. London, 1766.

Leonardo da Vinci. *The Literary Works of Leonardo da Vinci*. Edited by Jean Paul Richter. 2 vols. Berkeley and Los Angeles: University of California Press, 1977.

Lesur, Charles-Louis. *Annuaire historique universel pour 1825*. Paris, 1826.

Ley, Francis. *Le maréchal de Münnich et la Russie au XVIIIe siècle*. Paris: Plon, 1959.

Liebe, G. "Die soziale Wertung der Artillerie." *Zeitschrift für historische Waffenkunde* 2 (1900): 146–51.

Liechtenhan, Francine-Dominique. "Jacob von Stählin, academicien et courtisan." *Cahiers du monde russe* 43 (2002): 321–32.

Lignereux, Yann. *Lyon et le roi: De la "bonne ville" à l'absolutisme municipal (1594–1654)*. Seyssel: Champ Vallon, 2003.

Limon, Jerzy. *The Masque of Stuart Culture*. Newark: University of Delaware Press, 1990.

Ling, Wang. "On the Invention and Use of Gunpowder and Firearms in China." *Isis* 37 (1947): 160–78.

Livingstone, David N. *Putting Science in Its Place: Geographies of Scientific Knowledge*. Chicago: University of Chicago Press, 2003.

Locke, John. "Elements of Natural Philosophy." In *A Collection of Several Pieces of Mr. John Locke*, ed. John Hales, 179–230. London, 1720.

Loftis, John. "Richard Steele's Censorium." *Huntington Library Quarterly* 14 (1950): 53–62.

Lomonosov, Mikhail Vasil'evich. *Polnoe sobranie sochineniia*. 11 vols. Moscow-Leningrad: Izdatel'stvo Akademii nauk SSSR, 1950–83.

———. "Oration on the Origin of Light: A New Theory of Color." Translated by Henry M. Leicester. In *Lomonosov on the Corpuscular Theory*, ed. Henry M. Leicester, 247–69. Cambridge, MA: Harvard University Press, 1970.

———. "Oration on the Use of Chemistry: Presented in a Public Session of the Imperial Academy of Sciences, September 6th 1751." Translated by Henry M. Leicester. In *Lomonosov on the Corpuscular Theory*, ed. Henry M. Leicester, 186–202. Cambridge, MA: Harvard University Press, 1970.

Long, Pamela O. "Invention, 'Intellectual Property,' and the Origin of Patents." *Technology and Culture* 32 (1991): 846–84.

———. "Power, Patronage, and the Authorship of Ars: From Mechanical Know-How to Mechanical Knowledge in the Last Scribal Age." *Isis* 88 (1997): 1–41.

———. *Openness, Secrecy, Authorship: Technical Arts and the Culture of Knowledge from Antiquity to the Renaissance*. Baltimore: Johns Hopkins University Press, 2001.

*Lordonnance et ordre du tournoy, ioustes, combat, a pied, a cheual*. Paris, 1520.

Lorenzo, Claudio Di. *Il teatro del fuoco: Storie, vicende e architetture della pirotecnica*. Padua: F. Muzzio, 1990.

Lotz, Arthur. *Das Feuerwerk, seine Geschichte und Bibliographie*. Leipzig: Karl W. Hiersemann, 1941.

Lowman, R. *An Exact Narrative and Description of the Wonder-Full and Stupendious Fire-Works in Honour of Their Majesties Coronations, and for the High Entertainment of Their Majesties, the Nobility, and City of London; Made on the Thames, and Perform'd to the Admiration and Amazement of the Spectators on April the 24 1685*. London, 1685.

Lucar, Cyprian. "A Treatise Named Lucar Appendix." In Niccolò Tartaglia, *Three Books of Colloquies Concerning the Arte of Shooting*, trans. Cyprian Lucar, 80–120. London, 1588.

Luckett, Thomas Manley. "Hunting for Spies and Whores: A Parisian Riot on the Eve of the French Revolution." *Past and Present* 156 (1997): 116–43.

Luk'ianov, Pavel Mitrofanovich. *Istoriia khimicheskikh promyslov i khimicheskoi promyshlennosti Rossii do kontsa XIX veka*. Edited by S. I. Bol'fkovicha. 6 vols. Moscow: Izdatel'stvo Akademii nauk, 1948–65.

Luynes, Albert, Charles Philippe d', duc de. *Mémoires du duc de Luynes sur la cour de Louis XV (1735–58)*. 17 vols. Paris, 1860–65.

Lynn, Michael R. "Divining the Enlightenment: Public Opinion and Popular Science in Old Regime France." *Isis* 22 (2001): 34–54.

———. *Popular Science and Public Opinion in Eighteenth-Century France*. Manchester: Manchester University Press, 2006.

———. "Sparks for Sale: The Culture and Commerce of Fireworks in Early Modern France." *Eighteenth-Century Life* 30, no. 2 (2006): 74–97.

Lyon, John. *Experiments and Observations Made with a View to Point Out the Errors of the Present Received Theory of Electricity*. London, 1780.

Maccubbin, R., and M. Hamilton-Phillips, ed. *The Age of William III and Mary II: Power, Politics and Patronage, 1688–1702*. Williamsburg, VA: College of William and Mary, 1989.

MacCulloch, John. "Pyrotechny." In *Edinburgh Encyclopedia* (18 vols.), ed. David Brewster, 17:217–77. Edinburgh, 1830.

MacDougall, E. B., and N. Miller, eds. *Fons sapientiae: Renaissance Garden Fountains*. Washington, DC: Dumbarton Oaks, Trustees for Harvard University, 1978.

Macknight, Thomas. *The Life of Henry St. John, Viscount Bolingbroke, Secretary of State in the Reign of Queen Anne*. London, 1863.

Maggs, Barbara Widenor. "Firework Art and Literature: Eighteenth Century Pyrotechnical Tradition in Russia and Western Europe." *Slavonic and East European Review* 54 (1976): 24–40.

*The Magic Lantern, How to Buy and How to Use It, by "A Mere Phantom."* London, 1866.

*The Magnificent Marriage of the Two Great Princes, Fredericke Count-Palatine, &C: and the Lady Elizabeth, Daughter to the Imperiall Maiesties of King Iames, and Queene Anne to the Comfort of All Great Britaine*. London, 1613.

Makovskaia, A. K. "Melissino I. (1726–1797)." In *Zabytye imena i pamiatniki russkoi kul'tury: tezisy dokladov koferentsii k 60–letiiu Otdela istorii russkoi kul'tury*, ed. M. N. Diatlova, 38–43. St. Petersburg: State Hermitage, 2001.

Malcolm, James Peller. *Anecdotes of the Manners and Customs of London during the Eighteenth century. . . .* 2nd ed. 2 vols. London, 1810.

Malinovskii, K. V. "Iakob fon Shtelin i ego zapiski po istorii russkoi zhivopisi XVIII veka." In *Russkoe iskusstvo barokko: Materialy i issledovaniia*, ed. T. V. Alekseeva, 173–79. Moscow: Nauka, 1977.

Malthe, François de [Francis Malthus]. *Practiqve de la gverre: Contenant l'vsage [usage] de l'artillerie, bombes & mortiers, feux artificiels & petards, sappes & mines, ponts & pontons, tranchées & trauaux*. Paris, 1646. 3rd ed., Paris, 1668.

Malthus, Francis. *A Treatise of Artificial Fire-Works Both for Warres and Recreation. With Divers Pleasant Geometricall Obseruations, Fortifications, and Arithmeticall Examples. In Fauour of Mathematicall Students. Newly Written in French, and Englished by the Authour*. London, 1629.

Manstein, Christopher Hermann von. *Contemporary Memoirs of Russia, from the Year 1727 to 1744*. London, 1856.

Manzini, Luigi. *Applausi festivi fatti in Roma per l'elezzione di Ferdinando III*. Rome, 1637.

Marcet, J. Halimand. *Conversations on Political Economy, in Which the Elements of That Science Are Familiarly Explained*. 4th ed. London, 1821.

Marcigliano, Alessandro. "The Development of the Fireworks Display and Its Contribution to Dramatic Art in Renaissance Ferrara." *Theatre Research International* 14 (1989): 1–12.

*The Mariage of Prince Fredericke, and the Kings Daughter, the Lady Elizabeth, Vpon Shrouesunday Last*. 2nd ed. London, 1613.

Marin, Louis. *Portrait of the King*. Translated by Martha M. Houle. London: Macmillan, 1988.

Mariotte, Edme. *The Motion of Water, and Other Fluids, Being a Treatise of Hydroctaticks*. Translated by J. T. Desaguliers. London, 1718.

Marker, Gary. *Publishing, Printing and the Origins of Intellectual Life in Russia, 1700–1800*. Princeton, NJ: Princeton University Press, 1985.

Marlowe, Christopher. *The Tragicall History of Dr. Faustus*. London, 1604.

Marsden, Christopher. *Palmyra of the North: The First Days of St. Petersburg*. London: Faber & Faber, 1943.

Martin, D. C. "Former Homes of the Royal Society." *Notes and Records of the Royal Society of London* 22 (1967): 12–19.

Massuet, Pierre. *Suite de la science des personnes de cour, d'epée et de robe contenant les élémens de la philosophie moderne*. 2 vols. Amsterdam, 1752.

Mauskopf, Seymour H. "Gunpowder and the Chemical Revolution." In "The Chemical Revolution: Essays in Reinterpretation," special issue, *Osiris* 4 (1988): 93–118.

Mayow, John. *Medico-Physical Works: A Translation of the Tractatus quinque medico-physici Printed in the Sheldonian Theatre in 1674*. Oxford: Ashmolean Museum, 1926.

Mayr, Otto. *Authority, Liberty, and Automatic Machinery in Early Modern Europe*. Baltimore: Johns Hopkins University Press, 1986.

McClellan, James E., III. *Science Reorganized: Scientific Societies in the Eighteenth Century*. New York: Columbia University Press, 1985.

McConnell, Anita. "Hengler, Sarah (c. 1765–1845)." In *Oxford Dictionary of National Biography* (61 vols.), ed. H. C. G. Matthew and Brian Howard Harrison, 26:353. Oxford: Oxford University Press, 2004.

McGowan, Margaret M. *L'art du ballet de cour en France, 1581–1643*. Paris: Ecole du centre national de la recherche scientifique, 1963.

———. *Ideal Forms in the Age of Ronsard*. Berkeley and Los Angeles: University of California Press, 1985.

McKie, Douglas. "Fire and the Flamma Vitalis: Boyle, Hooke and Mayow." In *Science, Medicine, and History: Essays in Honour of Charles Singer* (2 vols.), ed. E. Ashworth Underwood, 1:469–88. Oxford: Oxford University Press, 1953.

Meehan-Waters, Brenda. *Autocracy and Aristocracy: The Russian Service Elite of 1730*. New Brunswick, NJ: Rutgers University Press, 1982.

Melissino, Peter Ivanovich. *Opisanie feierverka v Sanktpeterburge na Tsaritsinskom lugu 1 sentiabria 1796 goda*. St. Petersburg, 1796.

Melton, James Van Horn. *The Rise of the Public in Enlightenment Europe*. Cambridge: Cambridge University Press, 2001.

*Mémoire sur les feux d'air inflammable par M. Diller: Extrait des registres de l'Académie royale des sciences, du 4 juillet 1787*. Paris, 1787.

Mendelsohn, J. Andrew. "Alchemy and Politics in England, 1649–1665." *Past and Present* 135 (1992): 30–78.

Menestrier, Claude-François. *Les reiovissances de la paix, avec vn recveil de diuerses pieces sur ce sujet: Dedié a messievrs les preuost des marchands & escheuins de la ville de Lyon par le P.C.F.M. de la Compagnie de Iesvs. Advis necessaires pour la conduite des feux d'artifice*. Lyon, 1660.

Menshutkin, Boris Nikolaevich. *Russia's Lomonosov: Chemist, Courtier, Physicist, Poet*. Princeton, NJ: Princeton University Press, 1952.

Mersenne, Marin. *Correspondance du P. Marin Mersenne: Religieux minime*. Edited by Paul Tannery and Cornelius de Waard. 17 vols. Paris: G. Beauchesne/Presses Universitaires de France/Editions du Centre National de la Recherche Scientifique, 1933–88.

Merton, E. S. "Sir Thomas Browne's Theories of Respiration and Combustion." *Osiris* 10 (1952): 206–23.

Meya, Jörg, and Heinz Otto Sibum. *Das fünfte Element: Wirkungen und Deutungen der Elektrizität*. Hamburg: Deutsches Museum, Rowohlt, 1987.

Miller, John. *Popery and Politics in England, 1660–1688*. Cambridge: Cambridge University Press, 1973.

Mills, John. *An Essay on the Weather*. 2nd ed. London, 1773.

Mitchell, Trent. "The Politics of Experiment in the Eighteenth Century: The Pursuit of Audience and the Manipulation of Consensus in the Debate over Lightning Rods." *Eighteenth-Century Studies* 31 (1998): 307–31.

Mitra, Haridas. *Fire Works and Fire Festivals in Ancient India*. Calcutta: Abhedananda Academy of Culture, 1963.

Money, John. "Joseph Priestley in Cultural Context: Philosophic Spectacle, Popular Belief, and Popular Politics in 18th-Century Birmingham." *Enlightenment and Dissent* 7 (1988): 57–81; 8 (1989): 69–89.

Montucla, Jean-Etienne, ed. *Récréations mathématiques et physiques*. 4 vols. Paris, 1778.

Moore, David T. "Sir William Hamilton's Volcanology and His Involvement in Campi Phlegraei." *Archives of Natural History* 21 (1994): 169–93.

Moore, John E. "The Chinea, a Festival in Eighteenth Century Rome." Ph.D. thesis, Harvard University, 1992.

———. "Prints, Salami, and Cheese: Savoring the Roman Festival of the Chinea." *Art Bulletin* 77 (1995): 584–608.

Moore, William. *A Treatise on the Motion of Rockets and an Essay on Naval Gunnery*. London, 1813.

Moran, Bruce T. "German Prince-Practitioners: Aspects in the Development of Courtly Science, Technology, and Procedures in the Renaissance." *Technology and Culture* 22 (1981): 253–74.

———, ed. *Patronage and Institutions: Science, Technology, and Medicine at the European Court, 1500–1750*. New York: Boydell, 1991.

Morel, A. M. Th[omas]. *Traité pratique des feux d'artifice pour le spectacle et pour la guerre, avec les petits feux de table, et l'artifice à l'usage des théâtres*. Paris, 1800.

[Morel, Horace]. *Sviet dv fev d'artifice svr la prise de La Rochelle, que Morel doit faire pour l'arrivée du roy, sur la Seine, deuant le Louure*. Lyon, 1629.

Morse, William Northrop. "Lectures on Electricity in Colonial Times." *New England Quarterly* 7 (1934): 364–74.

Morton, Alan Q. "Concepts of Power: Natural Philosophy and the Uses of Machines in Mid-Eighteenth Century London." *British Journal for the History of Science* 28 (1995): 63–78.

———, ed. "Science Lecturing in the 18th Century." Special issue, *British Journal for the History of Science* 28 (1995): 1–99.

Morus, Iwan Rhys. *Frankenstein's Children: Electricity, Exhibition, and Experiment in Early-Nineteenth-Century London*. Princeton, NJ: Princeton University Press, 1998.

*Mr. Flockton's Theatre. At the [handwritten: White Lyen Highgate] in This Town. This Present Evening, Will Be Exhibited His Grand Exhibition, in the Same Manner as Performed Before the Royal Family and Most of the Nobility in the Kingdom*. London, ca. 1780.

Muir, Lynette R. *The Biblical Drama of Medieval Europe*. Cambridge: Cambridge University Press, 1995.

Müller, Regina. *Das Berliner Zeughaus: Die Baugeschichte*. Berlin: Deutsches Historisches Museum, Brandenburgisches Verl.-Haus, 1994.

Mulryne, J. R., Helen Watanabe-O'Kelly, and Margaret Shewring, eds. *Europa Triumphans: Court and Civic Festivals in Early Modern Europe*. 2 vols. Aldershot: Ashgate, 2004.

Myrone, Martin. "Vertue, George (1684–1756)." In *Oxford Dictionary of National Biography* (61 vols.), ed. H. C. G. Matthew and Brian Howard Harrison, 56:382–84. Oxford: Oxford University Press, 2004.

Napoleon, Emperor of France. *Memoirs of the History of France during the Reign of Napoleon*. 7 vols. London, 1823–24.

Needham, Joseph. *Military Technology: The Gunpowder Epic*. Pt. 7 of *Chemistry and Chemical Technology*. Vol. 5 of *Science and Civilisation in China*. Cambridge: Cambridge University Press, 1986.

Newman, William R. *Promethean Ambitions: Alchemy and the Quest to Perfect Nature*. Chicago: University of Chicago Press, 2004.

Newman, William R., and Lawrence M. Principe. *Alchemy Tried in the Fire: Starkey, Boyle, and the Fate of Helmontian Chymistry*. Chicago: University of Chicago Press, 2002.

Newton, Isaac. "Of Natures Obvious Laws and Processes in Vegetation." In *The Janus Face of Genius: The Role of Alchemy in Newton's Thought*, by Betty Jo Teeter Dobbs, 262–70. Cambridge: Cambridge University Press, 1991.

———. *Opticks; or, A Treatise of the Reflections, Refractions, Inflections and Colours of Light. The Second Edition, with Additions*. London, 1718.

———. *The Mathematical Principles of Natural Philosophy . . . Translated into English by Andrew Motte*. 2 vols. London, 1727.

Nichols, John. *The Progresses and Public Processions of Queen Elizabeth; among Which Are Interspersed Other Solemnities, Public Expenditures, and Remarkable Events during the Reign of That Illustrious Princess*. 3 vols. London, 1823.

———. *The Progresses, Processions, and Magnificent Festivities, of King James the First, His Royal Consort, Family, and Court, Etc.* 4 vols. London, 1828.

Nicolson, Benedict. *Joseph Wright of Derby, Painter of Light*. 2 vols. London, 1968.

Nieuhof, Johannes. *An Embassy from the East-India Company of the United Provinces, to the Grand Tartar Cham, Emperor of China Deliver'd by Their Excellencies, Peter De Goyer and Jacob De Keyzer, at His Imperial City of Peking*. London, 1673.

Nieuwentyt, Bernard. *The Religious Philosopher; or, The Right Use of Contemplating the Works of the Creator*. 3rd ed. 2 vols. London, 1724.

N.N., Gentleman formerly of Queen's College Oxon. *Will-with-a-Wisp; or, The Grand Ignis Fatuus of London. Being a Lay-Man's Letter to a Country-Gentleman, Concerning the Articles Lately Exhibited*. London, 1714.

Nollet, Jean-Antoine. *Essai sur l'électricité des corps*. Paris, 1746.

———. *Lectures in Experimental Philosophy . . . Translated from the French by John Colson*. London, 1748.

———. *Lettres sur l'électricité: Dans lesquelles on examine les dernières découvertes qui ont été faites sur cette matière, & les conséquences qu' l'on peut en tirer. Première partie*. Paris, 1753.

———. *Lettres sur l'électricité: Dans lesquelles on trouvera les principaux phénomènes qui ont été découverts depuis 1760, avec des discussions sur les conséquences qu'on peut en tirer . . . Troisième partie*. Paris, 1767.

Norton, Robert. *The Gunner Shewing the Whole Practise of Artillerie: With All the Appurtenances therevnto Belonging*. London, 1628.

———. *The Gunners Dialogue, with the Art of Great Artillery*. London, 1628.

*Nouvelle biographie générale depuis les temps les plus reculés jusqu'à nos jours.* Edited by J. C. F. Hoefer. 46 vols. Paris, 1852–66.

Ochs, Kathleen H. "The Royal Society of London's History of Trades Programme: An Early Episode in Applied Science." *Notes and Records of the Royal Society of London* 39 (1985): 129–58.

O'Connell, Sheila, Roy Porter, Celina Fox, and Ralph Hyde. *London, 1753.* London: British Museum Press; Boston: David R. Godine, 2003.

*An Ode for the Thanksgiving-Day, to the Tune of Derry Down.* By Titus Antigallicus. London, 1749.

Ogborn, Miles. *Spaces of Modernity: London's Geographies, 1680–1780.* New York: Guilford, 1998.

Ogborn, Miles, and Charles W. J. Withers. *Georgian Geographies: Essays on Space, Place and Landscape in the Eighteenth Century.* Manchester: Manchester University Press, 2004.

Ogden, Dunbar H. *The Staging of Drama in the Medieval Church.* Newark: University of Delaware Press; London: Associated Universities Press, 2002.

Olivier, Marc. "Jean-Nicolas Servandoni's Spectacles of Nature and Technology." *French Forum* 30 (2005): 31–47.

Ophir, Adi, Steven Shapin, and Simon Schaffer, eds. "The Place of Knowledge: The Spatial Setting and Its Relations to the Production of Knowledge." Special issue, *Science in Context,* vol. 4, 1991.

*Opisanie allegoricheskago izobrazheniia v zakliucheniu velikol'epnykh torzhestv, po sovershenii koronatsii Eiia Velichestva Ekateriny vtoryia . . . V imperatorskom rezidentsii Moskve protiv Kremlia, sentiabr 1762 goda.* Moscow, 1762.

*Opisanie koronatsii Eia Velichestva Imperatritsy, i Samoderzhitsy Vserossiiskoi, Anny Ioannovny, torzhestvenno otpravlennoi v tsarstvuiushchem gradie Moskvie, 28 aprielia, 1730 godu.* Moscow, 1730.

*Opisanie Velikoi illuminatsii v 28 Genvaria 1733 goda iako v vysokii den' rozhdeniia vsepresvetleishiia derzhavneishiia velikaia gosudaryni Anny ioannovny imperatritsy i samoderzhitsy vserossiiskiia pri feierverke v Sanktpeterburge predstavlennyia.* St. Petersburg, 1733.

Orden, Kate van. *Music, Discipline, and Arms in Early Modern France.* Chicago: University of Chicago Press, 2005.

Orgel, Stephen. *The Illusion of Power: Political Theater in the English Renaissance.* Berkeley and Los Angeles: University of California Press, 1975.

Ostrovitianov, K. V., ed. *Istoriia Akademii nauk SSSR: Tom pervyi (1724–1803).* Moscow-Leningrad, 1958.

*Oxford Dictionary of National Biography.* Edited by H. C. G. Matthew and Brian Howard Harrison. 61 vols. Oxford: Oxford University Press, 2004.

Ozouf, Mona. *Festivals and the French Revolution.* Translated by Alan Sheridan. Cambridge, MA: Harvard University Press, 1988.

Pallen, Thomas A., ed. *Vasari on Theater.* Carbondale: Southern Illinois University Press, 1999.

Pannabecker, John R. "Representing Mechanical Arts in Diderot's *Encyclopédie.*" *Technology and Culture* 39 (1998): 33–73.

Paracelsus. *Opera Bücher und Schrifften . . . durch J. Huserum . . . in Truck gegeben; Jetzt von newem mit Fleisz ubersehen, auch mit etlichen biszhero unbekandten Tractaten gemehrt . . . in zwen underschiedliche Tomos . . . gebracht.* Edited by Johann Huser. 2 vols. Strasbourg, 1616.

Parfaict, Claude, and François Parfaict. *Dictionnaire des théâtres de Paris.* 7 vols. Paris, 1767.

Partington, J. R. "The Life and Work of John Mayow (1641–79)." Pt. 2. *Isis* 47 (1956): 405–17.

———. *A History of Greek Fire and Gunpowder*. Cambridge: W. Heffer, 1960.

Paul, Michael C. "The Military Revolution in Russia, 1550–1682." *Journal of Military History* 68 (2004): 9–45.

Pavlova, G. E. "Proekty illiuminatsii Lomonosova." In *Lomonosov: Sbornik statei i materialov* (8 vols.), ed. A. I. Andreev and L. B. Modzalevskii, 4:219–37. Moscow-Leningrad: Izdatel'stvo Akademii nauk SSSR/Nauka, 1940–83.

Peacham, Henry. *The Compleat Gentleman Fashioning Him Absolute in the Most Necessary & Commendable Qualities Concerning Minde or Bodie That May Be Required in a Noble Gentleman*. London, 1622.

Pearson, Robin. *Insuring the Industrial Revolution: Fire Insurance in Great Britain, 1700–1850*. Aldershot: Ashgate, 2004.

Pekarskii, Petr. *Istoriia Imperatorskoi Akademii nauk v Peterburge*. 2 vols. St. Petersburg, 1870–73.

Pepys, Samuel. *Diary and Correspondence of Samuel Pepys, F.R.S.* Edited by Richard Lord Braybrooke. 4 vols. London, 1854.

Perrin, Claude. *La vie rocambolesque d'André Garnerin, pionnier du parachute*. Paris: Messene, 2000.

Perrinet d'Orval, Jean-Charles. *Essay sur les feux d'artifice pour le spectacle et pour la guerre*. Paris, 1745.

———. *Traité des feux d'artifice pour le spectacle et pour la guerre*. Berne, 1750.

———. "Feux d'artifice (artificier)." In *Encyclopédie; ou, Dictionnaire raisonné des science, des arts et des métiers* (35 vols.), ed. Denis Diderot and Jean d'Alembert, 6:640–46. Geneva, Paris, and Neufchastel, 1751–80.

———. *Manuel de l'artificier*. Neufchatel, 1755.

———. *Manuel de l'artificier: Seconde édition, revue, corrigée & augmentée*. Paris, 1757.

Petersburg Academy. *Relatsiia o illuminatsii . . . 3 dnia marta 1728 goda. Relation von der Illumination. . . .* St. Petersburg, 1728.

Petrov, A. N. "Palat' fel'dmarshala B. P. Sheremeteva i F. M. Apraksina v Moskve." *Arkhitekturnoe nasledstvo* 6 (1956): 138–46.

Philip, Chris. *A Bibliography of Firework Books: Works on Recreative Fireworks from the 16th to the 20th Century*. Winchester: St. Paul's Bibliographies, 1985.

Phillips, Edward. *The New World of English Words; or, A General Dictionary Containing the Interpretations of Such Hard Words as Are Derived from Other Languages*. 1st ed. London, 1658.

———. *The New World of Words; or, A Universal English Dictionary Containing the Proper Significations and Derivations of All Words from Other Languages . . . as Now Made Use of in Our English Tongue*. 5th ed. London, 1696.

Picinelli, Filippo. *Mundus symbolicus*. Cologne, 1695.

"A Pindaric Ode upon Oddities. Extempore. The Thanksgiving-Day! . . ." In *The Foundling Hospital for Wit, Number VI*, 90–91. London, 1749.

Pinto, J. A. "Nicola Michetti and Ephemeral Design in Eighteenth-century Rome." *Memoirs of the American Academy in Rome* or *Studies in Italian Art History* 35 (1980): 289–323.

Poirier, Jean-Pierre. *Lavoisier, Chemist, Biologist, Economist*. Translated by Rebecca Balinski. Philadelphia: University of Pennsylvania Press, 1996.

Polinière, Pierre. *Experiences de physique*. 2nd ed. 2 vols. Paris, 1734.

Porte, Joseph de la, Abbé de Fontenai. *Le voyageur françois; ou, Le connoissance de l'ancien et du nouveau monde*. 42 vols. Paris, 1769–95.

Priestley, Joseph. *The History and Present State of Electricity. . . .* 4th ed. London, 1775.

Prosper-Hardy, Simeon. *Mes loisirs: Journal d'événements tels qu'ils parviennent à ma connaissance (1764–1789)*. Edited by Maurice Tourneux and Maurice Vitrac. Paris: Picard, 1912.

*Protokoly zasedanii konferentsii Imperatorskoi Akademii nauk s 1725 do 1803 g.* 4 vols. St. Petersburg, 1897–1911.

Proudhomme J. C., and E. de Crauzat. *Les menus-plaisirs du roi: L'Ecole royale et la Conservatoire de musique*. Paris: Delagrave, 1929.

Pyliaev, M. I. *Zabytoe proshloe okrestnostei Peterburga*. St. Petersburg, 1889; reprint, St. Petersburg: Paritet, 2002.

Pyne, William Henry. "The Eidophusikon." In *Wine and Walnuts; or, After Dinner Chit-Chat* (2nd ed., 2 vols.), by Ephraim Hardcastle, 1:281–304. London, 1824.

*Pyrotechnica Loyolana, Ignatian Fire-Works; or, The Fiery Jesuits Temper and Behaviour*. London, 1667.

Quersitanus, Iosephus [Joseph Du Chesne]. *The Practise of Chymicall, and Hermeticall Physicke, for the Preservation of Health*. Translated by Thomas Tymme. London, 1605.

Rabiqueau, Charles. *Le spectacle du feu élémentaire; ou, Cours d'électricité expérimentale*. Paris, 1753.

Raeff, Marc. *The Well-Ordered Police State: Social and Institutional Change through Law in the Germanies and Russia, 1600–1800*. New Haven, CT: Yale University Press, 1983.

Raggio, Olga. "The Myth of Prometheus: Its Survival and Metamorphoses Up to the Eighteenth Century." *Journal of the Warburg and Courtauld Institutes* 21 (1958): 44–62.

*Ranelagh. A Magnificent Fire-Work Will Be Exhibited This Present Monday, May 20, 1793, in Honour of Her Majesty's Birthday. By Monsieur Caillot*. London, 1793.

Ransel, D. L. "The Government Crisis of 1730." In *Reform in Russia and the U.S.S.R.: Past and Prospects*, ed. Robert O. Crummey, 45–71. Champaign: University of Illinois Press, 1989.

Ravel, Jeffrey. *The Contested Parterre: Public Theater and French Political Culture, 1680–1791*. Ithaca, NY: Cornell University Press, 1999.

Ray, John. *The Wisdom of God Manifested in the Works of Creation*. London, 1691.

*Réception de tres-chrestien, tres-iuste, et tres-victorieux monarque Louys XIII. roy de France & de Nauarre . . . et de tres-chrestienne, tres-auguste, & tres-vertueuse Royne Anne d'Austriche: Par messieurs les doyen, chanoines, & comtes de Lyon . . . le XI. Decembre, 1622*. Lyon, 1623.

Reedy, Gerard. "Mystical Politics: The Imagery of Charles II's Coronation." In *Studies in Change and Revolution: Aspects of English Intellectual History, 1640–1800*, ed. Paul Korshin, 19–42. Menston: Scolar, 1972.

Regnault, [Père Noël]. *Philosophical Conversations; or, A New System of Physics*. Translated by Thomas Dale. 3 vols. London, 1731.

*Relatsiia o illuminatsii . . . 3 dnia marta 1728 goda/Relation von der Illumination. . . .* St. Petersburg, 1728.

Ricci, Matteo. *China in the Sixteenth Century: The Journals of Matthew Ricci, 1583–1610*. Edited by Louis J. Gallagher, S.J. New York: Random House, 1953.

Richardson, Thomas, and Henry Watts. *Chemical Technology; or, Chemistry in Its Applications to the Arts & Manufactures*. 2nd ed. 3 vols. London, 1863–67.

Riskin, Jessica. *Science in the Age of Sensibility: The Sentimental Empiricists of the French Enlightenment*. Chicago: University of Chicago Press, 2002.

Ritzarev, Marina. *Eighteenth-Century Russian Music*. Aldershot: Ashgate, 2006.

Roberts, Lissa. "The Death of the Sensuous Chemist: The 'New' Chemistry and the Transformation of Sensuous Technology." *Studies in History and Philosophy of Science* 26 (1995): 503–29.

Robinet, Jean Baptiste. *Dictionnaire universel des sciences morale, économique, politique et diplomatique; ou, Bibliotheque de l'homme-d'état et du citoyen*. Paris, 1783.

Robins, Benjamin. *New Principles of Gunnery*. London, 1747.

Roche, Daniel. *The Culture of Clothing: Dress and Fashion in the Ancien Régime*. Cambridge: Cambridge University Press, 1994.

Rodríguez-Camilloni, Humberto. "Utopia Realized in the New World: Form and Symbol of the City of Kings." In *Settlements in the Americas: Cross-Cultural Perspectives*, ed. Ralph Bennett, 28–52. Newark: University of Delaware Press; London: Associated University Presses, 1993.

Rodzevich, V. M. *Istoricheskoe opisanie Sanktpeterburgskogo Arsenala za 200 let ego sushchestvovaniia: 1712–1912 gg*. St. Petersburg: Tipo-litografiia S. Peterburgskoi Tiur'my, 1914.

Rogers, Nicholas. "Crowds and Political Festival in Georgian England." In *The Politics of the Excluded, c. 1500–1850*, ed. Tim Harris, 233–64. Basingstoke: Palgrave, 2001.

Röhling, Horst. "Illustrated Publications on Fireworks and Illuminations in Eighteenth Century Russia." In *Russia and the West in the Eighteenth Century: Proceedings of the Second Annual Conference Organized by the Study Group on Eighteenth-Century Russia . . . 17–22 July, 1981*, ed. A. G. Cross, 94–100. Newtonville, MA: Oriental Research Partners, 1983.

Rolle, Samuel. *Shilhavtiyah; or, The Burning of London in the Year 1666*. London, 1667.

Romano, Adriano. *Pyrotechnia, hoc est de ignibvs festivis, iocosis, artificialibvs et seriis*. Frankfurt a.M.: Officina Paltheniana, 1611.

Romocki, S. J. von. *Geschichte der Explosivstoffe: Sprengstoffchemie, Sprengtechnik und Torpedowesen*. Berlin, 1895; reprint, Hildesheim: Gerstenberg, 1983.

Roosevelt, Priscilla R. "Emerald Thrones and Living Statues: Theater and Theatricality on the Russian Estate." *Russian Review* 50 (1991): 1–23.

Rossi, Paolo. *Philosophy, Technology, and the Arts in the Early Modern Era*. Translated by Salvator Attanasio. New York: Harper & Row, 1970.

Rousseau, Jean-Jacques. *Julie; or, The New Heloise: Letters of Two Lovers Who Live in a Small Town at the Foot of the Alps*. Translated by Philip Stewart and Jean Vaché. Hanover, NH: Dartmouth College/University Press of New England, 1997.

Rovinskii, D. A. *Obozrenie ikonopisaniia v Rossii do kontsa XVII veka. Opisanie feierverkov i illuminatsii*. St. Petersburg: A. S. Suvorina, 1903.

Rowbottom, Margaret E. "John Theophilus Desaguliers (1683–1744)." *Proceedings of the Huguenot Society* 21 (1965–70): 196–218.

Ruggieri, Claude-Fortuné. *Elémens de pyrotechnie*. 1st ed. Paris, 1801.

———. *Pyrotechnie militaire; ou, Traité complet des feux de guerre et des bouches à feu*. Paris, 1812.

———. *Principles of Pyrotechnics*. Translated by Stuart Carlton. Paris, 1821; reprint, Buena Vista, CA: MP Associates, 1994.

———. *Notes explicites par Claude-Ruggiéri, artificier du roi, relativement à l'evénement du 8 juin 1825*. Paris, 1825.

———. *Précis historique sur les fêtes, les spectacles et les réjouissances publiques*. Paris, 1830.

Ruggieri, Gaetano. *A Description of the Machine for the Fireworks: With All Its Ornaments, and a Detail of the Manner in Which They Are to Be Exhibited in St. James's Park, Thursday, April 27, 1749, on Account of the General Peace, Signed at Aix La Chapelle, October 7, 1748. Published by Order of His Majesty's Board of Ordnance*. London, 1749.

Ruggieri, Michel-Marie [?]. *Précis pour Michel-Marie Ruggieri*. Paris, 1830.

Rykwert, Joseph. *The First Moderns: The Architects of the Eighteenth Century*. Cambridge, MA: MIT Press, 1980.

Sadoun-Goupil, Michelle. *Le chimiste Claude-Louis Berthollet, 1748–1822: Sa vie—son oeuvre*. Paris: Vrin, 1977.

Safranovskii, K. I. "Les salles de l'Académie des sciences de Saint-Petersbourg en 1741." *Cahiers du monde russe et soviétique* 8 (1967): 604–15.

Saint-Julien, Baron Louis Guillaume Baillet de. *L'art de composer et faire les fusées volantes et non volantes*. Paris, 1775.

Salatino, Kevin. *Incendiary Art: The Representation of Fireworks in Early Modern Europe*. Santa Monica, CA: Getty Research Institute for the History of Art and Humanities, 1997.

Sands, Mollie. *Invitation to Ranelagh, 1742–1803*. London: John Westhouse, 1946.

———. *The Eighteenth-Century Pleasure Gardens of Marylebone, 1773–1777*. London: Society for Theatre Research, 1987.

Sarton, George. "Montucla (1725–1799): His Life and Works." *Osiris* 1 (1936): 519–67.

Sassoli, Mario Gori. *Apparati architettonici per fuochi d'artificio a Roma nel Settecento: Della Chinea e di altre "Macchine di Gioia": Rome, Villa Farnesina, 24 marzo–28 maggio 1994*. Milan: Charta, 1994.

Say, Jean-Baptiste. *A Treatise on Political Economy; or, The Production, Distribution, and Consumption of Wealth*. Translated by Clement C. Biddle. Philadelphia, 1836.

Schaffer, Simon. "Natural Philosophy and Public Spectacle in the Eighteenth Century." *History of Science* 21 (1983): 1–43.

———. "Measuring Virtue: Eudiometry, Enlightenment and Pneumatic Medicine." In *The Medical Enlightenment of the Eighteenth Century*, ed. Andrew Cunningham and Roger French, 281–313. Cambridge: Cambridge University Press, 1990.

———. "The Show That Never Ends: Perpetual Motion in the Early Eighteenth Century." *British Journal for the History of Science* 28 (1995): 157–89.

———. "Experimenters' Techniques, Dyers' Hands and the Electric Planetarium." *Isis* 88 (1997): 456–83.

———. "Regeneration: The Body of Natural Philosophers in Restoration England." In *Science Incarnate: Historical Embodiments of Natural Knowledge*, ed. Christopher Lawrence and Steven Shapin, 83–120. Chicago: University of Chicago Press, 1998.

———. "Enlightened Automata." In *The Sciences in Enlightened Europe*, ed. William Clark, Jan Golinski, and Simon Schaffer, 126–65. Chicago: University of Chicago Press, 1999.

———. "Fish and Ships: Models in the Age of Reason." In *Models: The Third Dimension of Science*, ed. Soraya de Chadarevian and Nick Hopwood, 71–105. Stanford, CA: Stanford University Press, 2004.

Schaub, Owen W. "Pleasure Fires: Fireworks in the Court Festivals in Italy, Germany and Austria during the Baroque." Ph.D. thesis, Kent State University, 1978.

Schechner Genuth, Sara. *Comets, Popular Culture, and the Birth of Modern Cosmology*. Princeton, NJ: Princeton University Press, 1997.

Schiavo, Antonio di Pietro dello. *Il diario romano di Antonio di Pietro dello Schiavo dal 19 Ottobre 1404 al 25 Settembre 1417*. Edited by Francesco Isoldi. Città di Castello: Tipi della Casa editrice S. Lapi, 1917.

Schmidlap von Schorndorff, Johann. *Künstliche und rechtschaffene Feuerwerck zum schimpff*. Nuremberg, 1560.

———. *Künstliche und rechtschaffene Feuerwerck zum schimpff*. 4th ed. Nuremberg, 1591.

Schott, Gaspar. *Gasparis Schotti Magia universalis naturae et artis, sive, Recondita naturalium & artificialium rerum scientia . . . opus quadripartitum pars I, continet Optica, II. Acoustica, III. Mathematica, IV. Physica*. 4 vols. Würzburg, 1657–59.

Schulze, Ludmilla. "The Russification of the St. Petersburg Academy of Sciences and Arts in the Eighteenth Century." *British Journal for the History of Science* 18 (1985): 305–35.

Schwenter, Daniel. *Deliciae physico-mathematicae, oder, Mathemat: Und Philosophische Erquickstunden*. 3 vols. Nuremberg, 1651–53.

Schwoerer, Lois G. "The Glorious Revolution as Spectacle: A New Perspective." In *England's Rise to Greatness*, ed. Stephen B. Baxter, 109–49. Berkeley and Los Angeles: University of California Press, 1983.

Scott, W. S. *Green Retreats: The Story of Vauxhall Gardens, 1661–1859*. London: Odhams, 1955.

Secord, Anne. "Science in the Pub: Artisan Botanists in Early Nineteenth-Century Lancashire." *History of Science* 32 (1994): 269–315.

Semenova, L. N. *Byt naselenie Sankt-Peterburga (XVIII vek)*. St. Petersburg: Russko-Baltiiskii informatsionnyi tsentr, 1998.

Séris, Jean-Pierre. *Machine et communication: Du théâtre des machines à la méchanique industrielle*. Paris: Librarie philosophique J. Vrin, 1987.

Servandoni, Jean-Nicolas. *Description du spectacle de Pandore*. Paris, 1739.

Sewell, William H. "Visions of Labour: Illustrations of the Mechanical Arts before, in, and after Diderot's *Encyclopédie*." In *Work in France: Representations, Meaning, Organization, and Practice*, ed. Stephen Laurence Kaplan and Cynthia J. Koepp, 258–86. Ithaca, NY: Cornell University Press, 1986.

Shakespeare, William. *As You Like It*. Edited by Juliet Dusinberre. London: Arden, 2007.

Shapin, Steven. "The House of Experiment in Seventeenth-Century England." *Isis* 79 (1988): 373–404.

———. "The Invisible Technician." *American Scientist* 77 (1989): 554–63.

Shapin, Steven, and Simon Schaffer. *Leviathan and the Air Pump: Hobbes, Boyle and the Experimental Life*. Princeton, NJ: Princeton University Press, 1985.

Shapiro, Alan. *Fits, Passions and Paroxysms: Physics, Method, and Chemistry and Newton's Theories of Colored Bodies and Fits of Easy Reflection*. Cambridge: Cambridge University Press, 1993.

Sharpe, James. *Remember, Remember: A Cultural History of Guy Fawkes Day*. Cambridge, MA: Harvard University Press, 2005.

Shaw, Peter. *Chemical Lectures, Publickly Read at London, in the Years 1731, and 1732; and at Scarborough, in 1733, for the Improvement of Arts, Trades, and Natural Philosophy*. 2nd ed. London, 1755.

Shearman, John. *Mannerism*. Harmondsworth: Penguin, 1967.

Sibum, Heinz Otto. "Reworking the Mechanical Equivalent of Heat: Instruments of Precision and Gestures of Accuracy in Early Victorian England." *Studies in History and Philosophy of Science* 26 (1995): 73–106.

Siemienowicz, Kazimierz. *Artis magnae artilleriae pars prima*. Amsterdam, 1650.

———. *Grand arte d'artillerie par le sieur Casimir Siemienowicz chevalier Litvanien*. Translated by Pierre Noizet. Amsterdam, 1651.

———. *The Great Art of Artillery of Casimir Simienowicz*. Translated by George Shelvocke the Younger. London, 1729.

Sievernich, Gereon, and Hendrik Budde. *Das Buch der Feuerwerkskunst: Farbenfeuer am Himmel Asiens und Europas*. Nördlingen: Delphi, 1987.

Sigaud de la Fond, Joseph. *Description et usage d'un cabinet de physique expérimentale*. 2 vols. Paris, 1784.

Simms, D. L. "Archimedes and the Invention of Artillery and Gunpowder." *Technology and Culture* 28 (1987): 67–79.

Simon, Jonathan. *Chemistry, Pharmacy and Revolution in France, 1777–1809*. Aldershot: Ashgate, 2005.

Skalon, D. A., ed. *Glavnoe artilleriiskoe upravlenie*. Vol. 6, bk. 1, of *Stoletie Voennago ministerstsva, 1802–1902*. St. Petersburg, 1906.

Slare, Frederick. "An Account of Some Experiments Made with the Shining Substance of the Liquid and of the Solid Phosphorus." *Philosophical Collections*, no. 3 (December 10, 1681): 48–50.

Slaughter, Thomas, ed. *Ideology and Politics on the Eve of the Restoration: Newcastle's Advice to Charles II*. Philadelphia, 1984.

Smagina, G. I. "Publichnye lektsii Sankt-Peterburgskoi Akademii nauk vo vtoroi polovine XVIII v." *Voprosy istorii estestvoznaniia i tekhniki* 2 (1996): 16–26.

Smith, Adam. "Principles Which Lead and Direct Philosophical Enquiries, Illustrated by the History of Astronomy." In *Essays on Philosophical Subjects . . . to Which Is Prefixed, An Account of the Life and Writings of the Author*, ed. Dugald Stewart, 3–93. London, 1795.

Smith, Douglas. *Working the Rough Stone: Freemasonry and Society in Eighteenth-Century Russia*. DeKalb: Northern Illinois University Press, 1999.

Smith, Gil R. *Architectural Diplomacy: Rome and Paris in the Late Baroque*. Cambridge, MA: MIT Press, 1993.

Smith, Godfrey. *The Laboratory; or, School of Arts . . . Translated from the High Dutch*. 2nd ed. London, 1740.

Smith, Hannah. *Georgian Monarchy: Politics and Culture, 1714–1760*. Cambridge: Cambridge University Press, 2006.

Smith, Pamela H. *The Business of Alchemy: Science and Culture in the Holy Roman Empire*. Princeton, NJ: Princeton University Press, 1994.

———. *The Body of the Artisan: Art and Experience in the Scientific Revolution*. Chicago: University of Chicago Press, 2004.

———. "Art, Science, and Visual Culture in Early Modern Europe." *Isis* 97 (2006): 83–100.

———. "Laboratories." In *Early Modern Europe*, vol. 3 of *The Cambridge History of Science*, ed. Lorraine Daston and Katharine Park, 290–305. Cambridge: Cambridge University Press, 2006.

Smith, Pamela H., and Paula Findlen, eds. *Merchants and Marvels: Commerce, Science and Art in Early Modern Europe*. London: Routledge, 2001.

Smith, Thomas. *The Art of Gvnnery*. London, 1600.

Snobelen, S. D. "Caution, Conscience and the Newtonian Reformation: The Public and Private Heresies of Newton, Clarke and Whiston." *Enlightenment and Dissent* 16 (1997): 151–84.

———. "William Whiston: Natural Philosopher, Prophet, Primitive Christian." Ph.D. thesis, Cambridge University, 2000.

Sobel, Dava. *Longitude: The True Story of a Lone Genius Who Solved the Greatest Scientific Problem of His Time*. New York: Penguin, 1995.

Solov'ev, S. M. *Empress Anna: Favorites, Policies, Campaigns*. Translated by Walter J. Gleason Jr. Vol. 34 of *History of Russia from Earliest Times*. 48 vols. Gulf Breeze, FL: Academic International Press, 1976–2004.

———. *The Rule of Empress Anna*. Translated by Richard Hantula. Vol. 35 of *History of Russia from Earliest Times*. 48 vols. Gulf Breeze, FL: Academic International Press, 1976–2004.

Sorenson, Bent. "Sir William Hamilton's Vesuvian Apparatus." *Apollo* 159 (2004): 50–58.

Southworth, James Granville. *Vauxhall Gardens: A Chapter in the Social History of England*. New York: Columbia University Press, 1941.

"A Spectacle at Pentecost, Vicenza, 1379." In *The Medieval European Stage, 500–1550*, ed. William Tydeman, Glynne Wickham, John Northam, and W. D. Howarth, 432–33. Cambridge: Cambridge University Press, 2001.

*The Spectator*. 8 vols. London 1712–15.

Spedding, James. *The Letters and the Life of Francis Bacon, Including All His Occasional Works*. 7 vols. London, 1861–74.

Spence, Jonathan D. *The Memory Palace of Matteo Ricci*. London: Penguin, 1985.

Speranskii, Count M. M., ed. *Polnoe sobranie zakonov Rossiskoi Imperii, poveleniem Gosudaria Imperatora Nikolaia Pavlovicha sostavlennoe. Sobranie pervoe. S 1649 po 12 dekabria 1825 goda*. 45 vols. St. Petersburg, 1830–39.

Sprat, Thomas. *History of the Royal Society of London, for the Improving of Natural Knowledge*. London, 1667.

Stafford, Barbara Maria, Frances Terpak, and Isotta Poggi. *Devices of Wonder: From the World in a Box to Images on a Screen*. Los Angeles: Getty Research Institute, 2001.

Stählin, Jacob. "O pol'ze teatral'nykh deistve komedii; k vozderzhaniiu strastei chelovecheskikh." *Primechanii na vedemosti* 85 (1739): 337–44.

———. *Izobrazhenie feierverka i illuminatsii kotoryia v novyi 1748. god pred zimnim Eia Imperatorskago Velichestva domom predstavleny byli*. St. Petersburg, 1748.

———. *Original Anecdotes of Peter the Great*. London, 1788; reprint, New York: Arno Press/*New York Times*, 1970. As "Jacob Staehlin."

———. "Kratkaia istoriia iskusstva feierverkov v Rossii." In *Zapiski Iakoba Shtelina ob iziashchnykh iskusstvakh v Rossii* (2 vols.), 1:238–66. Moscow: Iskusstvo, 1990.

———. *Zapiski Iakoba Shtelina ob iziashchnykh iskusstvakh v Rossii*. Edited by K. V. Malinovskii. 2 vols. Moscow: Iskusstvo, 1990.

Stählin, Karl. *Aus den Papieren Jacob von Stählins*. Königsberg and Berlin: Ost-Europa, 1926.

Staniukovich, T. V. *Kunstkamera Peterburgskoi Akademii nauk*. Moscow-Leningrad: Izdatel'stvo Akademii nauk SSSR, 1953.

Starkey, George. *Pyrotechny Asserted and Illustrated to Be the Surest and Safest Means for Arts Triumph over Natures Infirmities Being a Full and Free Discovery of the Medicinal Mysteries Studiously Concealed by All Artists, and Onely Discoverable by Fire*. London, 1658.

States, Bert Olen, Jr. "Jean-Nicholas Servandoni: His Scenography and His Influence." Ph.D. thesis, Yale University, 1960.

Steele, Brett D. "Muskets and Pendulums: Benjamin Robins, Leonhard Euler, and the Ballistics Revolution." *Technology and Culture* 35 (1994): 348–82.

Stephens, H. M. "Desaguliers, Thomas (1721–1780)." In *Oxford Dictionary of National Biography* (61 vols.), ed. H. C. G. Matthew and Brian Howard Harrison, 15:893–94. Oxford: Oxford University Press, 2004.

Stern, Virginia F. "The Bibliotheca of Gabriel Harvey." *Renaissance Quarterly* 25 (1972): 1–62.

Stewart, Larry. *The Rise of Public Science: Rhetoric, Technology, and Natural Philosophy in Newtonian Britain, 1660–1750*. Cambridge: Cambridge University Press, 1992.

———. "A Meaning for Machines: Modernity, Utility, and the Eighteenth-Century British Public." *Journal of Modern History* 70 (1998): 259–94.

Strada, Famiano. *Prolusiones academicae academia secunda*. Leyden, 1627.

Strong, Roy. *Art and Power: Renaissance Festivals, 1450–1650*. Woodbridge: Boydell, 1984.

Stubbe, Henry. *Campanella Revived; or, An Enquiry into the History of the Royal Society, Whether the Virtuosi There Do Not Pursue the Projects of Campanella for the Reducing England unto Popery*. London, 1670.

———. "Animadversions upon the History of Making SALT-PETRE, Which Was Penned by Mr. Henshaw." In *Legends No Histories; or, A Specimen of Some Animadversions upon the History of the Royal Society Wherein, Besides the Several Errors against Common Literature, Sundry Mistakes about the Making of Salt-Petre and Gun-Powder Are Detected and Rectified*, 35–40. London, 1670.

———. *An Epistolary Discourse Concerning Phlebotomy: In Opposition to G. Thomson Pseudo-Chymist, a Pretended Disciple of the Lord Verulam*. London, 1671.

Stukeley, William. *The Philosophy of Earthquakes, Natural and Religious; or, An Inquiry into Their Cause, and Their Purpose*. 3rd ed. London, 1756.

Stukeley, William, W. C. Lukis, Roger Gale, and Samuel Gale. *The Family Memoirs of the Rev. William Stukeley, M.D., and the Antiquarian and Other Correspondence of William Stukeley, Roger & Samuel Gale, Etc.* 3 vols. Publications of the Surtees Society, vol. 76. Durham and London, 1882–87.

Sturges, W. Knight. "Jacques-François Blondel." *Journal of the Society of Architectural Historians* 11 (1952): 16–19.

Sukhomlinov, Mikhail Ivanov, ed. *Materialy dlia istorii Imperatorskoi Akademii Nauk*. 10 vols. St. Petersburg, 1885–1900.

Sumarokov, Alexandr. *Opisanie ognennago predstavleniia v pervyi vecher novago goda 1760*. St. Petersburg, 1759.

Sutton, Geoffrey V. *Science for a Polite Society: Gender, Culture, and the Demonstration of Enlightenment*. Boulder, CO: Westview, 1995.

Syndram, Dirk. "Princely Diversion and Courtly Display: The Kunstkammer and Dresden's Renaissance Collections." In *Princely Splendor: The Dresden Court, 1580–1620*, ed. Dirk Syndram and Antje Scherner, 54–69. Milan: Staatliche Kunstsammlungen Dresden/Metropolitan Museum of Art/Electa, 2004.

Tabournel, Raymond. "La catastrophe de la rue Royale, 30 mai 1770." *Revue des études historiques*, 1900, 414–18.

Targosz, Karolina. "'Le dragon volant' de Tito Livio Burattini." *Annali dell'Instituto e Museo di storia della scienza di Firenze* 2, no. 2 (1977): 67–85.

Tartaglia, Niccolò. *Three Books of Colloquies Concerning the Arte of Shooting*. Translated by Cyprian Lucar. London, 1588.

Tassoni, Estense. *L'Isola Beata, torneo fatto nella città di Ferrara per la venuta del Serenissimo Principe Carlo Arciduca d'Austria, a XXV Maggio MDLXIX*. Ferrara, 1569.

Taylor, John. *Heauens Blessing, and Earths Ioy; or, A True Relation, of the Supposed Sea-Fights & Fire-Workes, as Were Accomplished, Before the Royall Celebration, of the Al-Beloved Mariage, of the Two Peerlesse Paragons of Christendome, Fredericke & Elizabeth: With Triumphall Encomiasticke Verses, Consecrated to the Immortall Memory of Those Happy and Blessed Nuptials*. London, 1613.

Terrall, Mary. "Metaphysics, Mathematics, and the Gendering of Science in Eighteenth-Century France." In *The Sciences in Enlightened Europe*, ed. William Clark, Jan Golinski, and Simon Schaffer, 246–71. Chicago: Chicago University Press, 1999.

Tessier, Paul. *Chimie pyrotechnique; ou, Traité pratique des feux colorés*. Paris, 1859.

Thiery, Luc-Vincent. *Almanach du voyageur à Paris: Contenant une description intéressante de tous les monumens, chefs-d'oeuvre des arts, établissemens utiles. . . .* 7 vols. Paris, 1783–87.

Thompson, E. P. "Patrician Society, Plebeian Culture." *Journal of Social History* 6 (1974): 382–405.

Thompson, Edward Maunde, ed. *Correspondence of the Family of Hatton: Being Chiefly Letters Addressed to Christopher, First Viscount Hatton, A.D. 1601–1704*. 2 vols. London: Camden Society, 1878.

Thorndike, Lynn. *A History of Magic and Experimental Science*. 8 vols. New York: Columbia University Press, 1923–58.

Thorpe, W. N. "The Marquis of Worcester and Vauxhall." *Transactions of the Newcomen Society* 13 (1932–33): 75–88.

Thwaites, Reuben Gold, ed. *The Jesuit Relations and Allied Documents: Travels and Explorations of the Jesuit Missionaries in New France, 1610–1791: The Original French, Latin, and Italian Texts, with English Translations and Notes*. Edited by Reuben Gold Thwaites. 73 vols. Cleveland: Burrows, 1896–1901.

Tittmann, Wilfried. "Der Mythos vom 'Schwarzen Berthold.'" *Waffen- und Kostümkunde: Zeitschrift der Gesellschaft für Historische Waffen- und Kostümkunde* 25 (1983): 17–30.

Toone, William. *The Chronological Historian*. London, 1826.

Torlais, Jean. *Un physicien au siècle des lumières: L'Abbé Nollet, 1700–1770*. 1954; reprint, Elbeuf-sur-Andelle: Jonas, 1987.

Trapp, Joseph. *Proceedings of the French National Convention on the Trial of Louis XVI, Late King of France and Navarre*. 2nd ed. London, 1793.

*Die triumphirende Liebe*. Hamburg, 1653.

*The True and Particular Account of the Shocking and Dreadful Fire Which Happened on Wednesday Last, by the House of Mr Clithero, of Half Moon Alley, Bishopsgate-Street, Blowing Up, When Mrs. Clithero, Her Three Children, Another Four Persons, Perished in the Conflagration*. London, ca. 1791.

Turner, John. "Hengler's Fireworks—Famous for Sixty Years." *Fireworks Magazine*, no. 14 (September 1988): 13–15.

Tydeman, William, Glynne Wickham, John Northam, and W. D. Howarth, eds. *The Medieval European Stage, 500–1550*. Cambridge: Cambridge University Press, 2001.

Ufano, Diego. *Tratado dela artilleria y uso della platicado*. Brussels, 1612.

———. *Artillerie, c'est à dire vraye instruction de l'artillerie de toutes ses appartenances*. Translated by Jean Theodore de Bry. Frankfurt a.M., 1614.

Umbach, Maiken. "Visual Culture, Scientific Images and German Small-State Politics in the Late Enlightenment." *Past and Present* 158 (1998): 110–45.

Valturio, Roberto. *De re militari*. Verona, 1472.

Van Peursen, Cornelis-Anthonie. "Ars Inveniendi bei Leibniz." *Studia Leibnitiana: Zeitschrift für Geschichte der Philosophie und der Wissenschaften* 18 (1986): 183–94.

Vasil'iev, V. N. *Starinnye feierverki v Rossii (XVII–pervaia chetvert' XVIII veka)*. Leningrad, 1960.

Velimirovich, M. "Liturgical Drama in Byzantium and Russia." *Dumbarton Oaks Papers* 16 (1962): 351–85.

Venn, Captain Thomas. *The Compleat Gunner . . . Shewing the Art of Founding of Great Ordnance; Making Gun-Powder; the Taking of Heights and Distances*. London, 1672.

Vickers, Brian, ed. *Occult and Scientific Mentalities in the Renaissance*. Cambridge: Cambridge University Press, 1984.

Virgil. *The Aeneid of Virgil*. Translated by Alexander Strahan. 2 vols. London, 1767.

Vischer, Melchior. *Münnich: Ingenieur, Feldherr, Hochverräter*. Frankfurt am Main: Societäts, 1938.

Vockerodt, Johann Gotthilf. "Rossiia pri Petre Velikom." In *Chteniia v Imperatorskom obshchestve istoriia drevnostei rossiiskikh pri Moskovskom universitete*, 1874, pt. 2 (April–June), sec. 4:i–iv, 1–120.

Volkov, Vadim. "The Forms of Public Life: The Public Sphere and the Concept of Society in Imperial Russia." Ph.D. thesis, University of Cambridge, 1995.

Voltaire. *Letters Concerning the English Nation*. London, 1733.

Wackernagel, Rudolf H. *Das Münchner Zeughaus*. Munich: Schnell & Steiner, 1982.

Wade, Mara R. *Triumphus Nuptialis Danicus: German Court Culture and Denmark, the "Great Wedding" of 1634*. Wiesbaden: Harrassowitz, 1996.

Wairy, Louis Constant. *Mémoires de Constant, premier valet de chambre de l'empereur sur la vie privée de Napoléon, sa famille et sa cour*. 6 vols. Paris, 1830.

Waliszewski, K. *L'heritage de Pierre le Grand: Règne des femmes, gouvernement des favoris, 1725–1741*. Paris: Plon-Nourrit, 1900.

Walker, Adam. *Syllabus of a Course of Lectures on Natural and Experimental Philosophy*. York, 1772.

Wallace, Anthony F. C. *The Social Context of Innovation: Bureaucrats, Families, and Heroes in the Early Industrial Revolution, as Foreseen in Bacon's "New Atlantis."* Princeton, NJ: Princeton University Press, 1982.

Wallis, John. "A Letter of Dr. Wallis to Dr. Sloane, Concerning the Generation of Hail, and of Thunder and Lightning, and the Effects Thereof." In *Miscellanea Curiosa: Being a Collection of Some of the Principal Phaenomena in Nature, Accounted for by the Greatest Philosophers* (3 vols.), 2:315–20. London: Royal Society, 1705–7.

Walpole, Horace. *The Letters of Horace Walpole, Earl of Oxford*. Edited by Peter Cunningham. 9 vols. London, 1857–59.

Walton, Stephen. "The Art of Gunnery in Renaissance England." Ph.D. thesis, University of Toronto, 1999.

Warner, Deborah Jean. "Lightning Rods and Thunder Houses." *Rittenhouse* 11 (1997): 124–27.

Wauchope, Piers. "Beckman, Sir Martin (1634/5–1702)." In *Oxford Dictionary of National Biography* (61 vols.), ed. H. C. G. Matthew and Brian Howard Harrison, 4:740–42. Oxford: Oxford University Press, 2004.

Watanabe-O'Kelly, Helen. "Fireworks Displays, Firework Dramas and Illuminations: Precursors of Cinema?" *German Life and Letters* 48 (1995): 338–52.

Webster, Charles. *From Paracelsus to Newton: Magic and the Making of Modern Science*. Cambridge: Cambridge University Press, 1982.

Webster, John. *Academiarum examen; or, The Examination of Academies Wherein Is Discussed and Examined the Matter, Method and Customes of Academick and Scholastick Learning, and the Insufficiency Thereof Discovered and Laid Open*. London, 1654.

Wecker, Johann. *De secretis libri xvii*. Basel, 1582.

Weigert, Roger Armand. "Les feux d'artifice ordonnées par le bureau de la ville de Paris au XVIIe siècle." In *Paris et l'Ile-de-France, mémoires* (vol. 3), 173–99. Paris: Fédération des sociétés historiques et archéologiques de Paris et de l'Ile-de-France, 1951.

Wells, Maria Xenia Zevelechi. "A Flourish of Pyres." *FMR* 73 (1995): 116–26.

Werrett, Simon. "Healing the Nation's Wounds: Royal Ritual and Experimental Philosophy in Restoration England." *History of Science* 28 (2000): 377–99.

———. "An Odd Sort of Exhibition: The St. Petersburg Academy of Sciences in Enlightened Russia." Ph.D. thesis, Cambridge University, 2000.

———. "Das Feuer und die Höfe der Spätrenaissance." In *Feuer*, ed. Bernd Busch, 121–32. Bonn: Kunst- und Austellungshalle der Bundesrepublik Deutschland, 2001.

———. "Wonders Never Cease: Descartes's *Météores* and the Rainbow Fountain." *British Journal for the History of Science* 34 (2001): 129–47.

———. "From the Grand Whim to the Gasworks: Philosophical Fireworks in Georgian England." In *The Mindful Hand: Inquiry and Invention from the Late Renaissance to Early Industrialisation*, ed. Peter Dear, Lissa Roberts, and Simon Schaffer, 325–48. Amsterdam: Edita; Chicago: University of Chicago Press, 2007.

———. "The Techniques of Innovation: Historical Configurations of Art, Science, and Invention, from Galileo to GPS." In *Artists as Inventors: Inventors as Artists*, ed. Dieter Daniels and Barbara U. Schmidt, 54–69. Stuttgart: Hatje Cantz, 2008.

———. "Enlightenment in Russian Hands: The Inventions and Identity of Ivan Petrovich Kulibin in Eighteenth-Century St. Petersburg." *History and Technology* (in press).

Whiston, William. *A New Theory of the Earth from Its Original to the Consummation of All Things Wherein the Creation of the World in Six Days, the Universal Deluge, and the General Conflagration . . . Are Shewn to Be Perfectly Agreeable to Reason and Philosophy*. London, 1696.

Whiston, William, and Humphry Ditton. *A New Method for Discovering the Longitude Both at Sea and Land, Humbly Proposed to the Consideration of the Publick*. 1st ed. London, 1714.

———. *A New Method for Discovering the Longitude Both at Sea and Land, Humbly Proposed to the Consideration of the Publick*. 2nd ed. London, 1715.

Whitehorne, Peter. *Certaine Wayes for the Ordering of Soldiours in Battelray*. London, 1573.

Wiener, Philip. "Leibniz's Project of a Public Exhibition of Scientific Inventions." *Journal of the History of Ideas* 1 (1940): 232–40.

Wigelsworth, Jeffrey R. "Competing to Popularize Newtonian Philosophy: John Theophilus Desaguliers and the Preservation of Reputation." *Isis* 94 (2003): 435–55.

Williams, Sheila. "The Pope-Burning Processions of 1679, 1680 and 1681." *Journal of the Warburg and Courtauld Institutes* 21 (1958): 104–18.

Willis, Thomas. *An Essay of the Pathology of the Brain and Nervous Stock in Which Convulsive Diseases Are Treated Of*. London, 1681.

———. *Five Treatises, Viz. 1. Of Urines. 2. Of the Ascension of the Blood. 3. Of Musculary Motion. 4. The Anatomy of the Brain. 5. The Description and Use of the Nerves*. London, 1681.

———. "Of Convulsive Diseases." In *Dr. Willis's Practice of Physick*. London, 1684. Separately paginated.

Wilson, Alexander. "Biographical Account of Alexander Wilson, MD, Late Professor of Practical Astronomy in Glasgow." *Transactions of the Royal Society of Edinburgh* 10 (1824): 279–97.

Wilson, Benjamin. *An Account of Experiments Made at the Pantheon, on the Nature and Use of Conductors: To Which Are Added, Some New Experiments with the Leyden Phial*. London, 1778.

Wilson, Kathleen. "Empire, Trade and Popular Politics in Mid-Hanoverian Britain: The Case of Admiral Vernon." *Past and Present* 121 (1988): 74–109.

Wilson, Sir Robert Thomas. *Life of General Sir Robert Wilson . . . from Autobiographical Memoirs, Journals, Narratives, Correspondence, &c. Ed. by His Nephew and Son-in-Law, the Rev. Herbert Randolph*. 2 vols. London, 1862.

Winkler, Johann Heinrich. *Die Eigenschaften der electrischen Materie und des electrischen Feuers, aus verschiedenen neuen Versuchen erkläret, und, nebst etlichen neuen Maschinen zum Electrisiren beschrieben*. Leipzig, 1745.

———. *Elements of Natural Philosophy Delineated*. 1754. 2 vols. 1st English ed., translated from the 2nd German ed. London, 1757.

Winter, Frank H. *The First Golden Age of Rocketry: Congreve and Hale Rockets of the Nineteenth Century*. Washington, DC: Smithsonian Institution Press, 1990.

Winthrop, John. *A Lecture on Earthquakes; Read in the Chapel of Harvard-College in Cambridge, N.E. November 26th 1755. On Occasion of the Great Earthquake Which Shook New-England the Week Before*. Boston, 1755.

Withers, Charles W. J. *Placing the Enlightenment: Thinking Geographically about the Age of Reason*. Chicago: University of Chicago Press, 2007.

Wolff, Christian. *Cours de mathématique: Contenant toutes les parties de cette science . . . tome troisième, qui traite de la fortification, de l'attaque & la defence des places, de l'artillerie, des feux d'artifice, & de l'architecture*. Paris, 1757.

Wood, Anthony à. *The Life and Times of Anthony Wood, Antiquary of Oxford, 1632–1695, Described by Himself*. Edited by Andrew Clark. 5 vols. Oxford: Clarendon, 1891–1900.

Wood, Karen. "Making and Circulating Knowledge through Sir William Hamilton's *Campi Phlegraei*." *British Journal for the History of Science* 39 (2006): 67–96.

Wortman, Richard. *From Peter the Great to the Death of Nicholas I*. Vol. 1 of *Scenarios of Power: Myth and Ceremony in Russian Monarchy*. Princeton, NJ: Princeton University Press, 1995.

Wroth, Warwick. *The London Pleasure Gardens of the Eighteenth Century*. London: Macmillan, 1896.

Yates, Frances A. *The Rosicrucian Enlightenment*. London: Routledge, 2001.

Young, Davis A. *Mind over Magma: The Story of Igneous Petrology*. Princeton, NJ: Princeton University Press, 2003.

Young, Edward. *The Correspondence of Edward Young, 1683–1765*. Oxford: Clarendon, 1971.

Zanger, Abby E. *Scenes from the Marriage of Louis XIV*. Stanford, CA: Stanford University Press, 1997.

Zelov, D. D. *Ofitsial'nye svetskie prazdniki kak iavlenie russkoi kul'tury kontsa XVII- pervoi poloviny XVIII veka: Istoriia triumfov feierverkov ot Petra Velikogo do ego docheri Elizavety*. Moscow: URSS, 2002.

Zielinski, Siegfried. *Deep Time of the Media: Toward an Archeology of Hearing and Seeing by Technical Means*. Cambridge, MA: MIT Press, 2006.

Zilsel, Edgar. *The Social Origins of Modern Science*. Edited by D. Raven, W. Krohn, and R. S. Cohen. Boston Studies in the Philosophy of Science, vol. 200. Dordrecht: Kluwer Academic, 2003.

# INDEX

*The letter f following a page number indicates a figure.*